REMOTE SENSING OF THE CRYOSPHERE

Wiley-Blackwell Cryosphere Science Series

Permafrost, sea ice, snow, and ice masses ranging from continental ice sheets to mountain glaciers, are key components of the global environment, intersecting both physical and human systems. The study of the cryosphere is central to issues such as global climate change, regional water resources, and sea level change, and is at the forefront of research across a wide spectrum of disciplines, including glaciology, climatology, geology, environmental science, geography and planning.

The Wiley-Blackwell Cryosphere Science Series comprises volumes that are at the cutting edge of new research, or provide a focused interdisciplinary reviews of key aspects of the science.

Series Editor
Peter G Knight, Senior Lecturer in geography, Keele University

REMOTE SENSING OF THE CRYOSPHERE

EDITED BY

M. TEDESCO

CITY COLLEGE OF NEW YORK, NEW YORK, US

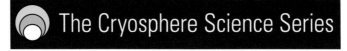

The Cryosphere Science Series

Series Editor: **Dr Peter Knight**, University of Keele

WILEY Blackwell

Library of Congress Cataloging-in-Publication Data

Remote sensing of the cryosphere / edited by M. Tedesco.
        pages cm
    Includes bibliographical references and index.
    ISBN 978-1-118-36885-5 (cloth)
    1. Cryosphere–Remote sensing–Textbooks. I. Tedesco, M., 1971- editor of compilation.
    GB2401.72.R42R47 2015
    551.31–dc23

                                                                    2014019575

A catalogue record for this book is available from the British Library.

Wiley also publishes its books in a variety of electronic formats. Some content that appears in print may not be available in electronic books.

Cover image: The Matusevich Glacier, East Antarctica. Courtesy of NASA Earth Observatory.

Set in 10/12.5pt, MinionPro by Laserwords Private Limited, Chennai, India
Printed and bound in Singapore by Markono Print Media Pte Ltd

1   2015

To my daughters, Olivia and Francesca and my wife, Luisa

Contents

Table of contents

List of contributors

Waleed Abdalati   University of Colorado, Boulder, CO, USA
Waleed.abdalati@colorado.edu

Liss M. Andreassen   Norwegian Water Resources and Energy Directorate, Oslo, Norway
lma@nve.no

A.A. Arendt   University of Alaska, Fairbanks, AK, USA
arendta@gi.alaska.edu

Monique Bernier   Centre Eau, Terre, Environnement, Québec, Québec G1K 9A9, Canada
monique.bernier@ete.inrs.ca

Suzanne Bevan   Swansea University, UK
S.L.Bevan@swansea.ac.uk

Tobias Bolch   University of Zurich, Zurich, Switzerland
tobias.bolch@geo.uzh.ch

Ludovic Brucker   NASA/Goddard Space Flight Center, Greenbelt, MD, and Universities Space Research Association, Columbia, USA
ludovic.brucker@nasa.gov

Jeffrey S. Deems   CIRES National Snow and Ice Data Center, and CIRES/NOAA Western Water Assessment – University of Colorado, Boulder, CO, USA
deems@nsidc.org

Chris Derksen   Environment Canada, Toronto, Canada
Chris.Derksen@ec.gc.ca

Stephen J. Déry   University of Northern British Columbia, Prince George, BC, Canada
sdery@unbc.ca

Claude R. Duguay   University of Waterloo, Waterloo, Ontario N2L 3G1, Canada
crduguay@uwaterloo.ca

Richard Forster   University of Utah, Salt Lake City, UT, USA
Rick.forster@geog.utah.edu

James L. Foster   NASA Goddard Space Flight Center, Greenbelt, MD, USA
James.l.foster@nasa.gov

Allan Frei   Hunter College, City University of New York, NY, USA
afrei@hunter.cuny.edu

Yves Gauthier   Centre Eau, Terre, Environnement, Québec, Québec G1K 9A9, Canada
yves.gauthier@ete.inrs.ca

Prasad Gogineni   Center for Remote Sensing of Ice Sheets, University of Kansas, Lawrence, KA, USA
pgogineni@ku.edu

Guido Grosse   University of Alaska, Fairbanks, AK, USA
ggrosse@gi.alaska.edu

Dorothy K. Hall   NASA / Goddard Space Flight Center, Greenbelt, MD, USA
dorothy.k.hall@nasa.gov

Robert L. Hawley   Dartmouth College, Hanover, NH, USA
Robert.L.Hawley@dartmouth.edu

M. Horwath   Technische Universitat Munchen, Munich, Germany
martin.horwath@bv.tum.de

Andreas Kääb   University of Oslo, Oslo, Norway
kaeaeb@geo.uio.no

Alexander A. Khokanovsky   Institute of Environmental Physics, University of Bremen, Bremen, Germany
alexk@iup.physik.uni-bremen.de

Lora Koenig   NASA/Goddard Space Flight Center, Greenbelt, MD, USA
Lora.s.koenig@nasa.gov

Alexei Kouraev   Laboratoire d'Etudes en Géophysique et Océanographie Spatiales (LEGOS), 31401, Toulouse Cedex 9, France
kouraev@notos.cst.cnes.fr

B.D. Loomis   SGT Inc., Science Division, Greenbelt, MD, USA
Bloomis@sgt-inc.com

Scott B. Luthcke   NASA Goddard Space Flight Center, Greenbelt, MD, USA
scott.b.luthcke@nasa.gov

Thorsten Markus   NASA Goddard Space Flight Center, Greenbelt, MD, USA
Thorsten.markus@nasa.gov

Hans-Peter Marshall   Boise State University, ID, USA
hpmarshall@boisestate.edu

Walter N. Meier   NASA Goddard Space Flight Center, Greenbelt, MD, USA
walt.meier@nasa.gov

Julie Miller   University of Utah, Salt Lake City, UT, USA
jjaimoe@gmail.com

Thomas Mote   University of Georgia, Athens, Georgia, USA
tmote@uga.edu

Tommaso Parrinello   European Space Agency, ESRIN, Frascati, Italy
tommaso.parrinello@esa.int

Bruce H. Raup   NSIDC, University of Colorado, Boulder, CO, USA
braup@nsidc.edu

D.D. Rowlands   NASA Goddard Space Flight Center, Greenbelt, MD, USA
david.d.rowlands@nasa.gov

T.J. Sabaka   NASA Goddard Space Flight Center, Greenbelt, MD, USA
terence.j.sabaka@nasa.gov

Konrad Steffen   Swiss Federal Institute for Forest, Snow and Landscape Research
WSL, Lausanne, Switzerland
Konrad.steffen@colorado.edu

Marco Tedesco   The City College of New York, City University of New York,
New York, NY, USA
mtedesco@ccny.cuny.edu

Charles Webb   NASA Headquarters, Washington DC, USA
charles.webb@nasa.gov

Sebastian Westermann   University of Oslo, Oslo, Norway
sebastian.westermann@geo.uio.no

Jie-Bang Yan   Center for Remote Sensing of Ice Sheets, University of Kansas,
Lawrence, KA, USA
syan@ku.edu

Cryosphere Science: Series Preface

Permafrost, sea ice, snow, and ice masses, ranging from continental ice sheets to mountain glaciers, are key components of the global environment, intersecting both physical and human systems. The scientific study of the cryosphere is central to issues such as global climate change, regional water resources, and sea-level change. The cryosphere is at the forefront of research across a wide spectrum of disciplinary interests, including glaciology, climatology, geology, environmental science, geography and planning.

The Wiley-Blackwell Cryosphere Science series serves as a framework for the publication of specialist volumes that are at the cutting edge of new research, or provide a benchmark statement in aspects of cryosphere science, where readers from a range of disciplines require a short, focused state-of-the-art text. These books lie at the boundary between research monographs and advanced text books, contributing to the development of the discipline, incorporating new approaches and ideas, but also providing a summary of the current state of knowledge in tightly focused topic areas. The books in this series are, therefore, intended to be suitable both as case studies for advanced undergraduates, and as specialist texts for postgraduate students, researchers and professionals.

Cryosphere science is in a period of rapid development, driven in part by an increasing urgency in our efforts to understand the global environmental system and the ways in which human activity impacts it. This rapid development is marked by the emergence of new techniques, concepts, approaches and attitudes.

Remote Sensing of the Cryosphere is an appropriate first volume in the series, as it clearly demonstrates this convergence of technological and theoretical developments in interdisciplinary efforts to address fundamental questions about the cryosphere.

Peter G Knight
February 2014.

Preface

The cryosphere, the region of the Earth where water is temporarily or permanently frozen, plays a key role on the climatological, hydrological, and energy cycles of our planet. Components of the cryosphere are snow on ground, terrestrial ice, such as glaciers and ice sheets, sea ice, river and lake ice, permafrost, and frozen soil. The harsh conditions, as well as the distribution and extent, characterizing the geographic areas where cryospheric components occur, are major impediments to data collection from the ground. In this context, remote sensing has provided a powerful and versatile tool to study the Earth's cryosphere, making "accessible" places that were otherwise inaccessible, or even unknown. It has been used to study, for example, the seasonal variability of snow cover, the advance and retreat of glaciers, the surface and internal properties of ice sheets, and the freezing and thawing of soil, to name a few examples. Because of the possibility of acquiring data over large areas, of the high number of observations available at high latitudes and, in some cases, of the independency of data acquisition from solar illumination or atmospheric conditions, remote sensing has been, and still is, among the major drivers (if not *the* major driver) for advancing our knowledge of the cryosphere.

The interdisciplinary nature of remote sensing, requiring people with engineering, science, geophysics, mathematics, physics and computer science background, is one of its characterizing aspects. This book is also the outcome of an interdisciplinary and collaborative effort (with 40 contributing authors over the 15 Chapters), stemming from several sessions that I co-organized and co-convened with many colleagues at the fall meeting of the American Geophysical Union (AGU) society in San Francisco and at the European Geophysical Union (EGU) meeting society in Vienna. The material submitted to the sessions led me to realize how much progress was being made, and how fast the community of researchers dealing with remote sensing of the cryosphere was expanding.

The scope of the book is to provide an overview of the methods, techniques and recent advances in applications of remote sensing of the cryosphere as well as a bibliographic source for those interested in deepening their understanding of the topics covered in the different chapters. These are: remote sensing and the cryosphere (Chapters 1 and 2); snow extent (Chapter 3); snow grain size and impurities (Chapter 4); snow depth and snow water equivalent (Chapter 5); surface and subsurface melting (Chapter 6); glaciers (Chapter 7); accumulation over the Greenland and Antarctica ice sheets (Chapter 8); ice thickness and velocities (Chapter 9); gravimetric measurements from space (Chapter 10); sea ice (Chapter 11); lake and river ice (Chapter 12); frozen ground and permafrost (Chapter 13); fieldwork activities (Chapter 14); and, lastly, recent and future cryosphere-oriented missions and experiments (Chapter 15).

Given the different styles adopted by the authors, and the broad spectrum of topics covered, the treatment throughout the book is technical in some places and more descriptive in others. The book is oriented towards readers with a limited or basic knowledge of the cryosphere and remote sensing methods, and who are willing to have an overview of the methods and techniques, such as senior undergraduate and master students. However, doctoral students can also use it as introductory textbook. For readers who are interested in specific topics, we have tried to keep the different chapters as self-contained as possible. Lastly, although it was never my intention to provide a complete anthology of recent results (sincere apologies for the unintentional omission of important works), I hope researchers will also find it helpful as a work of reference.

M. Tedesco
New York City
September, 2013

Acknowledgments

I would like to thank Wiley Publishing for providing me with the opportunity of publishing this book. Special thanks to the faculty, the chair, Dr. Jeff Steiner, and the students of the Department of Earth and Atmospheric Sciences at the City College of New York for understanding when my office door was locked from the inside. A special thanks goes also to the National Science Foundation, Office of Polar Programs (where I was serving as Program Director during the period when the book was finalized) for allowing me to work on the book within the framework of the Independent Research/Development Program.

My most sincere gratitude goes to the chapter leading authors and co-authors, for their contribution and the invaluable commitment to the publication of the book.

Special thanks to Dorothy Hall and W. Gareth Rees for their inspiring books *Remote Sensing of Ice and Snow* (University Press, Cambridge, 1985), by Dorothy Hall and Jaroslav Martinec and *Remote Sensing of Snow and Ice* (CRC Press, 2005) by W. Gareth Rees.

About the companion website

This book is accompanied by a companion website:

www.wiley.com/go/tedesco/cryosphere

The website includes:
- Powerpoints of all figures from the book for downloading
- PDFs of tables from the book

Remote sensing and the cryosphere

Marco Tedesco

The City College of New York, City University of New York, New York, USA

Summary

This introductory first chapter provides a general overview on remote sensing and an introduction to the cryosphere, exposing the reader to general concepts. The chapter is mainly oriented toward those readers with minimal or no experience on either of the two subjects. Remote sensing can be defined as to that ensemble of techniques, tools, data and sensors that allow us to study the Earth and its processes from airborne, spaceborne and *in situ* sensors without being in physical contact with the object under examination.

In the first part of this chapter, a brief history of remote sensing is introduced, describing early tools and applications (such as the pioneering work from air balloons and cameras attached to pigeons), followed by a basic introduction to the electromagnetic spectrum and electromagnetic radiation. The reader is then presented with a description of remote sensing systems, divided into the categories of *passive* (aerial photography, electro-optical sensors, thermal systems, microwave radiometers and gravimetric systems) and *active systems* (LiDAR, radar). Concepts such as spatial, temporal, spectral and radiometric resolutions are also introduced. In the second part of the chapter, the several elements of the cryosphere are introduced, together with a description of their basic physical properties and a general overview of their spatial distribution and the impact on other fields (such as biology, ecology, etc.).

1.1 Introduction

This chapter contains a general overview on both remote sensing and the cryosphere and briefly introduces the reader to their general concepts. Both topics are vast, and it is not possible to cover them in their entirety here. Nevertheless, it is helpful to provide an introductory overview of the two fields, with the references in this chapter (and throughout the book) suggesting reading material for those interested in more details.

1.2   Remote sensing

Remote sensing is the collection of information about an object or phenomenon without physical contact with the object. For practical applications, throughout this book we will refer to remote sensing as that ensemble of techniques, tools, data and sensors that allow us to study the Earth and its processes from airborne, spaceborne and *in situ* sensors without being in physical contact with the object under examination.

Remote sensing of the Earth began with the development of flight. The first photographs of Paris were taken from air balloons as early as 1858 by Gaspard-Félix Tournachon, a French photographer known as Nadar (http://www .papainternational.org/history.asp). In the 1880s, Arthur Batut attached cameras to kites to collect pictures over Labruguière, France. The apparatus also included an altimeter so that the scale of the images could be estimated. At the beginning of 1900, the Bavarian Pigeon Corps had cameras attached to pigeons, taking pictures every 30 seconds (http://www.sarracenia.com/astronomy/remotesensing /primer0120.html; Jensen, 2006).

Systematic aerial photography began with World War I and was improved during World War II. At the end of the Wars, the development of artificial satellites allowed remote sensing to begin performing measurements on a large scale, leading to the modern remote sensing era. More information on the history of remote sensing can be found, for example, in Jensen (2006).

1.2.1   *The electromagnetic spectrum and blackbody radiation*

Remote sensing of the Earth is based on the interaction between electromagnetic waves and matter, with the exception of those approaches based on gravimetry. The interaction between materials and electromagnetic waves depends on both the characteristics of the electromagnetic radiation (e.g., frequency) and on the chemical and physical properties of the targets. In many cases, the source of the electromagnetic radiation is the sun, which can be approximated as a black body (an idealized body that absorbs all incident electromagnetic radiation, regardless of frequency) at a temperature of about 5800 K. Though a large number of remote sensing applications deal with the visible portion (400–700 nm) of the electromagnetic spectrum (Figure 1.1), visible light occupies only a fraction of it. Indeed, a considerable portion of the incoming solar radiation is in form of ultraviolet and infrared radiation, and only a small portion is in form of microwave radiation.

Before reaching a spaceborne or airborne sensor, the electromagnetic radiation propagates through the atmosphere, hence interacting with the different atmospheric components. For example, as the sunlight enters the atmosphere, it interacts with gas molecules, suspended particles and aerosols. Because of the preferential scattering and absorption of particular wavelengths and elements, the radiation reaching the Earth is a combination of direct filtered solar radiation and diffused light scattered from the sky.

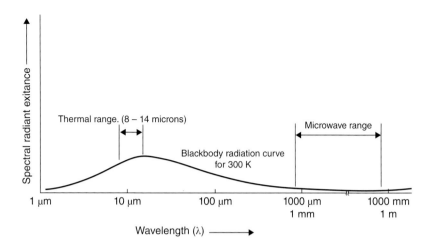

Figure 1.1 Spectral regions used for thermal and passive microwave sensing (Adapted from Lillesand *et al.*, 2007).

As the filtered and diffused solar radiation reaches the Earth, it interacts with surface targets (e.g., soil, snow, vegetation, ocean, etc.). Each of these materials interacts with the electromagnetic radiation through absorption, transmission and scattering, depending on its physical properties (e.g., leaves reflects back most of the radiation in the green regions, water reflects blue radiation more than red and green, etc.). Before reaching the sensors and being recorded by the instruments onboard, the upwelling radiation passes again through the atmosphere.

It follows that the atmospheric components (such as water vapor, carbon dioxide and ozone, Figure 1.2) drive the design of the sensors used to study the Earth from space. On the other hand, in the case of thermal infrared radiation (e.g., 800–1400 nm), the sensors will detect the radiation emitted by the surface as a result of the solar heating. In the case of passive microwave remote sensing, the instruments will record the naturally emitted radiation by the objects, because the incoming solar radiation in the microwave region is negligible.

Features characterizing the data collected by remote sensing platforms are spatial, temporal, spectral and radiometric resolutions. Spatial resolution is a measure of the spatial detail that can be distinguished in an image, being, in turn, a function of the sensor design and its altitude above the surface. Temporal resolution is the frequency of data acquisition (e.g., how many acquisitions are collected within a day) or the temporal interval separating successive data acquisitions. Spectral resolution is the ability of the system to distinguish different parts of the range of measured wavelengths (e.g., the number of measured bands and how narrow each band is).

For recording purposes, the energy received by an individual detector in a sensor must be "quantized" (e.g., divided into a number of discrete levels that are recorded as integer values). Radiometric resolution quantifies the number of levels: the more levels that are recorded, the greater is the radiometric resolution. Many current satellite systems quantize data into 256 levels (8 bits of data in a binary encoding system, $2^8 = 256$), but other systems can have higher radiometric resolution (e.g., 12 bits, $2^{12} = 4096$ levels).

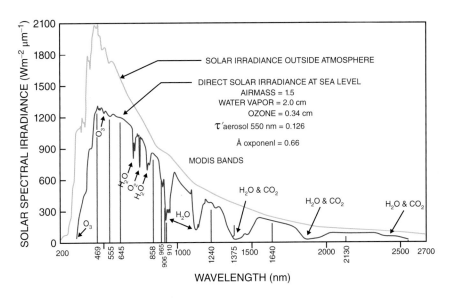

Figure 1.2 Solar spectral irradiance incident on the top of the atmosphere and transmitted through the atmosphere to the Earth's surface. Major absorption bands in the atmosphere are also shown (NASA).

Remote sensing instruments can be categorized into active and passive. Passive sensors measure the radiation that is naturally emitted or reflected by the target. For example, sensors operating in the visible range, measuring the solar radiation reflected by an object on Earth, are passive sensors. Other passive sensors are microwave radiometers, measuring the microwave radiation naturally emitted by the targets (in this case, as mentioned, the solar contribution is negligible). In this book, sensors collecting gravimetric data are also classified as passive. Active sensors emit energy and measure the amount of energy that is reflected or backscattered by the target. Examples of active sensors are radar (Radio Detection and Ranging) or LiDAR (Light Detection and Ranging).

1.2.2 Passive systems

1.2.2.1 Aerial photography

The first collection of aerial photography was performed by the French photographer and balloonist Gaspard-Félix Tournachon, known as Nadar, in 1858 over Paris (though it was destroyed, and the first surviving collection of aerial photography consists of a view of Boston taken from 630 m by James Wallace Black and Samuel Archer King in 1860). Aerial photography was crucial during World War I, supporting many strategic decisions for battlefronts, for example. After the wars, aerial photography became more and more accessible and was used by commercial companies and government agencies for many purposes. More information on the history of aerial photography can be found at http://www.papainternational.org/history.asp (History of Aerial Photography, Professional Aerial Photographers Association, accessed December 28, 2012). Aerial photography is nowadays used for cartography, land use planning and management, archeology and environmental studies, to name just a few applications.

Traditional photographic systems make use of photographic films to collect information in the visible and near-infrared regions. Obviously, intrinsic limitations exist due to the need of solar illumination and the impossibility of collecting data in the presence of clouds. The film consists of a suspension of small crystals of silver halides in a gelatin matrix, and its exposure to light converts the silver ions into thermodynamically stable metallic silver. The development process removes the crystals that have not been exposed, generating the so-called *negative*, which is then used to print the photograph. Panchromatic films have a relatively flat response in the visible, while color films are sensitive to the three bands of red, green and blue, and infrared films are sensitive to the near-infrared region. The spectral response of the photographic system can also be changed through the deployment of filters.

The parameters characterizing the performance of a photographic system are resolution, speed and contrast. Spatial resolution is driven by the size of the grains in the film (e.g., the smaller the grain, the higher the resolution). The spatial resolution on the ground (R_g) can be computed from the film spatial resolution (R_f) from the knowledge of the focal length f and the height of the sensor H (e.g., Rees, 2005), as:

$$R_g \approx R_f \cdot H/f \tag{1.1}$$

Film speed provides a measure of the exposure (defined as the illuminance at the film multiplied by the exposure time). Finally, the contrast refers to the range of exposures to which the film responds.

In aerial photography, data collection can be performed either through vertical photography (in which all points in the ground, assumed to be a planar surface, are assumed to be at the same distance H from the camera) or through oblique photography, in which the optical axis is not vertical. Oblique photography allows the coverage of larger areas, but has the disadvantage that the ground plane is no longer constant. Other techniques aiming at covering larger areas include the use of panoramic, strip and reconnaissance cameras. The assumption of a planar surface is often not valid (and consequently the assumption of the ground surface perpendicular to the optical axis). This translates into a phenomenon known as *relief displacement,* in which a point in an image is displaced from the position it would have if it had a zero elevation above the surface. This aspect can be solved by using two vertical images taken over the same scene from two different positions (stereophotography).

More details about aerial photography can be found in Jensen (2006), Lillesand *et al.* (2007), Rees (2001), and Rees (2005), or at http://www.nrcan.gc.ca/earth-sciences/products-services/satellite-photography-imagery/aerial-photos/about-aerial-photography/884 and http://www.fas.org/irp/imint/docs/rst/Sect10/Sect10_1.html.

1.2.2.2 Visible/near infrared electro-optical sensors

Electro-optical sensors operating in the visible and near-infrared regions are similar to the ones above described using films. The major difference between film-based and electro-optical sensors consists in the way the data is detected and stored. In the case of electro-optical sensors, indeed, the radiation is collected by means of electronic systems rather than photochemical processes. Electro-optical sensors are similar to modern digital cameras, making use of either charge-coupled device (CCD) or complementary metal-oxide-semiconductor (CMD) sensors.

There are several advantages in using electro-optical systems with respect to traditional photography. Large quantities of data can be stored or transmitted to receiving systems, hence allowing the mounting of such instruments on unmanned airborne and spaceborne platforms. Electronic systems can be calibrated and controlled from remote areas, hence providing a better performance. The collection of images from electro-optical systems is performed through different techniques than those used in traditional photography, especially from space. The approach analogous to traditional photography, indeed, using a two-dimensional array of CCD, would require a large number of detectors. Therefore, in the *pushbroom* method, a one-dimensional CCD is used and the two-dimensional image is obtained through the motion of the platform. An extremely common method is *whiskbroom* imaging, in which a single detector is used and the scene is scanned by means of a rotating mirror.

The spatial resolution, in the case of electro-optical sensors, is driven by the size of the detector projected onto the ground through the instrument's optics. The corresponding feature in an image is called a pixel. The spatial resolution can, therefore, be calculated once the pixel is known from the scaling factor, as already discussed above in the case of traditional photography. The spatial coverage of systems employing digital sensors is generally larger than that of systems using traditional photography, and it can range from tens to thousands of kilometers, depending on the sensor.

The collection of images at multiple bands (usually more than those available through traditional photography) allows performing classification based on multispectral data. In this case, the information in all bands is used to classify a pixel into a land cover or type through its spectral signature, this being a characteristic spectral variation of pixel values at the different available bands. Spectral mixture modeling can be performed on multispectral images. In this case, the area in the image is assumed to be composed of a number of different classes (with known spectral signatures called end-members), and the overall goal is to determine the fraction of each class within each pixel. The number of classes cannot exceed the number of bands, and spectral mixing is extremely helpful when many bands are available (hyperspectral images).

Hyperspectral sensors are instruments that acquire images in many, narrow spectral bands, from the visible through the thermal infrared, up to more than 200 bands. This enables the reconstruction of the reflectance (or emittance) spectrum for each pixel in the image, which can then be used to classify the features within the scene. The high number of points collected with hyperspectral sensors allows separation among different materials and targets. Applications include mineralogy, water quality, bathymetry, soil and vegetation types, and snow and ice properties.

1.2.2.3   Thermal infrared systems

Though the principles of detection in the case of thermal infrared (800 – 1400 nm) are similar to those of visible and near-infrared systems, the longer wavelength (and, consequently, the lower energy associated with the photons) in the case of the thermal infrared band imposes modifications on the detectors. Photon detectors employed in the 800 – 1400 nm range are the mercury cadmium telluride type (also known as MerCaT), or mercury-doped germanium sensors (e.g., Rees, 2005).

One crucial aspect is to minimize the sensors' own thermal emission. The sensors are, therefore, cooled down to temperatures that are as close as possible to the absolute zero and are surrounded by a dewar containing liquid nitrogen at a temperature of 77 K. Because of the larger size of detectors, the spatial resolution in the case of thermal infrared sensors is generally coarser than that of visible or near-infrared sensors. Spectral resolution of thermal sensors for earth surface studies range between 0.1 and 10 μm, depending on the application. Radiometric resolution characterizes the temperature resolution, in view of the possibility of discriminating between small changes in the brightness temperature of the incident radiation, and present-day systems can reach a temperature resolution of the order of 0.1°C.

1.2.2.4   Microwave radiometers

Microwave radiometers measure the energy naturally emitted by the Earth and its atmosphere in the microwave region (0.3 – 300 GHz). In this regard, passive microwave radiometers are similar to thermal radiometers and scanners. However, unlike the visible and thermal cases, the contribution from the sun is negligible with respect to that from the energy naturally emitted by the earth materials. The quantity measured by microwave radiometers is called brightness temperature (Tb), and is defined as the temperature that a black body in thermal equilibrium with its surrounding would have to be to match the intensity of a grey body at a given frequency. For the temperature range of objects on Earth and in the microwave band, it is possible to adopt the so-called Rayleigh-Jeans law, so that:

$$Tb = \varepsilon T \tag{1.2}$$

with ε being the emissivity and T the body's kinetic temperature.

Unlike the case of visible and near-infrared bands, it is also possible to acquire microwave data through clouds in many of the frequencies available on board spaceborne sensors. The possibility of collecting data in all-weather conditions, and without the need of solar illumination, allows spaceborne passive microwave remote sensors to collect data at high temporal resolution. The relatively coarse spatial resolution (of the order of several tens of kilometers) of data collected by spaceborne microwave sensors is compensated for by the large spatial coverage, due to the large swath.

As an example, Figure 1.3 shows spaceborne brightness temperatures collected at 19.35, 37.0 and 85.5 GHz horizontal polarization, over the northern hemisphere by the Special Sensor Microwave Imager (SSM/I) on November 1, 2006. Although the spatial patterns of the data at different frequencies are similar (e.g., land is warmer than ocean), differences exist due to the different interaction of electromagnetic waves with the earth materials at the different

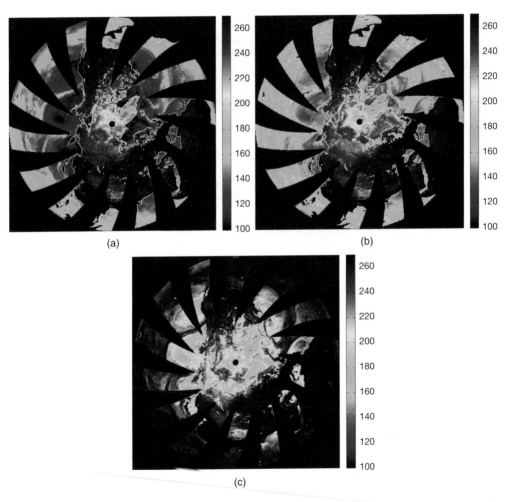

(a)

(b)

(c)

Figure 1.3 Spaceborne brightness temperatures [K] at (a) 19.35, (b) 37.0 and (c) 85.5 GHz horizontal polarization, collected over the Northern Hemisphere by SSM/I on November 1st, 2006 (Data from National Snow and Ice Data Center).

frequencies. This provides complementary information that can be use for retrieval of quantitative information on the physical properties of the different targets. Moreover, the large swath of passive microwave sensors allows covering large areas of the Earth on a daily basis.

1.2.2.5 Gravimetric systems

Gravimetry is the measurement of the strength of a gravitational field, and it has been recently used in remote sensing applications. In this regard, the Gravity Recovery And Climate Experiment (GRACE), a joint mission of NASA and the German Aerospace Center, has been measuring the Earth's gravity field and its changes since March 2002 (http://www.csr.utexas.edu/grace/). GRACE is the first earth-monitoring mission in the history of space flight whose key measurement is not derived from the interaction of electromagnetic waves with the target. The system is composed of two identical spacecraft flying in a polar orbit about 220 kilometers apart at an altitude of 500 kilometers. A microwave-ranging system is used to measure changes in the speed and distance between the two spacecraft with an accuracy as small as 10 μm.

As the twin satellites orbit over the Earth, they encounter gravity anomalies and, from the measurements of the distance between the two satellites and the support of precise positioning measurements from through Global Positioning System (GPS) instruments, scientists can construct a detailed map of Earth's gravity. Accelerometers are also used on each instrument to measure their own movement. Finally, GPS receivers, magnetometers and star cameras are also used to establish baseline positions.

GRACE data have been used for many applications, from the flow of water through aquifers to the hydrological cycle in the Amazon, to ocean circulation. In the case of the cryosphere, GRACE has provided an invaluable contribution in assessing the mass of both the Greenland and Antarctica ice sheets, for example, and seasonal accumulation of snow. A mission similar to GRACE, the Gravity Field and Steady-State Ocean Circulation Explorer (GOCE) by ESA, consisting of a satellite carrying a highly sensitive gravity gradiometer, was launched on March 17, 2009 (http://www.esa.int/Our_Activities/Observing_the_Earth/GOCE). A major difference between GRACE and GOCE is that GRACE measures time variations, while GOCE measures the static gravity field. More information on GRACE is reported in Chapters 8 and 15.

1.2.3 Active systems

1.2.3.1 LiDAR

LiDAR stands for LIght Detection And Ranging, and it makes use of transmitted pulses of laser light, measuring the time of flight to the return of the reflected laser energy. The recorded time is then used to calculate the distance between the sensor and the target. The use of LiDAR for elevation terrain determination and surface feature mapping began in the late 1970s. Modern LiDAR applications make use of aircraft equipped with GPS and sensors for measuring the angular orientation of the sensor with respect to the ground, called Inertial Measurement Unit (IMU), beside a pulsing laser (from 10 KHz to 800 KHz). Other components include an extremely accurate clock, a reliable data storage system and electronics.

Most airborne LiDAR systems record multiple returns per pulse, allowing discrimination not only between, for example, forest and bare ground, but also surfaces in between, and allowing the construction of digital elevation models of bare earth (e.g., when vegetation is removed). Some newer LiDAR systems record the time series of the return energy waveform rather than the time of several discrete peaks. This full waveform analysis provides an effectively infinite number of returns, and it allows an improved characterization of the physical properties of the medium and of the surface, as well as a more reliable retrieval of the ground surface in vegetated environments. Besides recording the three-dimensional coordinates, LiDAR systems also typically record the energy level of the returning pulse, referred to as *lidar intensity*. This information can be helpful for the identification of surface types, and it can be used as ancillary data or for visualization.

The primary data product from a LiDAR survey is a geolocated point cloud. The point cloud can be characterized by the spacing of points on the target surface, quantified by the ground point spacing (distance) or ground point density (number of points per unit area), which are analogous to the concept of resolution in other technologies. The point spacing is a function of the interaction of laser system properties and flight/orbit parameters with the properties of the target area. For airborne surveys, parameters such as *pulse repetition frequency* (PRF), scan angle, scan frequency, flight speed, and swath overlap are adjusted to achieve the desired ground point spacing as allowed by the terrain relief, target complexity, and surface reflectivity.

The first satellite carrying a LiDAR system for cryospheric applications was the NASA Ice, Cloud, and Land Elevation Satellite (ICESat). ICESat was launched on January 12, 2003 and carried the Geoscience Laser Altimeter System (GLAS), operating at 1.064 and 0.532 µm wavelengths. ICESat operated for seven years and was retired in February 2010. Its purpose was to collect precise altimetry measurements of the polar ice sheets in order to allow mass balance estimates and studies of the impact of Earth's climate on ice sheets and sea level. More information about ICESat, and its planned successor ICESat-2, is provided throughout the book and in Chapter 15.

Airborne LiDAR surveys are seeing increasing application for mapping of snow depth. In 2012, the NASA Airborne Snow Observatory (ASO) mapped snow depth distribution over an entire California river basin on a nominally weekly basis, providing unprecedented scientific opportunities and immediately valuable data products in support of operational water supply forecasting efforts. LiDAR applications such as ASO are likely to be a model for future snow mapping endeavors, providing high-resolution, multitemporal data products, with the flexibility to respond to rapidly changing conditions or extreme events.

1.2.3.2   Radar

Radar is an acronym for Radio Detection and Ranging. Unlike passive sensors, radar generates its own illumination by sending pulses of electromagnetic energy in the microwave region. A fraction of the energy reflected by the target returns to the radar's receiving antenna. The data recorded by the sensor can contain information about the target shape and its physical properties, both at the surface and below. The components of a radar system are a pulse generator, a transmitter, a duplexer (separating the outgoing and returned pulses), a directional antenna that focuses the pulses into a beam and records the returned pulses, a receiver and a recording device.

In the late 1940s, radar was an ideal reconnaissance system for military purposes, as it provides information independently from weather conditions and during both day and night times. The declassification of military technology led to civilian applications, which include terrain mapping, as well as forestry resources, water supplies, and monitoring of ocean surface to study winds and waves, for example.

Spaceborne radar remote sensing began in the late 1970s with the launch of Seasat, followed by the Shuttle Imaging Radar (SIR) and the Soviet Cosmos experiments during the 1980s. In the 1990s, the number of spaceborne radar missions increased, with agencies from different countries launching numerous satellites (such as ERS-1 by ESA, JERS-1 by Japan, Almaz-1 by the former Soviet Union and Radarsat by Canada). Following the 1990s, a number of spaceborne radar missions were launched (such as Envisat by ESA, PALSAR by Japan, the Shuttle Radar Topography mission, SRTM, Radarsat-2 by Canada). A summary of spaceborne radar missions is reported, for example, in Lillesand *et al.* (2007).

The *azimuthal direction* is defined as the forward direction of the aircraft at the flight altitude, while *range direction* is the one where the pulses spread out. Any line-of-sight from the radar to points on the ground defines the *slant range* to that point. The distance between the aircraft nadir (directly below) line and any ground target point is called *ground range*. The duration of a single pulse determines the resolution at a given slant range. The range resolution can be seen as the minimum distance between two reflecting points along the look direction, at that range at which these may be sensed as separate and distinct, and it can be written as:

$$R_r = \tau c / 2 \cos \beta \tag{1.3}$$

where:

τ is the pulse length (in microseconds)
c is the speed of light
β (beta) is the depression angle, defined as the angle between a horizontal plane and a given slant range direction.

The *azimuth resolution*, defined as the minimal size of an object along the direction of the flight path that can be resolved, is given by:

$$R_a = S\gamma \tag{1.4}$$

where:

S is the slant range

γ is the angular beamwidth, defined as the wavelength λ divided by the effective length of the antenna.

It follows that, for a given wavelength, the antenna beamwidth (and consequently the azimuth resolution) can be controlled by either the physical size of the antenna or by synthesizing a virtual antenna length. Systems in which the beamwidth is controlled by the physical size of the antenna are called *real aperture* or *non-coherent* radars.

Radar antennas on aircraft can be mounted below the platform and direct their beam to the side of the airplane in a direction normal to the flight path. For aircraft, this mode of operation is implied in the acronym SLAR, for Side Looking Airborne Radar. Antennas can reach a size up to 5–6 m. Unlike SLAR, the Synthetic Aperture Radar (SAR) uses small antennas to simulate a large one. This is achieved by sending pulses from different positions as the platform advances, and integrating the pulse echoes into a composite signal. With SAR, it is possible to simulate effective antenna lengths up to 100 m or more. The *Doppler* effect (apparent frequency shift due to the target's or the radar's velocity) is used in SAR. Indeed, as coherent pulses from the radar reflect from the ground to the moving platform, the target acts as if it were in apparent (relative) motion. This translates into changing frequencies, which give rise to variations in phase and amplitude in the returned pulses. This data is recorded by the radar system and is recombined to synthesize signals equivalent to those from a real-aperture system with a larger antenna.

1.3 The cryosphere

The cryosphere is the portion of the Earth where water is in its solid form, either seasonally or annually. The term comes from the fusion of the Greek words *cryos* (meaning cold, icy) and *sphaira* (ball, globe). According to Barry *et al.* (2011), the term was first introduced in 1923 by Antoni Boleslaw Dobrowolski, a Polish scientist, geophysicist and meteorologist. The components of the cryosphere are snow cover, glaciers, ice sheets and ice shelves, freshwater ice, sea ice, icebergs, permafrost and ground ice (Figure 1.4). In this book, we will study how remote sensing is applied to the components of the cryosphere, with the exception of icebergs (see, for example, Rees (2005) for remote sensing of icebergs).

The cryosphere plays a key role in the global climate system, with important linkages and feedbacks with the atmosphere and hydrosphere through its impact on surface energy and moisture fluxes, the release of large amounts of freshwater into oceans and the lockup of water during the freezing season. Atmospheric, geophysical, ecological, biological, chemical and geological processes, to name

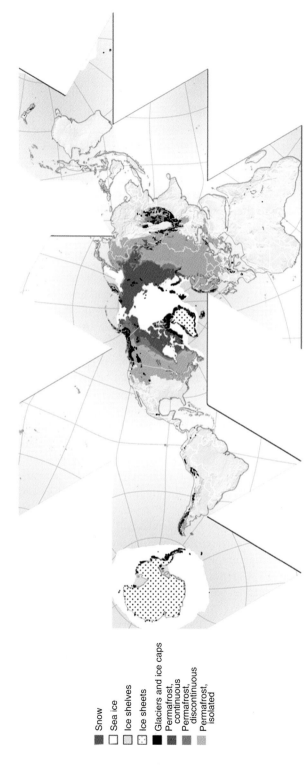

Snow
Sea ice
Ice shelves
Ice sheets
Glaciers and ice caps
Permafrost, continuous
Permafrost, discontinuous
Permafrost, isolated

Figure 1.4 Overview of the cryosphere and its larger components, from the UN Environment Program Global Outlook for Ice and Snow (Fraxen at en.wikipedia. CC-BY-SA-3.0/GFDL).

a few, are impacted by the cryosphere. A brief overview of the components of the cryosphere follows. Readers interested in a more detailed description can refer to Barry and Gan (2011), Marshall (2012) or http://en.wikipedia.org/wiki/cryosphere and related sources and links.

Frozen ground (permafrost and seasonally frozen ground) is the most extensive component of the cryosphere, with 55 million km^2 in the northern hemisphere (of which 22 million km^2 is due to permafrost; Barry and Gan, 2011). Permafrost might occur when the mean annual air temperature (MAAT) is lower than $-1°C$ and it is usually continuous when MAAT is below $-7°C$. The thickness of frozen ground can exceed ≈ 600 m along the Arctic coasts of Siberia and Alaska, becoming thinner toward the margins. The active layer (defined as the top layer of the soil that thaws during the summer and freezes again in autumn) plays a key role on the hydrologic and geomorphic regimes.

The second most extensive component of the cryosphere is snow, with a maximum cover extent of ≈ 47 million km^2 in January and a mean annual area of ≈ 26 million km^2 (Barry and Gan, 2011). Most of the snow cover is located in the northern hemisphere, with the maximum snow cover extent of ≈ 0.85 million km^2 in the southern hemisphere occurring in July. Though relatively limited in extent, snow stored in high mountain areas provides the major source of runoff for streamflow and groundwater recharge in many mid-latitude areas, and many regions of the Earth rely on snow accumulated during the winter for water resources and management. The high albedo of snow (e.g., the ratio between the incoming and outgoing solar radiation, which can be up to $0.8-0.9$ in the case of fresh snow) regulates the amount of solar energy that is absorbed by snow-covered areas in the northern hemisphere during winter.

Sea ice, formed by the freezing of the ocean water, is the third most extensive component of the cryosphere, with maximum winter extent of $\approx 14-16$ million km^2 in the northern hemisphere and $\approx 17-20$ million km^2 in the southern hemisphere. The extent in the northern hemisphere reduces down to less than $4-6$ million km^2 during the summer and $3-4$ million km^2 in the southern hemisphere.

The first sea ice is made of small (0.3 cm) separate disk-shaped crystals floating on the surface, then turning into crystals with a hexagonal shape and arms stretching out over the surface. As these arms break, the turbulence of the water causes a suspension of randomly shaped crystals of increasing density in the surface water, an ice type called *grease ice* or *frazil*. In quiet conditions, the crystals freeze together, forming a thin layer of ice, called *nilas* when it is still transparent. First-year ice (FYI) is then formed by the *congelation growth*, in which water freezes at the bottom of the ice layer. In rough waters, fresh new ice is formed by the cooling of the ocean as heat is released into the atmosphere. The upper layer of the ocean is supercooled and frazil ice forms, which then turns into grease ice, a mushy surface layer. Waves and wind then compress this ice into plates of several meters in diameter, called pancake ice, which will form upturned edges as the plates collide with one other while floating on the ocean surface. Eventually, the pancake ice plates will freeze together to form a consolidated pancake ice.

First-year sea ice is ice that is thicker than *young ice* but has no more than one year of growth. Simply put, first-year ice grows during the fall and winter, but melts during the summer. *Old sea ice* is sea ice that has survived at least one melting season and is generally divided into *second-year ice* (which has survived one melting season) and *multiyear ice* (which has survived more than one melting season). *Leads* and *polynyas* are areas of open water that occur even though air temperatures are below freezing. Leads are narrow and linear, while polynyas are larger, and both provide a direct interaction between the ocean and the atmosphere.

Glaciers and ice sheets are large bodies of ice that form where accumulation of snow exceeds its ablation (e.g., melting and sublimation) over time scales of many years (decades, centuries and millennia). Under the pressure of the layers above, snow and firn fuses into denser and denser firn. These layers undergo further compaction over many years and turn into glacial ice. By definition, ice sheets cover areas larger than $50,000$ km^2. At least 0.1 km^2 and 50 m thick, a glacier deforms and flows because of the stress due to its own weight.

Ice sheets hold 77% of the world's freshwater, of which Antarctica accounts for 90% and Greenland for almost 10% (and the remaining ice caps and glaciers ≈ 0.5%). The ice sheets contain an ice volume of 2.85 million km^3 (Greenland) and 24.7 million km^3 (Antarctica), compared to an estimated ice volume of 0.24 million km^3 in the case of glaciers and ice caps, of 0.66 million km^3 in the case of ice shelves and only 0.002 million km^3 in the case of snow on land in late January.

The ablation zone of a glacier is the region where there is a net loss in glacier mass (e.g., negative surface mass balance). The equilibrium line altitude (ELA) is the elevation where the surface mass balance turns from negative to positive, separating the ablation zone from the accumulation zone, where snowpack or superimposed ice accumulation persists. Within the accumulation zone, the dry snow zone is the region where no melt occurs (even during summer) and the percolation zone is where melt occurs, causing meltwater to percolate into the snowpack.

Ice forms over rivers and lakes during the winter. The effects of such components of the cryosphere are mostly local, in view of the small extent. However, because the freeze/thaw cycles respond to large-scale and local weather factors, measurements of lake and river ice can support climate studies. Because of the absence or low concentration of salt in lake and river waters, the ice occurring from their freezing can be considered as pure ice. In the case of lakes, the freezing begins on the surface with millimeter-size crystals (frazil). Because of their preferential growth along the crystallographic *c*-axis, a continuous layer of crystals with the *c*-axis horizontal forms. This ice is termed *black ice*, and it is represented by a smooth transparent slab of freshwater ice.

If snow deposits on the frozen lake, the weight of the snowpack might create depressions and, eventually, might crack the surface. Liquid water can then fill the cracks, saturating the snow overlying the ice. Once refrozen, this layer is called *white ice*. The formation on rivers differs slightly from that of lakes, because of the presence of river flow. In general, for slow-flowing rivers, the conditions are similar to those for a lake; but for fast-flowing rivers, the frazil can be distributed along the water column.

References

Barry, R.G. & Gan, T.Y. (2011). *The Global Cryosphere*. Cambridge University Press, 1st Edition, pp. 498, ISBN: 0521156858.

Barry, R.G., Jania, J., & Birkenmajer, K. (2011). A.B. Dobrowolski – the first Cryospheric scientist – and the subsequent development of Cryospheric science. *Hist. Geo-Space Sci.* **2**, 75–79.

Jensen, J. (2006). *Remote Sensing of the Environment: An Earth Resource Perspective*. Prentice Hall, pp. 608, ISBN-10: 0131889508.

Lillesand P., Kiefer, R.W., & Chipman, J. (2007). *Remote Sensing and image interpretation*. Wiley, 6th Edition, pp. 804.

Marshall, S.J. (2012). *The Cryosphere*. Princeton University Press, pp. 288, ISBN 9780691145259.

Rees, W.G. (2001). *Physical Principles of Remote Sensing*, Second edition. Cambridge, Cambridge University Press.

Rees, G.W. (2005). *Remote Sensing of Snow and Ice*. CRC Press. pp. 312.

2 Electromagnetic properties of components of the cryosphere

Marco Tedesco

The City College of New York, City University of New York, New York, USA

Summary

In this chapter, the electromagnetic properties of the components of the cryosphere are presented and discussed. For each component, the visible/infrared properties are first described, followed by those in the microwave spectrum. Snow properties are first presented, together with a section specifically dedicated to the ice permittivity in the microwave region. Then, sea ice and freshwater ice properties are introduced, followed by those concerning glaciers and ice sheets and, lastly, by those concerning frozen soil. This chapter is useful to both experienced and novel readers, as a quick reference, as a source for a more detailed bibliography, or as a starting point to study the electromagnetic properties of the different components of the cryosphere.

2.1 Electromagnetic properties of snow

2.1.1 Visible/near-infrared and thermal infrared

Dry snow appears bright to the human eye because of its high and relatively flat reflectance in the visible region (Figure 2.1). Snow reflectance is higher for fresh snow, and it decreases as grain size increases. This is mainly due to the reduction of the number of air-ice interfaces and the consequent reduction of scattering. Snow reflectance is also affected by other factors such as impurities (e.g., soot and dust). Fresh snow albedo can exceed 90% where metamorphosed, or dirty snow albedo can be as low as 20–40%. The presence of liquid water has little direct effect on snow albedo, mostly because of the fact that scattering in the visible and NIR bands continues to occur in presence of the liquid water. Moreover, the absorption of electromagnetic radiation in water is similar to that of ice in the visible and

Remote Sensing of the Cryosphere, First Edition. Edited by M. Tedesco.
© 2015 John Wiley & Sons, Ltd. Published 2015 by John Wiley & Sons, Ltd.
Companion Website: www.wiley.com/go/tedesco/cryosphere

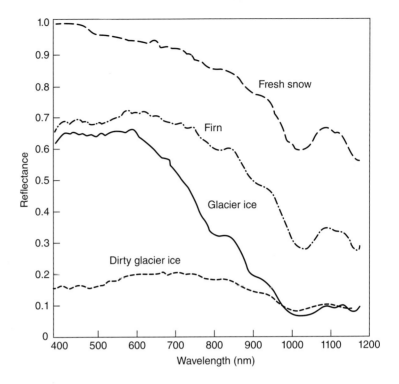

Figure 2.1 Reflectance spectra for different snow and ice surfaces (Hall & Martinec 1985. Adapted from Qunzhu *et al.*, 1984).

NIR regions. Nevertheless, the presence of liquid water within the snowpack impacts grain size growth and metamorphism (e.g., clustering), hence indirectly impacting albedo. Though reflection from a snowpack is anisotropic, with specular scattering due to reflection from ice crystals, most studies neglect this aspect.

Snow reflectance in the thermal infrared is relatively small, with values peaking at about 6–8% below 6 μm, and below or around 1% for wavelengths above 10 μm. Also in this case, the larger the grain, the lower the reflectance. Reflectance peaks are located between 4 and 6 μm, and around 8 μm (see, for example, Figure 4.3 and Rees, 2001).

Emissivity of snow in the thermal infrared ranges between 0.965 and 0.995, with a maximum near 10 μm. More details can be found in Rees (2007), Chapter 4.2.5.

2.1.2 Microwave region

2.1.2.1 Ice permittivity

For pure ice, there is a general agreement that the real part of the dielectric constant is independent of frequency between 100 MHz and 900 GHz, with weak dependence on temperature at microwave frequencies. The situation with respect to the imaginary part of the dielectric constant for ice at microwave frequencies is, on the other hand, different, and it depends on both temperature and frequency.

In this case, reported values may differ by almost half an order of magnitude. In Stogryn (1986), the following model has been developed by combining selected experimental data with certain physical assumptions to compute the real part of dielectric constant of ice:

$$Re(\varepsilon_{ice}) = (3.099T - 992.65)/(T - 318.896) \tag{2.1}$$

where T is the ice temperature.

In Mätzler & Wegmüller (1987), the following expression is also reported:

$$Re(\varepsilon_{ice}) = 3.1884 + 9.1 \cdot 10^{-4}(T - 273) \quad 243 \leq T \leq 273 \tag{2.2}$$

The imaginary part of pure ice in the microwave region can be written as:

$$Im(\varepsilon_{ice}) = \frac{A(T)}{f} + B(T)f \tag{2.3}$$

where f is the frequency in GHz and the coefficients A and B depend only on the ice temperature. Based on a fit of the data in Stogryn, (1986):

$$A(T) = \frac{e^{\left[12.50 - \frac{3.77 \cdot 10^3}{T}\right]}}{T} \tag{2.4}$$

$$B(T) = 10^{-4} Re(\varepsilon_{ice})(273.41 - T)^{-\frac{1}{2}} \tag{2.5}$$

In Hufford (1991), the values for $A(T)$ and $B(T)$ are computed using different data than the ones used in Stogryn (1986). Moreover, the Hufford model does not take into consideration the variation of real part with temperature. The relative formulae are the following:

$$\theta = \frac{300}{273.15 + T} - 1 \tag{2.6}$$

$$A(T) = (5.0415 + 6.1536 \cdot \theta) \cdot 1e^{-3} \cdot e^{.22.1..\theta} \tag{2.7}$$

$$B(T) = \frac{0.502 - 0.131 \cdot \theta}{1 + \theta} \cdot 0.0001 + 0.542 \cdot 1e^{-6} \cdot \left(\frac{1 + \theta}{\theta + 0.0073}\right)^2 \tag{2.8}$$

Among the factors influencing the values of ice permittivity is the presence of impurities. Indeed, it is known that impurities in ice can significantly change the value of both the real and imaginary parts, with major effects on the latter. In particular, Fujita *et al.* (1992) showed that the complex permittivity of doped ice can be expressed as a function of concentration of acid and temperature, and that the possible mechanism enhancing the complex permittivity of acid-doped ice is the formation of liquid phase in ice.

2.1.2.2 Dry snow permittivity

From an electromagnetic point of view, dry snow can be seen as a mixture of ice and air, with permittivity depending on the permittivities of the single constituent materials and on their fractional volume. The real part of dry snow permittivity can be considered constant with frequency and temperature, and it strongly depends on the fractional volume. The empirical relationships between the permittivity and density of dry snow have been widely studied in literature. For example, Hallikainen *et al.* (1986) propose Equations 2.9 and 2.10 for the real part of the dry snow permittivity (ε'_{ds}):

$$\varepsilon'_{ds} = 1 + 1.83\rho_{ds} \qquad\qquad \rho_{ds} \leq 0.5 \text{ g/cm}^3 \qquad\qquad (2.9)$$
$$\varepsilon'_{ds} = 0.51 + 2.88\rho_{ds} \qquad\qquad \rho_{ds} \geq 0.5 \text{ g/cm}^3 \qquad\qquad (2.10)$$

and Mätzler (1987) proposes Equation 2.11:

$$\varepsilon'_{ds} = 1 + (1.58\rho_{ds})/(1 - 0.365\rho_{ds}) \qquad\qquad (2.11)$$

Several empirical and theoretical mixing formulas have also been reported for the imaginary part of the dry snow permittivity (ε''_{ds}) such as, for example Equation 2.12 (Tiuri *et al.*, 1984):

$$\varepsilon_{ds}'' = \varepsilon_{ice}''(0.52\rho_{ds} + 0.62_{ds}) \qquad\qquad (2.12)$$

Sihvola (1999) proposes a mixing formula, generalizing the existing mixing formulas for granular media. Depending on the parameters' choice, the formula can express the Maxwell-Garnett formula, the effective formula for the Polder-Van Santen formula. In Mätzler (1996), several formulae fitting the experimental data are examined, with particular attention to the Polder-Van Santen formula.

2.1.2.3 Wet snow permittivity

Wet snow can be seen as a mixture of dry ice and liquid water, which can appear as free or bounded and is therefore more difficult to characterize electromagnetically than dry snow. Small changes in the distribution and small-scale structure of the water phase can cause large deviations on the wet snow permittivity, because dielectric contrasts are very large in the mixture. In addition, the practical difficulty of measuring the liquid water content with good accuracy is another factor influencing the test of performances of wet snow models.

In simple terms the snow permittivity can be written as:

$$\varepsilon = \varepsilon_{ds} + \Delta\varepsilon \qquad\qquad (2.13)$$

where ε_{ds} is the permittivity of the dry snow and the second term takes into account the effects of the presence of liquid water.

Table 2.1 Empirical relations between $\Delta\varepsilon$ or ε'' and snow wetness (Denoth 1989. Reproduced with permission of Elsevier).

Empirical relation	Remarks
$\Delta\varepsilon = 0.206 * Wv + 0.0046Wv^2$	$0.01 \leq freq \leq 1$ GHz
$\Delta\varepsilon = 0.02 * Wv + (0.06-3.1 * 10^{-4}(freq-4)^2)Wv^{1.5}$	$4 \leq freq \leq 12$ GHz
$\Delta\varepsilon = 0.089Wv + 0.0072Wv^2$	$freq = 1$GHz
$e'' = 0.073 * (e')^{1/2}Wv/freq$	$4 \leq freq \leq 12$ GHz
$e'' = 0.009 * Wv + 0.0007Wv^2$	$freq = 1$ GHz
$e'' = c(freq) * Wv^{1.5}$	

The term $\Delta\varepsilon$ can be, for example, expressed as a combination of powers of the wetness, with coefficients to be fitted from experimental data. As an example, Tiuri *et al.* (1984) obtained the following expression:

$$\Delta\varepsilon = (0.10Wv + 0.80Wv^2)\varepsilon_w \tag{2.14}$$

where Wv is the snow wetness and ε_w is the water permittivity.

Other expressions for $\Delta\varepsilon$ can be found in the literature (see, for example, Frolov and Macheret (1999), Ambach *et al.*, (1972), Sihvola, (1999)). Table 2.1 shows selected formulas relating the increment of permittivity or the imaginary part to snow wetness, as reported in Denoth (1989).

An important observation on the wet snow microstructure regards the separation of *pendular* and *funicular* regimes (Colbeck, 1982). The pendular regime, for low wetness, means that liquid forms separate bubbles and the air is continuous throughout the medium. In the high wetness range – the funicular regime – the liquid is continuous through the pore space and the water can be no longer be approximated as separate inclusions.

Relationships between the transition pendular-funicular regime and wet snow permittivity have been observed (Hallikainen *et al.*, 1986; Sihvola, 1999). The results obtained by Hallikainen (1986) shows that the Debye-like empirical model best describes the dielectric behavior of wet snow as a function of its physical parameters and frequency. Furthermore, if non-symmetry is assumed for the depolarization factors of water inclusions, both two-phase and three-phase Polder-Van Santen models describe the dielectric behavior of snow equally well. Finally, as water appears in snow, the water inclusions become needle-like shaped. A transition in shape occurs around liquid water volume content of 3%, at which point the inclusions become disk-like.

The values of the depolarization factors can be computed by using three approaches:

- first, water particles are assumed to be symmetric and their shape independent from liquid water content;

- second, water particles are assumed to be non symmetric and their shape independent from liquid water content;
- third, water particles are assumed to be non-symmetric and their shape dependent from liquid water content.

Other models such as the Maxwell-Garnett formula (Sihvola, 1999) may be used when modeling the water in wet snow as a thin film surrounding the ice particles.

2.2 Electromagnetic properties of sea ice

2.2.1 Visible/near-infrared and thermal infrared

The optical properties of young sea ice are distinct from other ice types and are mostly governed by the included brine (this being the major driver), air bubbles, growth rate and temperature. Thicker first year ice has a higher reflectance than thinner young gray ice. For the same thickness, full salinity first year ice at very low temperature has the highest reflectance (between ≈0.8 and ≈0.9), which decreases with temperature. Fresh bubbly ice has the lowest reflectance, with values around 0.1 over the entire spectrum. As an example, Figure 2.2a shows a spectral albedo sequence that first-year ice might follow through a melt cycle (Perovich, 1996).

Snow albedo is high and a 0.1 m thick layer of wind-packed snow is sufficient to eliminate any contribution to the albedo from the underlying ice. Albedo decreases as the surface changes from snow-covered ice to cold, bare multiyear ice (curve b in Figure 2.2a) and continues to do so as the ice melts and some of the air voids are filled with water. A more evident wavelength dependency is also observable as the snow is removed. The presence of melt ponds considerable reduces the albedo, with a steep decrease to values below 0.1 for wavelengths above 500 nm.

Spectral albedos for young sea ice grown off of East Antarctica are plotted in Figure 2.2b (Perovich, 1996). Also in this case, the presence of 30 cm of snow on the ice surface masks the optical properties of the ice. The presence of a thin (2 cm) snow layer over young gray ice still increases the albedo but only by a relatively small amount. Albedo values for nilas and open water are also reported in the figure and are generally around 0.1 over the entire spectrum.

The thermal infrared properties of sea ice are similar to those of freshwater ice, with values of about 0.98 (e.g. Rees, 2005).

2.2.2 Microwave region

The electromagnetic properties of sea ice in the microwave region depend on many factors – for example, salinity, ice type, frequency and temperature. Moreover, snow covering the sea ice can alter the microwave electromagnetic response of the target observed by the sensor, with the overall impact depending on frequency

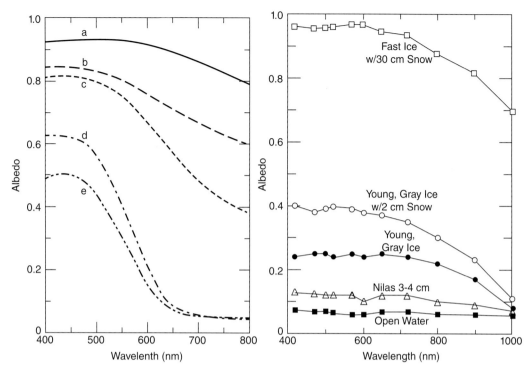

Figure 2.2 (a) Spectral albedos for a possible evolutionary sequence of multiyear ice (Grenfell & Maykut, 1977): a. snow-covered ice; b. cold bare ice; c. melting bare ice; d. early-season melt pond; and e. mature melt pond.

(b) Spectral albedos of Antarctic sea ice (Allison, Brandt & Warren 1993. Reproduced with permission of John Wiley & Sons Ltd).

and snowpack properties (e.g., thickness, grain size, density, temperature, etc.). Because of the dependency on many factors, modeling of permittivity of sea ice in the microwave region is a complex task. In simple terms, the permittivity of sea ice can be approximated by the following linear relationship with the relative brine volume V_b (for $V_b \leq 70\%$; Carsey, 1992):

$$\varepsilon'_{sea\_ice} = A + B \cdot V_b \tag{2.15}$$

$$\varepsilon''_{sea\_ice} = C + D \cdot V_b \tag{2.16}$$

with the coefficients having the following values: A = 3.12, B = 0.009, C = 0.04 and D = 0.005 at 1 GHz, A = 3.05, B = 0.0072, C = 0.02 and D = 0.0033 at 4 GHz, and A = 3, B = 0.012, C = 0.0 and D = 0.01 at 10 GHz. Emissivity values of sea ice and other surfaces at 10.7, 18 and 37 GHz, vertical polarization, are shown in Figure 2.3. Microwave backscattering coefficients for different sea ice conditions are plotted in Figure 2.4.

A detailed treatise of microwave remote sensing of sea ice is reported in Carsey (1992); http://www.agu.org/books/gm/v068/.

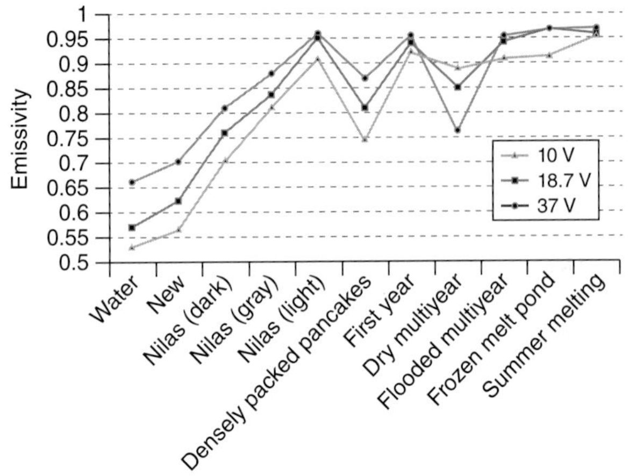

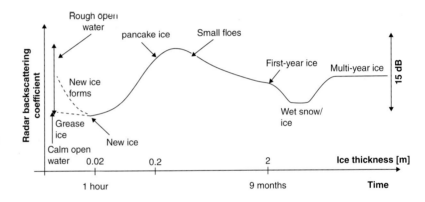

Figure 2.3 Emissivity at 10, 18.7 and 37 GHz for different sea ice and surface conditions. (Data from Carsey 1992).

Figure 2.4 Backscattering coefficient for different sea ice conditions. Modified from: http://Earth.esa.int/applications /data_util/hrisk/ice/ice.htm (Nansen Environmental and Remote Sensing Center).

2.3 Electromagnetic properties of freshwater ice

The electromagnetic properties in the visible and NIR regions of black ice are essentially the same as those of pure ice, where those of white ice are similar to those of snow.

Similarly, thermal infrared emissivity of freshwater ice is similar to that of snow. In the case of microwaves, the real part of the permittivity for lake and river ice is generally assumed to be that of ice. In the case of the imaginary part, Rees (1987) reports the following formula (Equation 2.17) originally proposed by Nyfors (1982):

$$\varepsilon'' = 57.34(f^{-1} + 2.48 \times 10^{-14}f^{0.5})e^{0.0362T} \tag{2.17}$$

where f is the frequency in Hz and T is the temperature in K.

The presence of roughness, and inhomogeneity such as air bubbles, will deviate the values of permittivity from that of ice. For example, ice with air bubbles has a generally lower dielectric constant than ice with no bubbles. More on the electromagnetic and physical properties of lake and river ice are provided in Chapter 12.

2.4 Electromagnetic properties of glaciers and ice sheets

2.4.1 Visible/near-infrared and thermal infrared

During winter, snow covers glaciers and ice sheets. Therefore the spectral properties in the visible and near-infrared regions are driven by the presence of the snowpack overlying the ice. As snow melts during the spring and summer, constructive metamorphism reduces the albedo, and ice or firn can be exposed when the seasonal snow is removed. Spectral properties of exposed ice depend on the presence of impurities and darker materials (e.g., moraine, cryoconite) or surface lakes, which reduce the albedo in the visible and near-infrared regions. Figure 2.1 shows spectral reflectance curves of fresh snow, firn, glacier ice, and dirty glacier ice. Reflectance reduces as snow ages and, for example, goes through several melting/refreezing cycles. Generally, the reflectance of a glacier increases as one moves from its terminus to higher elevations, in view of the reduced presence of darker materials, of the persistency of snow and because of the reduced number of melting/refreezing cycles, if any.

Thermal infrared emissivity of glaciers is driven by the emissivity of snow and ice, as well as that of liquid water.

2.4.2 Microwave region

Microwave properties of glaciers are controlled by snow and ice microwave properties. During the winter, snow covers the glaciated surface and, consequently, microwave properties are driven either by the snow surface or/and by the interaction of the snow cover with the underlying ice layers, depending on the frequency. For relatively high frequency and deep snow, the electromagnetic signal will be mostly due to the snow cover; but, as frequency decreases, penetration depth increases and the microwave response is due to the interaction of the electromagnetic waves with snow and ice and their internal layers. During summer, the glacier surface is either covered by wet snow or melting ice. The presence of liquid water drives the electromagnetic response into the microwave region, in view of the increased imaginary part of wet snow permittivity with respect to that of dry snow and ice.

Microwave brightness temperature during winter is relatively low, and it is impacted by the physical properties of the ice layers and by the snowpack. As snow melts during summer, the recorded brightness temperature increases, as a consequence of the presence of liquid water, which behaves similarly to a black body. This is, among other things, also the foundation of melt detection algorithms from space. In the active case, the snow covering the glacier during the winter has relatively small impact on the backscattering coefficient at the frequencies available from most spaceborne platforms (e.g., C-, X-band).

Recent research has shown the potential of studying snow properties using Ku-band data collected by the QuikSCAT satellite. In general, the presence of wet snow drastically and suddenly decreases the recorded backscattering values, in view of the relatively smooth surface of wet snow and of the dominating surface scattering over the volumetric one. Surface scattering also dominates in the ablation area, and it is mostly controlled by the surface roughness. Over the accumulation area, volumetric scattering dominates, controlled by snow grain dimension, ice layers, lenses, and other internal structures.

2.5 Electromagnetic properties of frozen soil

2.5.1 Visible/near-infrared and thermal infrared

The spectral properties of frozen soil in the visible/near-infrared are driven by the soil composition, and there exist many references in the literature discussing the dependence of spectral signature of soil on moisture content and soil composition. The thermal/infrared spectrum depends on the temperature of the material. Of course, one can expect a change in thermal infrared properties between frozen and unfrozen soil, but this is not a straightforward issue if, for example, a (seasonally) unfrozen active layer covers the permanently frozen ground (permafrost). The intrinsic vertical structure sub-surface nature of the frozen soil properties do not favor the use of surface reflectance studies for direct interpretation of soil properties. Nevertheless, surface reflectance in the visible/near-infrared and thermal regions is used to study features that can be related to the presence or evolution of permafrost or frozen soil.

2.5.2 Microwave region

At microwave frequencies, the dielectric constant of soil-water mixture depends mainly on liquid water fraction. Not all liquid water in soil freezes at or below 0°C, with the fraction of unfrozen water or ice in frozen soil depending on the temperature, salinity, initial moisture, and soil texture. Because of the large dielectric contrast between liquid water and ice permittivity, the amount of liquid water in frozen soil plays a major role in the dielectric constant of frozen soil. Zhang *et al.* (2003) propose an expression for calculating the dielectric constant of frozen soil (ε_{fs}) as:

$$\varepsilon^{\alpha}_{fs} = 1 + (\rho_b/\rho_s)(\varepsilon^{\alpha}_{s} - 1) + m^{\beta}_{vu} \cdot \varepsilon^{\alpha}_{w} - m_{vu} + m_i \varepsilon^{\alpha}_{i} \tag{2.18}$$

where:

m_{vu} is the unfrozen volumetric moisture content
m_i is the volumetric ice content
ρ_b and ρ_s are, respectively, the soil bulk and specific density
ε_s, ε_i and ε_w are, respectively, the soil, ice and water dielectric constants
$\alpha = 0.65$
β is a soil-texture-dependent coefficient.

Mironov & Lukin (2009) report the results of measurement of complex dielectric permittivity of loam soil in the range of temperatures from 25°C to −20°C and frequencies from 0.5 GHz to 15 GHz. Tikhonov (1994) describes the soil permittivity at microwave frequencies as function of its physical parameters. Here, frozen soil is modeled by an air medium containing spherical ice particles and spherical quartz particles covered with bound water film.

Zuerndorfer et al. (1990) use Nimbus 7 SMMR data to map daily freeze/thaw patterns in the upper Midwest for the fall of 1984, while Toll et al. (1999) use passive-microwave satellite data and simulation modeling to assess seasonally frozen soils for north central US and Canada. More recently, Zhang et al. (2003b) use an approach combining a passive microwave remote sensing algorithm and a one-dimensional numerical heat transfer model with phase change to detect the near-surface soil freeze/thaw cycle over snow-free and snow-covered land areas in the contiguous United States.

References

Allison, I., Brandt, R.E. & Warren, S.G. (1993). East Antarctic sea ice: Albedo, thickness distribution and snow cover. *Journal of Geophysical Research* **98**(C7), 12417–12429.

Ambach, W. and Denoth, A. (1972). Studies on the dielectric properties of snow. *Zeitschrift fur Gletscherkunde und Glazialgeologie* **8**, 113–123.

Carsey, F.D. (Ed.) (1992). Microwave Remote Sensing of Sea Ice. *Geophysical Monograph Series* vol. 68, 462 pp., AGU, Washington, DC, doi:10.1029/GM068.

Colbeck, S.C. (1982). The geometry and permittivity of snow at high frequencies, *Journal of Applied Physics* **53**, 4495–4500.

Denoth, A. (1989). Snow dielectric measurements. *Advances in Space Research* **9**(1), 233–243.

Frolov, A.D. & Macheret, Y.Y. (1999). On dielectric properties of dry and wet snow. *Hydrological processes* **13**, 1755–1760.

Fujita, S., Shiraishi, M. & Mae, S. (1992). Measurement on the dielectric properties of acid-doped ice at 9.7 GHz. *IEEE Transactions on Geoscience and Remote Sensing* **30**, 799–803.

Grenfell, T.C. & Maykut, G.A. (1977). The optical properties of ice and snow in the Arctic Basin. *Journal of Glaciology* **18**, 445–463.

Hall & Martinec (1985). *Remote Sensing of Ice and Snow*. Chapman and Hall Ltd., London.

Hallikainen, M., Ulaby, F. & Abdelrazik, M. (1986). Dielectric properties of snow in 3 to 37 GHz range. *IEEE Trans. on Antennas and Propagation* **34**(11), 1329–1340.

Hufford, G. (1991). A model for the complex permittivity of ice at frequencies below 1 THz. *International Journal Of Infrared and MM Waves* **12**(7), 677–680.

Mätzler, C. (1987). Applications of the interaction of microwaves with the natural snow cover. *Remote Sensing Reviews* **2**, 259–387.

Mätzler, C. (1996). Microwave permittivity of dry snow. *IEEE Transactions on Geoscience and Remote Sensing* **34**, 573–581.

Mätzler, C. and Wegmüller, U. (1988). Dielectric Properties of freshwater ice at microwave frequencies. *Journal of Physics D: Applied Physics* **20**, 1623–1630; Errata, Vol. 21, pp. 1660.

Mironov, V.L. and Lukin, Y.I. (2009). Temperature Dependable Microwave Dielectric Model for Frozen Soils. *PIERS Online* **5**(5), 406–410. doi:10.2529/PIERS090220040026.

Nyfors, E. (1982). *On the dielectric properties of dry snow in the 800 MHz to 13 GHz range*. Helsinki: Radio Laboratory, Helsinki University of Technology.

Perovich, D. (1996). The optical properties of sea ice. U.S. CRREL Monography, 96-1, pp. 33, available at http://www.dtic.mil/cgi-bin/GetTRDoc?AD= ADA310586, accessed January 5th, 2013.

Rees, G.W. (2001). *Physical principles of remote sensing*. Cambridge University Press, 360 pp.

Rees, G.W. (2005). *Remote Sensing of Snow and Ice*. CRC Press, 312 pp.

Sihvola, A. (1999). *Electromagnetic mixing formulas and applications*. IEE Publishing, London.

Stogryn, A. (1986). A study of the microwave brightness temperature of snow from the point of view of strong fluctuation theory. *IEEE Transactions On Geoscience and Remote Sensing* **GE-24**(2), 220–231.

Tikhonov, V.V. (1994). *Model of complex dielectric constant of wet and frozen soil in the 1–40 GHz frequency range*. IGARSS '94 Proceedings. Surface and Atmospheric Remote Sensing: Technologies, Data Analysis and Interpretation, Volume 3, pp. 1576–1578.

Tiuri, M.E., Sihvola, A., Nyfors, E.G. & Hallikainen, M.T. (1984). The complex dielectric constant of snow at microwave frequencies. *IEEE Journal of Oceanic Engineering* **9**, 377–382.

Toll, D.L., Owe, M., Foster, J. & Levine, E. (1999). *Monitoring Seasonally Frozen Soils Using Passive Microwave Satellite Data and Simulation Modeling*. IGARSS '99 Proceedings, IEEE 1999, Volume: 2, pp. 1149–115.

Zhang, L., Shi, J., Zhang, Z. & Zhao, K. (2003a). The estimation of dielectric constant of frozen soil-water mixture at microwave bands. Proceedings of IGARSS '03. 2903–2905 vol. 4.

Zhang, T., Armstrong, R.L. & Smith, J. (2003b). Investigation of the near-surface soil freeze-thaw cycle in the contiguous United States: Algorithm development and validation. *Journal of Geophysical Research.* **108**, NO. D22, 8860. doi:10.1029/2003JD003530.

Zuerndorfer, B.W., England, A.W., Dobson, M.C. & Ulaby, F.T. (1990). Mapping freeze/thaw boundaries with SMMR data. *Agricultural and Forest Meteorology* **52**, 199–225.

Acronyms

CCD	Charge-coupled device
CMD	Complementary metal-oxide-semiconductor sensors
ELA	Equilibrium line altitude
FYI	First-year ice
GLAS	Geoscience Laser Altimeter System
GOCE	Gravity Field and Steady-State Ocean Circulation Explorer
GPS	Global Positioning System
GRACE	Gravity Recovery And Climate Experiment
ICESat NASA	Ice, Cloud, and Land Elevation Satellite
IMU	Inertial Measurement Unit
LiDAR	Light detection and ranging
MAAT	Mean annual air temperature
MerCaT	Mercury cadmium telluride type
PRF	Pulse repetition frequency
Radar	Radio Detection and Ranging
SAR	Synthetic Aperture Radar
SIR	Shuttle Imaging Radar
SLAR	Side Looking Airborne Radar
SRTM	Shuttle Radar Topography mission
SSM/I	Special Sensor Microwave Imager
Tb	Brightness temperature

Websites cited

http://www.papainternational.org/history.asp
http://www.sarracenia.com/astronomy/remotesensing/primer0120.html
http://www.sarracenia.com/astronomy/remotesensing/primer0120.html
http://www.papainternational.org/history.asp
http://www.nrcan.gc.ca/Earth-sciences/products-services/satellite-photography-imagery/aerial-photos/about-aerial-photography/884

http://www.fas.org/irp/imint/docs/rst/Sect10/Sect10_1.html
http://www.csr.utexas.edu/grace/
http://www.esa.int/Our_Activities/Observing_the_Earth/GOCE
http://en.wikipedia.org/wiki/Cryosphere
http://www.agu.org/books/gm/v068/
http://www.agu.org/books/gm/v068/

Remote sensing of snow extent

Dorothy K. Hall[1], Allan Frei[2] and Stephen J. Déry[3]

[1] NASA/Goddard Space Flight Center, Greenbelt, USA
[2] Hunter College, City University of New York, USA
[3] University of Northern British Columbia, Prince George, Canada

Summary

Snow plays an important role in the Earth's energy balance because of its high albedo and low thermal conductivity, and seasonal snow cover is a critical water resource in many mountainous regions all over the globe. The high albedo of snow presents a good contrast with most other natural surfaces, often making it readily visible in satellite images.

Since the advent of satellite snow-cover mapping in 1966, many advances have been made, particularly in the mapping of snow-cover extent (SCE) using visible/near-infrared (VNIR) data. Advances in imaging technology have led to improvements in spatial and temporal resolution of snow maps, and in our ability to measure snow albedo. Advancements in both SCE mapping, and measuring snow albedo, continued with the development and launch of the Moderate-resolution Imaging Spectroradiometer (MODIS), first launched in 1999, with its 36 channels and daily or near-daily global coverage. Daily SCE maps are produced from MODIS by an automated algorithm at a spatial resolution of 500 m, with fractional snow cover and daily snow albedo.

Passive-microwave (PM)-derived snow cover and snow water equivalent (SWE) data have been available since the late 1970s, providing daily or near-daily maps irrespective of cloud cover and darkness, though at a rather coarse resolution of ≈25 km.

Progress has been made in recent years by combining VNIR and PM data, allowing improvements in mapping snow cover and SWE, even in mountain environments. Use of both VNIR and PM data in models has also permitted improvements in model predictions of snowmelt runoff.

Continued improvements in both remotely sensed snow cover products and numerical models ensure the expanding use of both to accurately depict land surface and hydrological processes at the watershed to continent scale. This will provide enhanced realism in predicting flooding events caused by rapid snowmelt.

Remote Sensing of the Cryosphere, First Edition. Edited by M. Tedesco.
© 2015 John Wiley & Sons, Ltd. Published 2015 by John Wiley & Sons, Ltd.
Companion Website: www.wiley.com/go/tedesco/cryosphere

3.1 Introduction

Snow was easily identified in the first image obtained from the Television Infrared Operational Satellite-1 (TIROS-1) weather satellite in 1960, because the high albedo of snow presents a good contrast with most other natural surfaces. Subsequently, the National Oceanic and Atmospheric Administration (NOAA) began to map snow using satellite-borne instruments in 1966. Snow plays an important role in the Earth's energy balance, causing more solar radiation to be reflected back into space compared to most snow-free surfaces. Seasonal snow cover also provides a critical water resource, through meltwater emanating from rivers that originate from high-mountain areas such as the Tibetan Plateau. Meltwater from mountain snow packs flows to some of the world's most densely populated areas such as Southeast Asia, benefiting over one billion people (Immerzeel *et al.*, 2010).

In this section, we provide a brief overview of the remote sensing of snow cover using visible and near-infrared (VNIR) and passive microwave (PM) data. Snow can be mapped using the microwave part of the electromagnetic spectrum, even in darkness and through cloud cover, although at a coarser spatial resolution than when using VNIR data. Fusing VNIR and PM algorithms to produce a blended product offers synergistic benefits. Snow-water equivalent (SWE), snow extent, and melt onset are important parameters for climate models and for the initialization of atmospheric forecasts at daily and seasonal time scales. Snowmelt data are also needed as input to hydrological models to improve flood control and irrigation management.

3.2 Visible/near-infrared snow products

In 1966 NOAA began producing weekly snow and ice cover analysis charts for the northern hemisphere that were created manually using a combination of *in situ* and satellite data (Matson and Wiesnet, 1981; Matson *et al.*, 1986). Since 1997, the Interactive Multi-Sensor Snow and Ice Mapping System (IMS) (http://www.natice .noaa.gov/ims/) has produced snow products daily at a spatial resolution of up to 4 km, still utilizing data from various satellites and ground stations (Ramsay, 1998; Helfrich *et al.*, 2007). The key feature that distinguishes the IMS maps from other standard snow products is human involvement in the analysis. IMS analysts map snow extent across the globe every day, regardless of the presence of clouds, for input as boundary conditions into weather forecasting models.

Snow-cover maps generated by the Rutgers University Global Snow Lab (http://climate.rutgers.edu/snowcover/) are considered a climate-data record (CDR). The dataset consisting of the Rutgers snow maps, based on NOAA satellite data, constitutes a >45-year record of snow-cover extent for the Northern Hemisphere, and is used in many climate studies (e.g. Robinson *et al.*, 1993; Frei and Robinson, 1999; Frei *et al.*, 1999; Brown, 2000; Robinson and Frei, 2000;

Déry and Brown, 2007; Brown *et al.*, 2010; Brown and Robinson, 2011; Frei *et al.*, 2012). Some inconsistencies that were discovered in the early part of the NOAA data record (before 1972) were corrected before producing the Rutgers CDR.

The advent of the Landsat series of sensors, beginning with the launch of Landsat-1 in 1972, heralded a new era of producing multispectral images from which snow maps could be created at 80 m spatial resolution (Rango and Martinec, 1979). With Landsat data came the ability to create detailed basin-scale snow-cover maps when cloud cover permitted. Landsats-4 and -5, launched on 16 July 1982 and 1 March 1984, respectively, each carried a Thematic Mapper (TM) sensor with a spatial resolution of 30 m, and Landsat-7 carries an Enhanced Thematic Mapper Plus (ETM+) with spatial resolution of 30 m and the panchromatic band with a spatial resolution of 15 m. Landsat-8, launched on 11 February 2013, carries the Operational Land Imager (OLI) and the Thermal Infrared Sensor (TIRS), with a spatial resolution up to 15 m and 100 m resolution, respectively. Although the Landsat series has provided high-quality, scene-based snow maps, the 16- or 18-day repeat-pass interval of the Landsat satellites is not adequate for most snow-mapping requirements, especially during spring snowmelt, when changes can occur over daily timescales.

In December 1999, NASA launched the Terra satellite as part of the Earth Observing System (EOS). The first of two Moderate Resolution Imaging Spectroradiometers (MODIS) was the flagship instrument of the five-instrument Terra payload. A second MODIS instrument was launched on the Aqua satellite in May of 2002. Since early 2000, snow cover has been mapped daily using MODIS data (Figure 3.1), with automated algorithms developed at Goddard Space Flight Center (GSFC) in Greenbelt, Maryland (http://modis-snow-ice.gsfc.nasa.gov); Riggs *et al.*, 2006; Hall and Riggs, 2007).

The standard MODIS snow map products provide global, daily coverage at 500 m and 0.05° resolution, and include fractional snow cover (Salomonson and Appel, 2004) and a daily cloud-gap filled product (Hall *et al.*, 2010). Daily snow albedo is also provided (Klein and Stroeve, 2002). Improvements to the algorithm are incorporated each time the entire dataset is reprocessed (Riggs *et al.*, 2012).

With more than 12 years of MODIS snow-cover maps now available, many time-series studies are being undertaken. For example, focusing on Southeast Asia, Immerzeel *et al.* (2009) used MODIS standard snow-cover maps to show a significant negative winter snow-cover trend in the Indus River basin, for the period 2000–2008; a similar trend for 2000–2006 was found in the western Himalaya region by Prasad and Singh (2007). Meltwater from snow in the mountains is extremely important in the Indus River basin for water resources. Using only seven years of MODIS snow-cover data, Pu *et al.* (2007) found large spatial variability between years in the amount of snow cover over the Tibetan Plateau, with a possible slight decrease in snow cover over the time period from 2000–2006.

For the reader's convenience, we report in the following a summary of an integral part of the algorithm that is used to generate the operational snow cover product from MODIS, the so-called Normalized Difference Snow Index (NDSI).

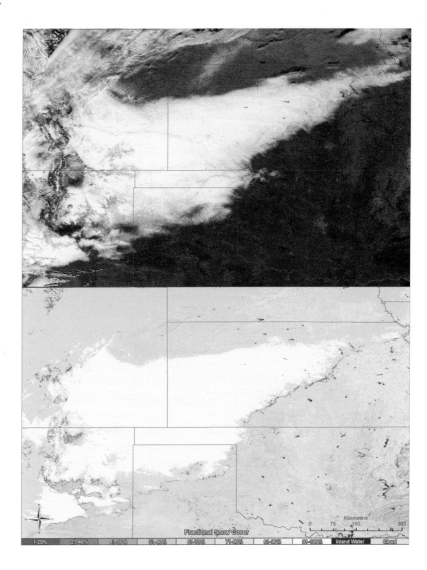

Figure 3.1 MODIS Aqua image (top) and snow map (bottom), acquired on 23 December 2011.

3.2.1 The normalized difference snow index (NDSI)

Snow is highly reflective in the visible part of the electromagnetic spectrum and highly absorptive in the near-infrared or short-wave infrared part of the spectrum, whereas the reflectance of most clouds remains high, allowing good separation of most clouds and snow. The NDSI is the normalized difference of a visible and a NIR or short-wave infrared (SWIR) band. The NDSI has a long history, as described in Hall and Riggs (2011). The use of band ratios of visible and NIR or short-wave infrared (SWIR) channels to separate snow and clouds was documented in the literature beginning in the1970s and early 1980s (e.g., Valovcin, 1976; Kyle *et al.*, 1978; Bunting and d'Entremont, 1982). Band ratios as applied to Landsat Thematic Mapper (TM) data for mapping snow was developed by Dozier (1987) and Dozier and Marks (1987). Later, methods that relied on the VIS/NIR ratio were refined substantially using satellite data, by Crane and Anderson (1984), Dozier (1989), and Rosenthal and Dozier (1996) for regional scales, and by Riggs *et al.* (1993), Hall *et al.* (1995, 2002), and Hall and Riggs (2007) for automated global snow cover mapping.

With the launch of MODIS at the end of the 1990s, a computationally-conservative global snow-mapping algorithm needed to be developed that would perform automatically. Using the heritage algorithms discussed above, Riggs *et al.* (1993) and Hall *et al.* (1995) introduced a snow-mapping algorithm that serves as the basis of the current MODIS standard snow-mapping product. The MODIS snow-mapping algorithm uses a normalized difference between MODIS band 4 and 6, and employs several spectral tests. A planetary reflectance below or equal to 11% is a threshold test in which values smaller than 11% were determined not to be snow.

The prototype MODIS snow-mapping algorithm was improved with additional spectral tests. One key modification is that the NDSI threshold was changed in forested areas, based on results of a canopy reflectance model (Klein *et al.*, 1998), using both the Normalized Difference Vegetation Index (NDVI) (Tucker, 1979) and NDSI in densely-forested areas, as determined from the NDVI test. Following the 1999 launch of the MODIS on the Terra spacecraft, the snow algorithm was modified several times, but the NDSI has remained the basis of the algorithm (see Riggs *et al.*, 2006). The automated MODIS snow-mapping algorithm uses at-satellite reflectance in MODIS bands 4 (0.545 – 0.565 μm) and 6 (1.628 – 1.652 μm) to calculate the NDSI (Hall *et al.*, 1995):

$$\text{NDSI} = \frac{(\text{band4} - \text{band6})}{(\text{band4} + \text{band6})} \qquad (3.1)$$

A pixel in a non-densely forested region will be mapped as snow if the NDSI is ≥0.4 and reflectance in MODIS band 2 (0.841 μm – 0.876 μm) is >11%. However, if the MODIS band 4 reflectance is <10%, then the pixel will not be mapped as snow, even if the other criteria are met. This is required because very low reflectances cause the denominator in the NDSI to be quite small, and only small increases in the visible wavelengths are required to make the NDSI value high enough to classify a pixel, erroneously, as snow. In dense forests, snow cover tends to lower the NDVI, so MODIS bands 1 (0.620 – 0.670 μm) and 2 (0.841 – 0.876 μm) are used to calculate the NDVI.

$$\text{NDVI} = \frac{(\text{band2} - \text{band1})}{(\text{band2} + \text{band1})} \qquad (3.2)$$

Several important refinements of the original algorithm have been implemented. Please see Riggs and Hall (in press) and the MODIS snow and ice mapping project website (http://modis-snow-ice.gsfc.nasa.gov/) for the most current algorithm refinements.

3.3 Passive microwave products

Historical PM measurements are available from the Electrically-Scanning Microwave Radiometer (ESMR) beginning in December 1972, the Scanning Multichannel Microwave Radiometer (SMMR) instrument (1978 through 1987), and the Special Sensor Microwave / Imager (SSM/I) instrument (1987 through to the present), although some compatibility issues between snow products from the two instruments exist (Armstrong and Brodzik, 2001; Derksen and Walker, 2003; Brodzik *et al.*, 2007). The NASA EOS Aqua platform also provided science data from the Advanced Microwave Scanning Radiometer – Earth Observing System (AMSR-E) from 2002 until the instrument failed in October 2011 (Kelly *et al.*, 2003; Derksen *et al.*, 2005; Tedesco *et al.*, 2011), after outliving its expected lifetime of three years.

PM imaging does not depend on the presence of sunlight. Radiation is largely (but not completely) transmitted through non-precipitating clouds, offering the potential to estimate snow cover under conditions that preclude the acquisition of VNIR observations.

Snow is efficient at scattering microwave radiation emitted from the Earth's surface, since snow grain dimensions are often similar to microwave wavelengths, resulting in a diminished signal measured at the satellite from snow-covered surfaces (Chang *et al.*, 1976; Schanda *et al.*, 1983; Matzler, 1994). Microwave scattering by ice crystals is frequency-dependent, enabling the use of two or more bands to estimate SWE (Chang *et al.*, 1987; Grody and Basist, 1996), although other methods have been evaluated such as one based on the inversion of a snow emission model (e.g., Pulliainen and Hallikainen, 2001). A thorough discussion of SWE retrieval from space appears in Chapter 5 of this book.

Limitations to the monitoring of snow extent and SWE using PM sensors result from a variety of factors, including the presence of liquid water, ice lenses, grain size variations such as those resulting from depth hoar (Hall *et al.*, 1986), and vertical heterogeneities within the snowpack (Chang *et al.*, 1996; Foster *et al.*, 1997 and 2005; Tedesco *et al.*, 2005).

Due to the inherent difficulties and regional variations in the interpretation of PM signals, the production of a data set that is consistently accurate across all Northern Hemisphere regions requires either:

1 a physical approach, which includes robust representations of snowpack processes and their parameterization in retrieval schemes (e.g., Pulliainen and Hallikainen, 2001); or

2 a regional approach, which includes regionally-tuned algorithms (Foster *et al.*, 1997) that statistically represent regional snowpack processes but are not applicable in different snow accumulation regimes.

3.4 Blended VNIR/PM products

Data from different satellite instruments, ground observations and models can be combined to enhance information content about snow cover. A recent example of a blended product is the Air Force Weather Agency (AFWA)/National Aeronautics and Space Administration (NASA) Snow Algorithm (ANSA) blended, global snow product (Foster *et al.*, 2011), which utilizes MODIS and AMSR-E (Kelly *et al.*, 2003; Tedesco *et al.*, 2011) standard data products (Figure 3.2).

This blended product improves mapping of snow cover extent, fractional snow cover, SWE, the onset of snowmelt, and actively melting snow covers (Foster *et al.*, 2011). The blended snow products have been produced at a spatial resolution of 5 or 25 km, but the inherent resolution of the AMSR-E precludes the SWE or snow cover (during cloudy periods), being finer than 25 km, though snow cover can be mapped by MODIS at 5 km (or finer) resolution under clear skies. The product accuracy has been assessed in the lower Great Lakes region of the US (Hall *et al.*, 2012), in the mountains in Turkey (Akyurek *et al.*, 2011), and in Finland (Casey *et al.*, 2008). The confidence for mapping snow cover extent is greater with the MODIS product than with the microwave product when cloud-free MODIS observations are available, so therefore the MODIS product is used as the default for detecting snow cover. Other examples of combined products include the Canadian Meteorological Centre (CMC) snow product (Brasnett, 1999) and GlobSnow (Pulliainen, 2006; Takala *et al.*, 2011; Tait *et al.*, 2000; Derksen *et al.*, 2004; Biancamaria *et al.*, 2011).

3.5 Satellite snow extent as input to hydrological models

Over the past decade, there have been rapid developments in land surface and hydrological modeling. With the amplification of climate change in polar (Serreze and Francis, 2006; Déry and Brown, 2007) and alpine (Bradley *et al.*, 2004) regions driven in part by the snow/ice-albedo feedback, there is renewed interest in improving the representation of snow in land surface and hydrological models. To this end, the assimilation of remotely-sensed snow cover extent by numerical models provides a strong constraint in retrospective or (near) real-time simulations of land surface and hydrological processes at the watershed to continental scale.

For example, Déry *et al.* (2005) improved their snow and streamflow simulations for the Kuparuk River Basin of Alaska's North Slope by constraining the Catchment-based Land Surface Model (CLSM) (Koster *et al.*, 2000) with snow areal depletion curves from MODIS. Andreadis and Lettenmaier (2006) used an Ensemble Kalman Filter (EnKF) to assimilate MODIS snow cover extent data into the Variable Infiltration Capacity (VIC) macroscale hydrological model with application to the Snake River Basin of the western United States. Sun *et al.* (2004) demonstrated the feasibility of using an EnKF to assimilate remotely-sensed snow cover extent data for simulations of continental-scale land surface conditions.

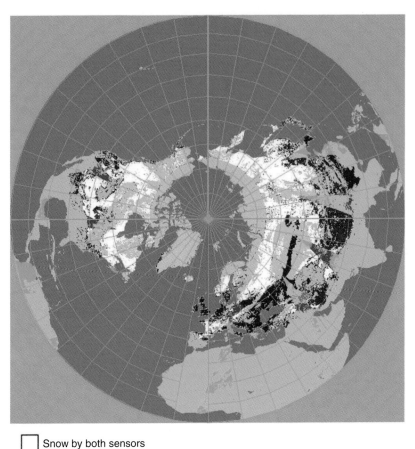

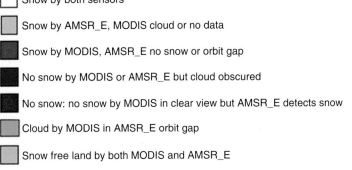

Figure 3.2 ANSA blended-snow product for 26 January 2007 in the Lambert Azimuthal polar projection (adapted from Foster *et al.*, 2011).

Snow by both sensors

Snow by AMSR_E, MODIS cloud or no data

Snow by MODIS, AMSR_E no snow or orbit gap

No snow by MODIS or AMSR_E but cloud obscured

No snow: no snow by MODIS in clear view but AMSR_E detects snow

Cloud by MODIS in AMSR_E orbit gap

Snow free land by both MODIS and AMSR_E

MODIS snow cover products have also been used to develop statistical models that predict the timing of the spring freshet for the Quesnel River of western Canada (Figure 3.3; Tong *et al.*, 2009) and for the Upper Euphrates River of Turkey (Akyurek *et al.*, 2011). Other recent efforts have explored the potential to assimilate fractional snow coverage information from MODIS to reconstruct SWE in Sierra Nevada watersheds (Rice *et al.*, 2011) and for snowmelt energy balance modeling in Alaska (Homan *et al.*, 2011).

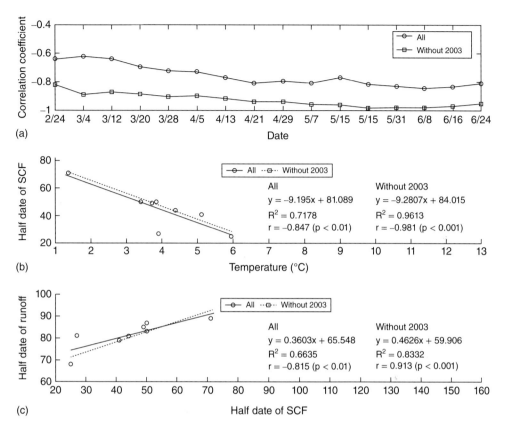

Figure 3.3 (a) Correlation coefficients and (b) scatter plots between average air temperature within different periods and Julian day of a snow cover fraction of 50% ($SCF_{50\%}$), and (c) between Julian day of $SCF_{50\%}$ and normalized accumulated runoff ($R_{50\%}$) during snow ablation seasons 2000–2007 in the Quesnel River Basin (QRB) of western Canada (Tong et al. 2009). The scatter plots reveal an inverse linear relationship between spring air temperatures and the date for which half of the QRB remains snow-covered, with the latter preceding the freshet by about a month.

Rodell and Houser (2004) used MODIS snow-cover maps in the Mosaic land surface model, driven by the NASA/NOAA Global Land Data Assimilation System (GLDAS), to directly correct SWE fields when there was a mismatch between observations and model predictions. A suite of data assimilation experiments using the Land Information System (LIS) framework of Kumar et al. (2008) was also performed. Output from this simulation was compared to that from a control (not updated) simulation, and both were assessed using a conventional snow-cover product and data from ground-based observation networks over the continental US. Output from the updated simulations generally displayed more accurate snow coverage and compared more favorably with *in situ* snow time series than did the models without assimilation of MODIS snow-cover maps, as reported in Hall et al. (2010).

3.6   Concluding remarks

Since the advent of satellite snow-cover mapping in the 1960s, many advances have been made, particularly in the mapping of snow-cover extent using VNIR data. Rapid advances in imaging technology have led to improvements in spatial and temporal resolution of snow maps. Furthermore, the ability to measure snow albedo has advanced, as increasingly sophisticated Landsat instruments with higher dynamic range (to capture reflectance changes within a snow cover) were launched. Advancements continued with the development and launch of the MODIS instrument, first launched in 2000, with its 36 channels and daily or near-daily global coverage of snow-prone areas. An important outcome of the Landsat heritage is daily snow-cover maps that are now produced by an automated algorithm at a spatial resolution of 500 m, with fractional snow cover and daily snow albedo derived from MODIS.

PM-derived snow-cover and SWE maps have been available since the late 1970s, providing daily or near-daily maps, irrespective of cloud cover and darkness at a rather coarse resolution of ≈ 25 km. The technology of satellite-borne PM sensors has not advanced as rapidly as has the technology of VNIR instruments, largely because of the difficulty in having a large enough microwave antenna in space to significantly improve the spatial resolution of PM-derived snow maps. Consequently, it has been challenging to effect major improvements in maps of snow cover and SWE derived from PM data. Until the spatial resolution of PM sensors flown on-board satellites can be improved significantly, our ability to map SWE will not make giant leaps forward.

Progress, however, has been made in recent years by combining VNIR and PM data, allowing improvements in mapping snow cover and SWE, even in mountain environments. The use of both VNIR and PM data in models has also permitted improvements in model predictions of snowmelt runoff.

Future prospects suggest that techniques for snow-cover data assimilation are likely to be improved and optimized, allowing (near) real-time land surface and hydrological simulations. This will provide enhanced modeling of potential flooding events caused by rapid snowmelt. Continued improvements in both remotely sensed snow-cover products and numerical models ensure the expanding use of both to accurately depict land surface and hydrological processes at the watershed-to-continent scale.

Acknowledgments

The authors thank Jody Hoon-Starr/SSAI for image processing of Figure 3.1, and Jinjun Tong/UCLA for production of Figure 3.3. This work was supported, in part, by NASA's Cryospheric Sciences Program.

References

Akyurek, Z., Surer, S. & Beser, Ö. (2011). Investigation of the snow-cover dynamics in the Upper Euphrates Basin of Turkey using remotely sensed snow-cover products and hydrometeorological data. *Hydrological Processes* **25**, 3637–3648.

Andreadis, K.M. & Lettenmaier, D.P. (2006). Assimilating remotely sensed snow observations into a macroscale hydrology model. *Advances in Water Resources* **29**, 872–886.

Armstrong, R.L. & Brodzik, M.J. (2001). Recent Northern Hemisphere Snow Extent, A Comparison of Data Derived from Visible and Microwave Satellite Sensors. *Geophysical Research Letters* **28**(19), 3673–3676.

Biancamaria, S., Cazevave, A., Mognard, N.M., Llovel, W. & Frappart, F. (2011). Satellite-based high latitude snow volume trend, variability and contribution to sea level over 1989/2006. *Global and Planetary Change* **75**, 99–107.

Bradley, R.S., Keimig, F.T. & Diaz, H.F. (2004). Projected temperature changes along the American cordillera and the planned GCOS network. *Geophysical Research Letters* **31**, L16210, doi:10.1029/2004GL020229.

Brasnett, B. (1999). A Global Analysis of Snow Depth for Numerical Weather Prediction. *Journal of Applied Meteorology* **38**(6), 726–740.

Brodzik, M.J., Armstrong, R.A. & Savoie, M. (2007). Global EASE-Grid 8-day Blended SSM/I and MODIS Snow Cover. From http://nsidc.org/data/docs/daac/nsidc0321_8day_ssmi_modis_blend/index.html.

Brown, R., Derksen, C. & Wang, L. (2010). A multi-data set analysis of variability and change in Arctic spring snow cover extent, 1967–2008. *Geophysical Research Letters* **115**(D16111), 16pp, doi:10.1029/2010JD013975.

Brown, R.D. (2000). Northern Hemisphere snow cover variability and change, 1915–1997. *Journal of Climate* **13**, 2339–2355.

Brown, R.D. & Robinson, D.A. (2011). Northern Hemisphere spring snow cover variability and change over 1922–2010 including an assessment of uncertainty. *The Cryosphere* **5**, 219–229.

Bunting, J.T. & d'Entremont, R.P. (1982). *Improved cloud detection utilizing Defense Meteorological Satellite Program near infrared measurements.* Air Force Geophysics Laboratory, Environmental Research Papers No. 765, AFGL-TR-82-0027, 27 January 1982, 91.

Casey, K.A., Kim, E., Hallikainen, M.T., Foster, J.L., Hall, D.K. & Riggs, G.A. (2008). *Validation of the AFWA-NASA blended snow-cover product in Finland, 2006–2007.* Proceedings of the 65th Eastern Snow Conference, Fairlee, VT, 28–30 May 2008.

Chang, A.T.C., Gloersen, P., Schmugge, T.J., Wilheit, T. & Zwally, H.J. (1976). Microwave emission from snow and glacier ice. *Journal of Glaciology* **16**(74), 23–39.

Chang, A.T.C., Foster, D.R. & Hall, D.K. (1987). Nimbus-7 SMMR derived global snow cover parameters. *Annals of Glaciology* **9**, 39–44.

Chang, A.T.C., Foster, D.R. & Hall, D.K. (1996). Effects of forest on the snow parameters derived from microwave measurements during the BOREAS winter field campaign. *Hydrological Processes* **10**, 1565–1574.

Crane, R.G. & Anderson, M.R. (1984). Satellite discrimination of snow/cloud surfaces. *International Journal of Remote Sensing* **5**(1), 213–223.

Derksen, C. & Walker, A.E. (2003). Identification of systematic bias in the cross-platform (SMMR and SMM/I) EASE-Grid brightness temperature time series. *IEEE Transactions on Geoscience and Remote Sensing* **41**(4), 910–915.

Derksen, C., Brown, R. & Walker, A.E. (2004). Merging conventional (1915–92) and passive microwave (1978–2002) estimates of snow extent and water equivalent over central North America. *Journal of Hydrometeorology* **5**, 850–861.

Derksen, C., Walker, A. & Goodison, B. (2005). Evaluation of passive microwave snow water equivalent retrievals across the boreal forest/tundra transition of western Canada. *Remote Sensing of Environment* **96**, 315–327.

Déry, S.J. & Brown, R.D. (2007). Recent Northern Hemisphere snow cover extent trends and implications for the snow-albedo feedback. *Geophysical Research Letters* **34**, L22504. doi: 10.1029/2007GL031474.

Déry, S.J., Salomonson, V.V., Stieglitz, M., Hall, D.K. & Appel, I. (2005). An approach to using snow areal depletion curves inferred from MODIS and its application to land surface modelling in Alaska. *Hydrological Processes* **19**, 2755–2774. doi: 10.1002/hyp.5784.

Dozier, J. (1987). *Remote sensing of snow characteristics in the southern Sierra Nevada, Large Scale Effects of Seasonal Snow Cover.* Proceedings of the Vancouver Symposium August 1987, IAHS, 166, 305–314.

Dozier, J. (1989). Spectral signature of alpine snow cover from the LANDSAT Thematic Mapper. *Remote Sensing of Environment* **28**, 9–22.

Dozier, J. & Marks, D. (1987). Snow mapping and classification from LANDSAT Thematic Mapper data. *Annals of Glaciology* **9**, 1–7.

Foster, J.L., Chang A.T.C. & Hall D.K. (1997). Comparison of snow mass estimates from a prototype passive microwave snow algorithm, a revised algorithm and a snow depth climatology. *Remote Sensing of Environment* **62**, 132–142.

Foster, J.L., Sun, C., Walker, J.P., *et al.* (2005). Quantifying the uncertainty in passive microwave snow water equivalent observations. *Remote Sensing of Environment* **94**, 187–203.

Foster, J.L., Hall, D.K., Eylander, J.B., *et al.* (2011). A blended global snow product using visible, passive microwave and scatterometer data. *International Journal of Remote Sensing* **32**(5–6), 1371–1395.

Frei, A. & Robinson, D.A. (1999). Northern hemisphere snow extent, regional variability 1972–1994. *International Journal of Climatology* **19**, 1535–1560.

Frei, A., Robinson D.A. & Hughes, M.G. (1999). North American snow extent, 1900–1994. *International Journal of Climatology* **19**, 1517–1534.

Frei, A., Tedesco, M., Lee, S., *et al.* (2012). A review of current-generation satellite-based snow products. *Advances in Space Research* **50**(8), 1007–1029. doi: 10.1016/j.asr.2011.12.021.

Grody, N.C. & Basist, A.N. (1996). Global identification of snowcover using SSM/I measurements. *IEEE Transactions on Geoscience and Remote Sensing* **34**(1), 237–249.

Hall, D.K. & Riggs, G.A. (2007). Accuracy assessment of the MODIS snow-cover products. *Hydrological Processes* **21**(12), 1534–1547. doi:10.1002/hyp.6715.

Hall, D.K. & Riggs, G.A. (2011). Normalized-Difference Snow Index (NDSI). In: Singh, V.P., Singh, P. & Haritashya, U.K. (eds.). Encyclopedia of Earth Sciences Series, *Encyclopedia of Snow, Ice and Glaciers*. Springer Science, doi:10.1007/978-90-481-2642-2_376.

Hall, D.K., Chang, A.T.C. & Foster, J.L. (1986). Detection of the depth hoar layer in the snowpack of the Arctic Coastal Plain of Alaska, USA, using satellite data. *Journal of Glaciology* **32**(1), 87–94.

Hall, D.K., Riggs, G.A. & Salomonson, V.V. (1995). Development of methods for mapping global snow cover using Moderate Resolution Imaging Spectroradiometer (MODIS) data. *Remote Sensing of Environment* **54**, 127–140.

Hall, D.K., Riggs, G.A., Salomonson, V.V., DiGirolamo, N.E. & Bayr, K.J. (2002). MODIS snow-cover products. *Remote Sensing of Environment* **83**, 181–194.

Hall, D.K., Riggs, G.A., Foster, J.L. & Kumar, S. (2010). Development and validation of a cloud-gap filled MODIS daily snow-cover product. *Remote Sensing of Environment* **114**, 496–503. doi:10.1016/j.rse.2009.10.007.

Hall, D.K., Foster, J.L., Kumar, S., Chien, J.Y.L. & Riggs, G.A. (2012). Evaluation of the AFWA-NASA (ANSA) snow product in the Lower Great Lakes Region. *Hydrology and Earth Systems Sciences Discussions* **9**, 1141–1161.

Helfrich, S.R., McNamara, D., Ramsay, B.H., Baldwin, T. & Kasheta, T. (2007). Enhancements to, and forthcoming developments in the Interactive Multisensor Snow and Ice Mapping System (IMS). *Hydrological Processes* **21**, 1576–1586.

Homan, J.W., Luce, C.H., McNamara, J.P. & Glen, N.P. (2011). Improvement of distributed snowmelt energy balance modeling with MODIS-based NDSI-derived fractional snow-covered area data. *Hydrological Processes* **25**, 650–660, doi:10.1002/hyp.7857.

Immerzeel, W.W., Droogers, P., de Jong, S.M. & Bierkens, M.F.P. (2009). Large-scale monitoring of snow cover and runoff simulation in Himalayan river basins using remote sensing. *Remote Sensing of Environment* **113**, 40–49.

Immerzeel, W.W., van Beek, L.P.H. & Bierkens, M.F.P. (2010). Climate change will affect the Asian water towers. *Science* **328**, 1382–1385.

Kelly, R.E.J., Chang, A.T.C., Tsang, L. & Foster, J.L. (2003). Development of a prototype AMSR-E global snow area and depth algorithm. *IEEE Transactions on Geoscience and Remote Sensing* **41**(2), 230–242.

Klein, A.G. & Stroeve, J. (2002). Development and validation of a snow albedo algorithm for the MODIS instrument. *Annals of Glaciology* **34**, 45–52.

Klein, A.G., Hall, D.K. & Riggs, G.A. (1998). Improving snow-cover mapping in forests through the use of a canopy reflectance model. *Hydrological Processes* **12**, 1723–1744.

Koster, R.D., Suarez, M.J., Ducharne, A., Stieglitz, M. & Kumar, P. (2000). A catchment-based approach to modeling land surface processes in a general circulation model, 1. Model structure. *Journal of Geophysical Research* **105**(D20), 24809–24822.

Kyle, H.L., Curran, R.J., Barnes, W.L. & Escoe, D. (1978). *A cloud physics radiometer*. Third Conference on Atmospheric Radiation, American Meteorological Society, 28–30 June 1978, Davis, Calif., 107.

Matson, M. & Wiesnet, D.R. (1981). New data base for climate studies. *Nature* **289**, 451–456.

Matson, M., Roeplewski, C.F. & Varnadore, M.S. (1986). *An Atlas of Satellite-Derived Northern Hemisphere Snow Cover Frequency*. National Weather Service, Washington, DC, 75.

Matzler, C. (1994). Passive microwave signatures of landscapes in winter. *Meteorological and Atmospheric Physics* **54**, 241–260.

Prasad, A.K. & Singh, R.P. (2007). Changes in Himalayan Snow and Glacier Cover Between 1972 and 2000. *Eos Transactions of the American Geophysical Union* **88**(33), 326.

Pu, Z., Xu, L. & Salomonson, V.V. (2007). MODIS/Terra observed seasonal variations of snow cover over the Tibetan Plateau. *Geophysical Research Letters* **34**, L06706. doi:10.1029/2007GL029262.

Pulliainen, J. (2006). Mapping of snow water equivalent and snow depth in boreal and sub-Arctic zones by assimilating space-borne microwave radiometer data and ground-based observations. *Remote Sensing of Environment* **101**(2), 257–269.

Pulliainen, J. & Hallikainen, M. (2001). Retrieval of regional snow water equivalent from space-borne passive microwave observations. *Remote Sensing of Environment* **75**(1), 76–85.

Ramsay, B.H. (1998). The interactive multisensor snow and ice mapping system. *Hydrological Processes* **12**, 1537–1546.

Rango, A. & Martinec, J. (1979). Application of a snowmelt-runoff model using LANDSAT data. *Nordic Hydrology* **10**, 225–238.

Rice, R., Bales, R.C., Painter, T.H. & Dozier, J. (2011). Snow water equivalent along elevation gradients in the Merced and Tuolumne River basins of the Sierra Nevada. *Water Resources Research*, W08515. doi:10.1029/2010WR009278.

Riggs, G.A. & Hall, D.K. (in press). *MODIS Snow cover Algorithms and Products – Plans for Next Version*. Proceedings of the 68th Annual Eastern Snow Conference, Montreal, QC, Canada.

Riggs, G.A., Hall, D.K., Barker, J.L. & Salomonson, V.V. (1993). *The Developing Moderate Resolution Imaging Spectroradiometer (MODIS) Snow Cover Algorithm*. Proceedings of the 50th Annual Eastern Snow Conference. 8–10 June. Quebec, Canada: Quebec City, 51–58.

Riggs, G.A., Hall D.K. & Salomonson, V.V. (2006). *MODIS Snow Products Users' Guide*. from http://modis-snow-ice.gsfc.nasa.gov/sugkc2.html.

Riggs, G.A. & Hall, D.K. (2012). *MODIS snow-mapping algorithm changes for Collection*. Proceedings of the 69th Eastern Snow Conference, 5–7 June 2012, Claryville, NY.

Robinson, D.A., Dewey K.F. & Heim, R.R.J. (1993). Global Snow Cover Monitoring, An Update. *Bulletin of the American Meteorological Society* **74**(9), 1689–1696.

Robinson, D.A. & Frei, A. (2000). Seasonal variability of Northern Hemisphere snow extent using visible satellite data. *Professional Geographer* **51**, 307–314.

Rodell, M. & Houser, P.R. (2004). Updating a Land Surface Model with MODIS-Derived Snow Cover. *Journal of Hydrometeorology* **5**, 1064–1075.

Rosenthal, W. & Dozier, J. (1996). Automated mapping of montane snow cover at subpixel resolution from the LANDSAT Thematic Mapper. *Water Resources Research* **32**(1), 115–130.

Salomonson, V.V. & Appel, I. (2004). Estimating fractional snow coverage from MODIS using the Normalized Difference Snow Index (NDSI). *Remote Sensing of Environment* **89**, 351–360.

Schanda, E., Matzler, C. & Kunzi, K. (1983). Microwave remote sensing of snow cover. *International Journal of Remote Sensing* **4**, 149–158.

Scherer, D., Hall, D.K., Hochschild, V., *et al.* (2005). Remote sensing of snow cover. In: Duguay, C.R. & Pietroniro, A. (eds). *Remote Sensing in Northern Hydrology, Measuring Environmental Change*. Geophysical Monograph 163, American Geophysical Union, Washington, DC.

Serreze, M.C. & Francis, J.A. (2006). The Arctic amplification debate. *Climatic Change* **76**, 241–264. doi: 10.1007/s10584-005-9017-y.

Sun, D., Walker, J.P. & Houser, P.R. (2004). A methodology for snow data assimilation in a land surface model. *Journal of Geophysical Research* **109**, D08108. doi: 10.1029/2003JD003765.

Tait, A.B., Hall, D.K., Foster, J.L. & Armstrong, R.L. (2000). Utilizing multiple datasets for snow-cover mapping. *Remote Sensing of Environment* **72**, 111–126.

Takala, M., Luojus, K., Pulliainen, J., Derksen, C., Lemmetyinen, J., Kärnä, J.-P., Koskinen, J. & Bojkov, B. (2011). Estimating northern hemisphere snow water equivalent for climate research through assimilation of space-borne radiometer data and ground-based measurements. *Remote Sensing of Environment* **115**(12), 3517–3529.

Tedesco, M., Kim, E.J., Gasiewski, A. & Stankov, B. (2005). Analysis of multi-scale radiometric data collected during the Cold Land Processes Experiment-1 (CLPX-1). *Geophysical Research Letters* **32**(L18501), 4.

Tedesco, M., Kelly, R.E.J., Foster, J.L. & Chang, A.T.C. (2011 updated). AMSR-E/Aqua Daily L3 Global Snow Water Equivalent EASE-Grids V002. National Snow and Ice Data Center Digital Media, Boulder Colorado, USA.

Tong, J., Déry, S.J. & Jackson, P.L. (2009). Interrelationships between MODIS/Terra remotely sensed snow cover and the hydrometeorology of the Quesnel River Basin, British Columbia, Canada. *Hydrology and Earth System Sciences* **13**, 1439–1452.

Tucker, C.J. (1979). Red and photographic infrared linear combinations for monitoring vegetation. *Remote Sensing of Environment* **8**, 127–150.

Valovcin, F.R. (1976). *Snow/cloud discrimination*. AFGL-TR-76-0174, ADA 032385.

Acronyms

AFWA	Air Force Weather Agency
AMSR-E	Advanced Microwave Scanning Radiometer - Earth Observing System
ANSA	Air Force Weather Agency / National Aeronautics and Space Administration Snow Algorithm
CLSM	Catchment-based Land Surface Model
CMC	Canadian Meteorological Centre
EnKF	Ensemble Kalman Filter
EOS	Earth Observing System
ETM	Enhanced Thematic Mapper Plus
GLDAS	The NASA/NOAA Global Land Data Assimilation System
GSFC	Goddard Space Flight Center
IMS	Interactive Multi-Sensor Snow and Ice Mapping System
LIS	Land Information System
MODIS	Moderate Resolution Imaging Spectroradiometer
NASA	National Aeronautics and Space Administration
NASA EOS	National Aeronautics and Space Administration Earth Observing System
NDSI	Normalized Difference Snow Index
NDVI	Normalized Difference Vegetation Index
NIR	Near-infrared
NOAA	National Oceanic and Atmospheric Administration
PM	Passive microwave
SMMR	Scanning Multichannel Microwave Radiometer
SSM/I	Special Sensor Microwave / Imager
SWE	Snow-water equivalent
SWIR	Short-wave infrared
TIROS-1	Television Infrared Operational Satellite-1
TM	Thematic Mapper
VIC	Variable Infiltration Capacity
VIS	Visible
VNIR	Visible and near-infrared

Websites cited

http://www.natice.noaa.gov/ims/ (Interactive Multi-Sensor Snow and Ice Mapping System)

http://climate.rutgers.edu/snowcover/ (Rutgers University Global Snow Lab)

http://modis-snow-ice.gsfc.nasa.gov (MODIS)

4

Remote sensing of snow albedo, grain size, and pollution from space

Alexander A. Khokanovsky

Institute of Environmental Physics, University of Bremen, Germany

Summary

Snow reflectance in the visible region of the electromagnetic spectrum depends on the amount and type of pollutants. Fresh and clean snow is very reflective and its albedo is close to one, almost irrespective of grain size. On the other hand, snow with larger grains is less reflective in the near-infrared, compared with finely grained snow. This is due to the enhanced light absorption by large ice grains compared with the small ones. This difference in reflectance can be actually measured, and it provides a means of determining the grain effective radius and, therefore, the separation of different types of snow (e.g., snow with very fine, fine, medium, coarse, very coarse, and extreme grains (see, e.g., the classification http://www.crrel.usace.army.mil/library/booksnongovernment/Seasonal_Snow.pdf).

The size of snow grains is important for determining snow albedo, snow thickness using microwave techniques, the thermal environment within the snowpack and, also, for detecting signs of snow melt and snowfall. In this chapter we have summarized the techniques used for estimating snow grain size, snow albedo and level of pollution, using spaceborne observations. The main focus of the chapter is on the theoretical background of snow grain size and albedo/pollution monitoring using spaceborne observations. The presented results can be also used for the interpretation of optical airborne and ground-based observations of snowfields and the cryosphere in general.

4.1 Introduction

Optical remote sensing techniques are used for the determination of various snow parameters, including the concentration of pollutants in snow, the average ice grain size and the snow spectral albedo (Dozier and Painter, 2004). The extent of global snow cover and its temporal behaviour can be also studied using optical remote sensing methods (Seidel and Martinec (2004); see Chapter 3 of this book). Generally,

Remote Sensing of the Cryosphere, First Edition. Edited by M. Tedesco.
© 2015 John Wiley & Sons, Ltd. Published 2015 by John Wiley & Sons, Ltd.
Companion Website: www.wiley.com/go/tedesco/cryosphere

fresh snow is a very bright object in the visible range. However, the brightness decreases considerably if snow is contaminated by soot, dust or algae (Warren, 1982). The decrease in snow brightness can be easily detected by a radiometer or spectrometer operated in the UV-visible spectral range (see Table 4.1).

The snow reflectance is hardly influenced by the grain size in the UV-visible region of the electromagnetic spectrum. However, ice grains absorb solar light in the near-IR spectral range and the amount of absorbed electromagnetic energy depends on the grain size being larger for larger grains. This reduces light reflectance in the near-IR spectral range (e.g., at the wavelength 1.24 μm) and can be also easily detected using optical measurements.

As a matter of fact, snow darkening is proportional to the square root of the snow grain size in the visible (Kokhanovsky et al., 2011). Both snow grain size and the amount of impurities determine snow spectral albedo, which can be easily calculated using radiative transfer theory for the defined levels of snow pollution and snow grain size (see, for example, the online calculator at http://snow.engin .umich.edu/). The focus of this chapter is on the theoretical background of snow grain size and albedo/pollution monitoring using spaceborne observations. The presented results can be also used for the interpretation of optical airborne and ground-based observations of snow fields and the cryosphere in general.

There are several approaches to retrieve the grain size from the spectral reflectance of snow in the near-infrared region of the electromagnetic spectrum (Bourdelles and Fily, 1993; Fily et al., 1997; Green et al., 2002; Scambos et al., 2007; Stamnes et al., 2007; Zege et al., 2011; Kokhanovsky and Rozanov, 2012). The processing software is based either on the look-up-tables calculated using the exact solution of radiative transfer equation (Stamnes et al., 2007) or the asymptotic radiative transfer theory (Kokhanovsky and Rozanov, 2012). In particular, the spectral reflectance measurements at the wavelengths 0.46 and 0.865 μm were used by Stamnes et al. (2007) for the retrieval of snow impurity concentration and grain size. The subsurface snow grain size was retrieved using measurements at 1.6 μm.

It should be pointed out that snow has a layered structure (Colbeck, 1991), with different grain sizes along different layers. Therefore, the retrieved particle size depends on the channel used, because of different penetration depths occurring at different wavelengths. This enables a profiling of snow grain size using spectral reflectance measurements (Li et al., 2001). Kokhanovsky et al. (2011) used the asymptotic radiative transfer theory and measurements at 0.44 and 0.865 μm for the retrieval of snow grain size. Lyapustin et al. (2009) used the same theory, except the band ratio (0.645 and 1.24 μm) was used. Zege et al. (2011) used the MODIS measurements at the wavelengths 0.47, 0.86, and 1.24 μm.

The use of multiple channels (or their ratio) offers the possibility of improving the accuracy of the algorithm (at least for vertically homogeneous snow layers). Nolin and Dozier (2000) proposed a hyperspectral method for the determination of snow grain size. They used the snow spectral reflectance centered at 1.03 μm, the wavelength of a prominent ice absorption feature. The interpretation is based on the area – not the depth – of the absorption feature scaled to absolute reflectance. Therefore, the method does not require topographic correction and is insensitive to the instrument noise.

Table 4.1 Channels of relevant remote sensing instruments (new channels as compared to heritage instruments are given in bold) used for snow remote sensing. Airborne and ground-based instrumentation is described by Dozier and Painter (2004), Green *et al.* (2006), and Aoki *et al.* (2007).

Instrument	Channels	Comments
MODIS	0.4125, 0.443, 0.469, 0.488, 0.531, 0.551, 0.555, 0.645, 0.667, 0.678, 0.748, 0.858, 0.8695, 0.905, 0.936, 0.940, 1.24, 1.375, 1.64, 2.13, 3.75, 3.959, 4.05, 4.465, 4.516, 6.715, 7.325, 8.55, 9.73, 11.03, 12.02, 13.335, 13.635, 13.935, 14.235 μm	Bandwidth: 10–50 nm ($\lambda \leq 2.2$ μm Spatial resolution: 0.25 –1 km Swath: 2330 km
MISR	0.446, 0.558, 0.672, 0.866 μm	Bandwidth: 20–40 nm Spatial resolution: 0.275 km Swath: 360 km Viewing zenith angles (for the same target): 0, 26.1, 45.6, 60, 70.5 degrees in forward backward views
MERIS, 01.03.2002–08.04.2012	0.4125, 0.4425, 0.49, 0.51, 0.56, 0.62, 0.665, 0.68125, 0.705, 0.75375, 0.76, 0.775, 0.865, 0.89, 0.9 μm	Bandwidth: 7.5–20 nm (3.75 nm at $\lambda = 760.625$ nm) Spatial resolution: 0.3 km Swath: 1150 km
SeaWiFS, 18.09.1997–11.12.2010	0.412, 0.443, 0.49, 0.51, 0.555, 0.67, 0.765, 0.865 μm	Bandwidth: 20 nm Spatial resolution: 1.1 km Swath: 2801 km
SEVIRI	0.635, 0.81, 1.64, 3.92, 6.25, 7.35, 8.7, 9.66, 10.8, 12, 13.4 μm	Bandwidth: 15–30 nm ($\lambda \leq 1.64$ μm) Spatial resolution: 0.25–1.0 km for most of channels Geostationary (15 min repetition time from 0° longitude)
GLI, 14.12.02–24.10.2003	0.38, 0.4, 0.412, 0.443, 0.46, 0.49, 0.52, 0.545, 0.565, 0.625, 0.666, 0.68, 0.678, 0.71, 0.749, 0.763, 0.825, 0.865, 1.05, 1.135, 1.24, 1.38, 1.64, 2.21, 3.715, 6.7, 7.3, 8.6, 10.8, 12 μm	Bandwidth: 10 nm ($\lambda \leq 0.865$ μm) Spatial resolution: 3.0 km for most of channels Swath: 1600 km
MSI (S-2) 2015 ESA, Europe	0.443, 0.490, 0.560, 0.665, 0.705, 0.740, 0.775, 0.842, 0.865, 0.940, 1. 375, 1.61, 2.19 μm	13 bands 4 bands – 10 m 6 bands – 20 m 3 bands – 60 m Bandwidth: 20–115 nm Swath: 290 km
OLCI (S-3) 2015 ESA, Europe	0.400, 0.412, 0.443, 0.490, 0.510, 0.560, 0.620, 0.665, **0.67375**, 0.68125, 0.70875, 0.75375, 0.76125, **0.76435**, **0.7675**, 0.77875, 0.865, 0.885, 0.900, **0.94**, **1.02** μm	21 bands All bands – 300 m Bandwidth: 2.5–40nm Swath: 1300 km (5 cameras)
SLSTR (S-3) 2015 ESA, Europe	0.555, 0.659, 0.865, **1.375**, 1.61, **2.25**, 3.74, 10.95, 12.0 μm	9 bands (**dual view**) 6 bands – 500 m 3 bands – 1000 m (TIR) Bandwidth: 15–60 nm (0.38–1 μm in TIR) Swath: 1700 km (750 km for backward view)

Table 4.1 *(continued)*

Instrument	Channels	Comments
MSI (Earthcare) 2016 ESA-JAXA, Europe-Japan	0.659, 0.865, 1.61, 2.2, 8.8, 10.8, 12 µm	Bandwidth: 0.02, 0.02, 0.06, 0.1, 0.9, 0.9, 0.9 µm, respectively Spatial resolution: 500 m Swath: 150 km
VIIRS (NPP) 2011 NASA, USA	0.412, 0. 445, 0.488, 0.555, 0.64, 0.672, 0.7, 0.746, 0.865, 1.24, 1.378, 1.61, 2.25, 3.7, 4.05, 8.55, 10.673, 11.45, 12.013 µm	Bandwidth: 15–60 nm (0.18–1.0 µm in TIR) Spatial resolution: 400–800 m Swath: 3000 km
S-GLI 2016 JAXA, Japan	0.38, 0.412, 0.443, 0.490, 0.530, 0.565, 0.670, 0.763, 0.865, 1.05, 1.38, 1.64, 2.21, 10.8, 12.0 µm	Bandwidth: 10–20 nm (0.05–0.7 µm in TIR) Spatial resolution: 250–1000 m Swath: 1400 km, channels 0.67 and 0.865 µm will also provide information on polarization characteristics of reflected light and perform multi-angular observations

The method was applied to the data of Airborne Visible/Infrared Imaging Spectrometer (AVIRIS). Painter *et al.* (2003) developed a method for the retrieval of sub-pixel snow grain size also using AVIRIS. The measured AVIRIS spectral reflectance was matched with the theoretical one, iterating the grain size and snow cover for a ground scene under study. Green *et al.* (2002) used AVIRIS data to map the solid, liquid, and vapor phases of water in snow analyzing spectral reflectance at 0.94, 0.98, and 1.03 µm, and using the fact that the absorption feature in reflectance moves to somewhat longer wavelengths while going from the vapor to the crystalline state of the same matter.

4.2 Forward modeling

The optical instrument measures the specific intensity I of the reflected, transmitted or internal light field. A recent review on the directional radiometry, the notion of the specific intensity, and radiative transfer is given by Mishchenko (2011). In satellite applications, the reflection function:

$$R(\mu_0, \mu, \varphi) = I^{\uparrow}(\mu_0, \mu, \varphi)/I_L^{\uparrow}(\mu_0) \tag{4.1}$$

is used. Here, $I^{\uparrow}(\mu_0, \mu, \varphi)$ is the specific intensity of reflected light for a given target, $I_L^{\uparrow}(\mu_0)$ is the specific intensity of reflected light for a perfectly reflecting Lambertian white screen (with albedo 1.0), μ_0 is the cosine of the solar zenith angle (SZA), μ is the cosine of the viewing zenith angle (VZA), and φ is the difference of the solar and observations azimuthal angles.

The relative azimuth angle is equal to zero in the glint region (SZA = VZA) and is equal to π in the exact backscattering direction. The value of $I_L^\uparrow(\mu_0)$ can be easily calculated if the irradiance of the incident solar radiation E_0 on an area perpendicular to the light beam is known (Liou, 2002; Kokhanovsky, 2006):

$$I_L^\uparrow(\mu_0) = \mu_0 E_0/\pi \tag{4.2}$$

Both E_0 and $I^\uparrow(\mu_0, \mu, \varphi)$ can be measured by sensors mounted on satellites. This enables the calculation of the reflection function:

$$R(\mu_0, \mu, \varphi) = \pi I^\uparrow(\mu_0, \mu, \varphi)/\mu_0 E_0 \tag{4.3}$$

or of the bidirectional reflectance distribution function, BRDF = R/π.

It follows that, to retrieve snow properties from optical satellite data, it is important to provide some means for the calculation of the reflection function for different observation geometries. This is usually done in the framework of the radiative transfer theory (Chandrasekhar, 1960; Liou, 2002). In particular, one needs to specify the boundary conditions, the spectral single scattering albedo ω_0, the optical thickness τ and phase function $p(\theta)$ (θ is the scattering angle) for a given turbid medium (e.g., snow, ice), and solve the radiative transfer equation for the specific intensity of reflected light. In most cases, snow can be considered as a semi-infinite medium. Then only parameters ω_0 and $p(\theta)$ must be specified if the effects of vertical snow inhomogeneity are ignored. Advanced snow radiative transfer models account for the vertical variation of snow single scattering albedo, phase function, and extinction coefficient.

The radiative transfer equation is usually solved numerically. However, in the case of weakly absorbing media such as snow, the approximate solution of the radiative transfer equation is possible. In particular, the reflection function (for a vertically homogeneous snow) can be presented as (Zege $et\ al.$, 1991; Kokhanovsky, 2006):

$$R(\mu_0, \mu, \varphi) = R(\mu_0, \mu, \varphi)A^n, A = \exp[-\alpha], \quad n = K_0(\mu_0)K_0(\mu)/R_0(\mu_0, \mu, \varphi) \tag{4.4}$$

where $R_0(\mu_0, \mu, \varphi)$ is the reflection function of a semi-infinite non-absorbing snow layer, A is the spherical albedo of snow, and

$$\alpha = 4\sqrt{\frac{\beta}{3(1-g)}}, \qquad K_0(\mu) = \frac{3}{7}(1 + 2\mu) \tag{4.5}$$

Here, $\beta = 1 - \omega_0$ is the probability of photon absorption in a unit volume of snow, g is the asymmetry parameter, $K_0(\mu)$ is the so-called escape function (van de Hulst, 1980). Kokhanovsky (2006) proposed the following parameterization for $R_0(\mu_0, \mu, \varphi)$:

$$R_0(\mu_0, \mu, \varphi) = \frac{a + b(\mu_0 + \mu) + c\mu_0\mu + p(\theta)}{4(\mu_0 + \mu)} \qquad (4.6)$$

$$p(\theta) = 11.1 \exp(-0.087\theta) + 1.1 \exp(-0.014\theta) \qquad (4.7)$$

$$\cos\theta = -\mu_0\mu + \sqrt{(1 - \mu_0^2)(1 - \mu^2)} \cos\varphi, \quad a = 1.247, \quad b = 1.186, \quad c = 5.157 \qquad (4.8)$$

Here θ is the scattering angle in degrees.

This parameterization was performed on the assumption that the snow particles can be presented as rough fractal ice crystals. The results for rough hexagonal particles produce similar phase functions (see Figure 4.1).

In the case of a two-layered snow model, the following approximation can be used:

$$R_0(\mu_0, \mu, \varphi) = R(\mu_0, \mu, \varphi, \tau) + \frac{At^2 K_0(\mu_0)K_0(\mu)}{1 - Ar} \qquad (4.9)$$

where:

$R(\mu_0, \mu, \varphi, \tau)$ is the reflection function of the upper layer (for the case of an underlying black surface)

A is the spherical albedo of the second layer.

Figure 4.1 The dependence of the phase function on the shape of particles. The following shapes were considered: hexagons with the length to the diameter ratios equal to 100/25, 100/50, 100/100 μm and fractal particles (Macke *et al.*, 1996) at wavelength 550 nm. The diameter is defined as the distance between opposite sides of the hexagon. For all particles, the surface was assumed to be rough in the calculations (Khokanovsky 2011. Reproduced with permission of Taylor & Francis).

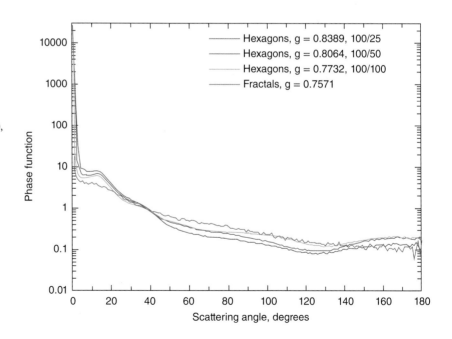

The functions $R(\mu_0, \mu, \varphi, \tau)$, $r(\tau)$ and $t(\tau)$ for the upper layer can be calculated either using the radiative transfer equation solution or an approximation (Zege et al., 1991; Kokhanovsky, 2006):

$$R(\mu_0, \mu, \varphi, \tau) = R_0(\mu_0, \mu, \varphi, \tau) \exp(-an) - t \exp(-x - \alpha) K_0(\mu_0) K_0(\mu),$$

$$t(\tau) = \frac{\sinh \alpha}{\sinh(\varepsilon x + \alpha)}, r(\tau) = \exp(-\alpha) - t(\tau) \exp(-x - \alpha), \tag{4.10}$$

$$\varepsilon = 1.072, x = \gamma\tau, \alpha = 4q\gamma, q = \frac{1}{3(1-g)}, \gamma = \sqrt{3\beta(1-g)}$$

Here, local optical parameters (e.g., β, g) correspond to that of an upper snow layer. The case of arbitrary number of layers was considered by Kokhanovsky (2006). Equation 4.9 is often used also in the case of a single snow layer deposited on the surface (e.g., grass) with the spherical albedo A. The equations given above can be used to estimate the snow spectral single scattering albedo (or the parameter β) from spectral snow reflectance measurements for an assumed phase function (and, therefore, value of g) of a snow layer (see Figure 4.1). Look-up-tables based on the exact radiative transfer equation solution can be also used (Stamnes et al., 2007). The determination of the snow grain size and level of pollution requires the derivation of the relationships between local optical and microphysical snow characteristics. This is discussed in the next section.

4.3   Local optical properties of a snow layer

We will assume that snow consists of ice grains and other scatterers (e.g., soot, dust, algae, etc.) suspended in air. The dense media effects (Tsang et al., 2000) are ignored. Then the snow extinction coefficient k_{ext}, absorption coefficient k_{abs}, and phase function $p(\theta)$ can be presented as:

$$k_{ext} = k_{ext}^{ice} + \sum_{j=1}^{N} k_{ext,j} \tag{4.11}$$

$$k_{abs} = k_{abs}^{ice} + \sum_{j=1}^{N} k_{abs,j} \tag{4.12}$$

$$p(\theta) = \frac{k_{sca} p^{ice}(\theta) + \sum_{j=1}^{N} k_{sca,j} p_j(\theta)}{k_{sca} + \sum_{j=1}^{N} k_{sca,j}} \tag{4.13}$$

Here, "ice" means that the parameter corresponds to the pure snow (no impurities, only ice phase), $k_{sca} = k_{ext} - k_{abs}$ is the scattering coefficient, and the account for N impurities is given in the second terms of corresponding expressions. The probability of photon absorption is calculated as $\beta = k_{abs}/k_{ext}$ and the asymmetry parameter $g = \frac{1}{2} \int\limits_{0}^{\pi} p(\theta) \sin \theta \cos \theta d\theta$. All local optical characteristics given above can be computed using Mie theory for spherical particles (Mie, 1908) or geometrical optics approaches (van de Hulst, 1981; Liou, 2002; Kokhanovsky, 2006) for particles of different shapes. Also, approximations, look-up-tables, and parameterizations of results of exact calculations are possible. The selected approximations are discussed below.

The extinction and absorption coefficients of the pure ice can be presented as:

$$k_{ext}^{ice} = NC_{ext}, k_{abs}^{ice} = NC_{abs} \qquad (4.14)$$

where:

C_{ext} and C_{abs} are average extinction and absorption cross-sections of snow grains
$N = c_i/V$ is the number concentration of snow grains (c_i is their volumetric concentration and V is the average volume of grains).

In the framework of the geometrical optics approximation, it follows that (Kokhanovsky, 2006):

$$C_{ext} = 2\Sigma, C_{abs} = \xi \gamma V \qquad (4.15)$$

for large (as compared to the wavelength), randomly oriented ice grains of arbitrary shape. The expression for C_{abs} presented here is valid in the case of weakly absorbing large particles such as ice grains in the visible and near IR regions of the electromagnetic spectrum (Kokhanovsky and Zege, 2004). Here, Σ is the average projection area of the particles and $\gamma = \frac{4\pi\kappa_i(\lambda)}{\lambda}$, κ_i is the imaginary part of the ice refractive index at the wavelength λ. The parameter ξ depends on the shape of particles (being close to 1.28 for ice spheres; Kokhanovsky and Zege, 2004) and the real part of the ice refractive index. Its dependence on the size of particles can be ignored in the visible and near-IR regions of the electromagnetic spectrum. Let us introduce the effective grain size (EGS) as:

$$a_{ef} = \frac{3V}{4\Sigma} \qquad (4.16)$$

The value of a_{ef} coincides with the radius of particles in the case of monodispersed spheres. Then one easily derives for the probability of photon absorption:

$$\beta = \sigma \gamma a_{ef} \qquad (4.17)$$

where $\sigma = 2\xi/3$ (≈ 0.85 for spheres) (Kokhanovsky and Zege, 2004).

The phase function of ice grains depends on the assumed shapes of scatterers. It is often assumed that snow grains can be presented as fractal particles (Macke et al., 1996). The corresponding phase function does not differ considerably from the phase function measured in situ in ice clouds (Kokhanovsky et al., 2011). The asymmetry parameter for this phase function is equal approximately to 0.76 in the visible (see Figure 4.1).

The parameter α (see Equation 4.5) in the case of pure snow(no pollutants) is as follows:

$$\alpha = \sqrt{\gamma a_{ef}} \mathbf{P} \tag{4.18}$$

where $\mathbf{P} = 4\sqrt{\sigma/3(1-g)}$. The parameter α and, therefore, the grain size a_{ef}, can be derived from snow reflectance measurements, assuming the shape of particles.

In particular, the shape of particles determines the value of $\mathbf{P}$. The geometric optics calculations for spherical particles give $\mathbf{P} = 6.3$. It is difficult to calculate this parameter from first principles, due to the irregular shape of snow grains. Therefore, it is important to report both values $\mathbf{P}$ and a_{ef} in outputs of corresponding retrieval algorithms. One can also introduce the optically equivalent radius $\alpha_{opt} = \mathbf{P}^2 a_{ef}$. This radius can be directly derived from the reflectance measurements and is less influenced by assumptions on the shape of particles. It can be also used to find the spectral snow albedo using a simple approximation:

$$A(\lambda) = \exp\left\{-\sqrt{\gamma(\lambda)\alpha_{opt}}\right\} \tag{4.19}$$

Let us consider the polluted snow now. If we assume that soot is the only pollutant, then we may assume that light scattering and extinction in snow is dominated by ice grains. The equation for the parameter β must be modified as:

$$\beta = \beta_{ice} + \beta_{soot} \tag{4.20}$$

It follows (see Equation 4.17) that:

$$\beta_{ice} = \sigma\gamma a_{ef} \tag{4.21}$$

Let us introduce the soot mass absorption coefficient ε:

$$k_{abs}^{soot} = \varepsilon\rho_s c_s \tag{4.22}$$

where ρ_s is the soot density and c_s is the volumetric concentration of snow grains. Taking into account that $k_{ext} = 1.5c_i/a_{ef}$ (Kokhanovsky, 2006), one derives:

$$\beta_s = \frac{2c_s}{3c_i}\varepsilon\rho_s a_{ef} \tag{4.23}$$

We will assume that $\varepsilon : \lambda^{-1}$ (as for Rayleigh scatterers) and, therefore, $\varepsilon(\lambda) = \varepsilon(\lambda_0)\lambda_0/\lambda$. Taking into account the formulae given above, we arrive to the following equation:

$$u(\lambda) = \sqrt{(B_1\chi_i(\lambda) + B_s c)x_{ef}} \tag{4.24}$$

where:

$$B_i = \frac{32\sigma}{3(1-g)}, \quad B_s = \frac{16\rho_i\varepsilon(\lambda_0)\lambda_0}{9\pi(1-g)}, \quad c = \frac{\rho_s c_s}{\rho_i c_i}, \quad x_{ef} = \frac{2\pi a_{ef}}{\lambda} \tag{4.25}$$

and $\rho_i = 0.9167$ g/cm^3 is the ice density.

Figure 4.2 Measured spectral dependence of plane albedo (Hadley and Kirchstetter, 2012) at the nadir illumination and various levels of soot absorption (soot concentrations are 1.1e-7(filled circles), 4.5e-7(triangles), 8.6e-7(stars), 1.68e-6 (boxes)). The effective radius of particles is 55 microns. The results of calculations are given by the solid lines. The imaginary part of ice refractive index was taken from Warren and Brandt (2008).

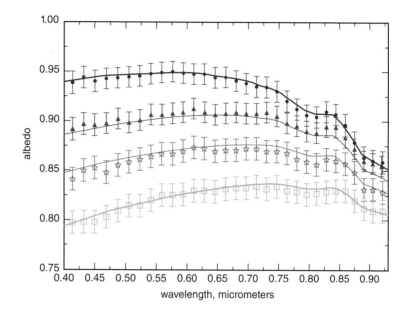

The function $\alpha(\lambda)$ given in Equation 4.24 can be used to find the snow spherical albedo $A(\lambda) = \exp(-\alpha(\lambda))$, the plane albedo $A_p = A^{K_0(\mu_0)}$ and also the snow BRDF, as discussed above.

A comparison of calculations according to these equations with experimental results for the plane albedo is given in Figure 4.2. The excellent agreement of theory with the experiment performed by Hadley and Kirchstetter (2012) is found, assuming that $B_i = 77.02$, $B_s = 22.15$. These parameters were derived using the minimization procedure at the largest concentration of pollutants (see Figure 4.2). Theoretical calculations for spherical particles give $\xi = 1.28$, $g = 0.884$ (Kokhanovsky and Zege, 2004) and $B_i = 78.47$. It follows that theoretical and experimental values of B_i are in a good agreement. The theoretical value of B_s cannot be derived, because the value of $\varepsilon(\lambda_0)$ for soot in snow was not determined experimentally. The experimental value of $B_s = 22.15$ is consistent with the assumption that $\varepsilon(0.435\ \mu m) = 9.27\ g/cm^3$ (see Equation 4.25), which is plausible for the laboratory experiment performed by Hadley and Kirchstetter (2012).

4.4 Inverse problem

The soot concentration and snow grain size are usually found using two approaches: 1) look up table (LUT) and 2) asymptotic theory. In the case of the LUT approach (Stamnes *et al.*, 2007), the radiances are pre-calculated at several channels and then the parameters a_{ef} and c are sought, by minimizing the difference of observed and calculated spectra of light reflected from the snow.

The asymptotic theory enables the analytical inversion. Therefore it is described in more detail here for the case of dual-channel retrievals:

$$R_1 = R_0 \exp\left[-\sqrt{\left(B_i \chi_{i,1} + B_s c\right) x_{ef}}\right], \qquad R_2 = R_0 \exp\left[-\sqrt{\left(B_i \chi_{i,2} + B_s c\right) x_{ef}}\right] \tag{4.26}$$

The value of a_{ef} from the second equation in 4.26 (the measurements in the near IR) can be represented via R_2 and c. Namely, it follows that:

$$a_{ef} = \frac{\ln^2(R_2/R_0)}{\kappa(B_i \chi_{i,2} + B_s c)} \tag{4.27}$$

where $\kappa = 2\pi/\lambda$. The corresponding equation is substituted to the first equation in 4.26. This enables the determination of c as:

$$c = \frac{\Psi_1 - \Xi \Psi_2}{(\Xi - 1)B_s} \tag{4.28}$$

where $\Psi_1 = B_i \chi_{i1}$, $\Psi_2 = B_i \chi_{i2}$, $\Xi = \ln^2(R_1/R_0)/\ln^2(R_2/R_0)$. The effective radius is then found from Equation 4.27. In principle, the measurements at several channels can be used simultaneously. This enables, for example, the illumination of constant R_0 in Equation 4.28 (Zege et al., 2011). One should avoid the application of the algorithm to directions close to the forward (glint) and backward scattering geometries, and also to cases with a low Sun (with SZA < 75°), where the accuracy of the underlying theory drops (Kokhanovsky and Breon, 2012; Wiebe et al., 2013).

The application of the algorithm described above to the MEdium Resolution Imaging Spectrometer (MERIS) data is shown in Figure 4.3 (Kokhanovsky et al., 2011). Currently, there are several algorithms based on the asymptotic theory (Tedesco and Kokhanovsky, 2007, 2010; Lyapustin et al., 2009; Zege et al., 2011; Wiebe et al., 2013). The principal differences between them are discussed by Kokhanovsky and Rozanov (2012).

An example of retrievals based on the LUT approach over Arctic from Global Imager (GLI) data is shown in Figure 4.4 (Hori et al., 2007). The retrievals based on the asymptotic radiative transfer theory for the Moderate-resolution Imaging Spectroradiometer (MODIS) data are shown in Figure 4.5 (Lyapustin et al., 2009). In particular, the results presented in Figure 4.5 quantify the temporal change of snow grain size over Greenland (for the time period April to August, 2004).

The soot concentration in snow is generally very small and, therefore, the accuracy of retrievals of the parameter c is low (Aoki et al., 2007; Zege et al., 2011; Warren, 2013). The dependence of the spherical albedo A on soot concentration is given in Figure 4.6.

Assuming combined measurement and forward modeling errors around 5%, we arrive at the conclusion that the retrievals of concentrations of soot below 50–100 ng/g are hardly possible from a satellite. Zege et al. (2011) have also reached this conclusion, suggesting the use of their method for remote sensing of

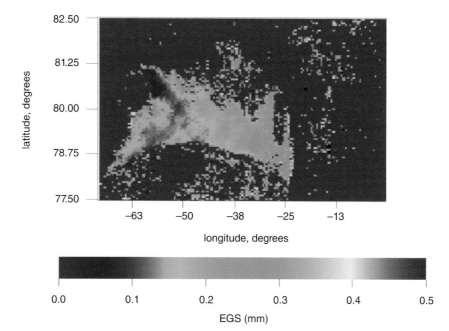

Figure 4.3 The snow effective radius derived from MERIS observations (Khokanovsky 2011. Reproduced with permission of Taylor & Francis).

soot only at high levels of snow pollution (above 1000 ng/g). On the other hand, the concentration of soot in the Arctic is usually in the range 3 – 30 ng/g (Warren, 2013). Values are even lower in Antarctica. Therefore, spaceborne retrievals are possible only close to rural areas and also for occasional pollution events. Therefore, we confirm the findings of Warren (2013) that optical remote sensing is not likely to be useful for quantifying the effect of black carbon on snow albedo (except for areas with a high concentration of pollutants).

However, the temporal changes of snow albedo (without attribution to specific causes) can be well determined from a satellite. In particular, Strove et al. (2005) have found that the mean difference between the MODIS algorithm albedo retrievals and the *in situ* data is less than 0.02 for all the stations in Greenland combined (RMSE = 0.07). Thus, the corresponding retrievals can be used not only for climate modeling but also for ice sheet mass balance studies.

Snow albedo can be easily calculated at any wavelength if the values of a_{ef} and c are retrieved. It is possible to retrieve the snow albedo directly from the reflectance (see Equation 4.4): $A = (R/R_0)^{1/n}$; the albedo derived in such a way is less influenced by *a priori* assumptions. Models of the bidirectional reflectance of snow, created using a discrete ordinate radiative transfer model, can be also used for the retrievals of snow albedo, as demonstrated by Klein and Stroeve (2002) and Stroeve et al. (2005).

The validation of snow grain size, snow albedo and concentration of impurities retrievals has been performed by Nolin and Dozier (2000), Stroeve et al. (2005), Aoki et al. (2007), Kokhanovsky et al. (2011), and Wiebe et al. (2013). The error

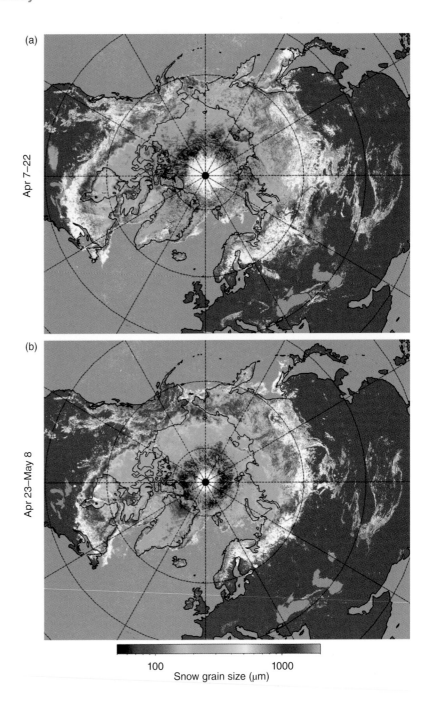

Figure 4.4 The retrieved snow grain size (16-day average, April 7 to May 8, 2003) based on the LUT approach (Hori *et al.*, 2007. Reproduced with permission of Elsevier).

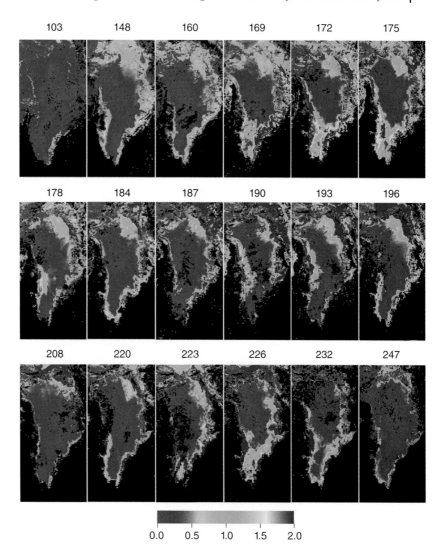

Figure 4.5 Snow grain size (mm) retrieved from MODIS TERRA data for the melting period of 2004 (April–August). Each image is obtained as a three-day composite to cover gaps caused by clouds. The numbers give a Julian day (Lyapustin *et al.*, 2009. Reproduced with permission of Elsevier).

on snow albedo is on average within 0.02, as stated above. The value of a_{ef} is difficult to measure on the ground (also in view of its spatial variability within a pixel), although a high correlation between the ground – measured and satellite – retrieved grain sizes has been found (Aoki *et al.*, 2007). Because of the lack of standard methods to measure the snow grain size on the ground, results depend on the experience of the observer and the selection of the grain size definition (largest or the smallest crystal size, etc.).

The results of the validation campaign for the satellite determination of soot concentration performed by Aoki *et al.* (2007) show that the coefficients of correlation between *in situ* and satellite-derived values of c are considerably lower (0.36–0.51), compared to the results for a_{ef} as derived from optical measurements

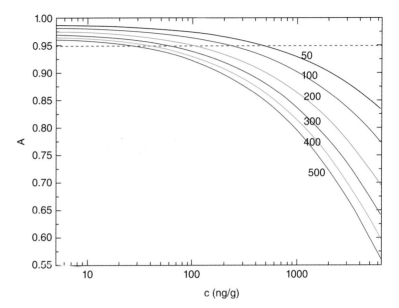

Figure 4.6 The dependence of spherical albedo on the soot concentration for various effective radii of snow grains (50, 100, 200, 300, 400, and 500 μm).

at the wavelength 0.9 μm (0.82). This is most probably due to the fact that the snow was contaminated not only by soot but also by other impurities (e.g., dust), with high mass concentrations in the range 70–60 000 ng/g.

4.5   Pitfalls of retrievals

The retrievals as described above look quite simple. However, there are complications due to atmospheric effects and, also, due to assumptions on the snow field properties under study (e.g., horizontally and vertically inhomogeneous snow layers, finite snow depth and influence of underlying surface, wet snow, snow on the slopes of mountains, snow in forested areas, blowing snow, slash, firn, etc.). Selected problems related to optical remote sensing of snow from space are discussed in this section.

Cloud screening is difficult over snow. Clouds look similar to snow if viewed from a satellite. Therefore, robust cloud screening algorithms (over snow) are needed ahead of retrievals. Cloud screening over snow is usually performed using the following:

a) the analysis of reflectance in gaseous absorption bands (oxygen A-band, water vapor, etc);
b) different time scales for snow and cloud variability (accumulation of images, reference image);
c) the analysis of differential Snow Index (the 865/885 nm ratio contrast is lower for cloudy scenes);

d) the study of the shape of spectral reflectance curves (UV-VIS-NIR);
e) the analysis of brightness temperature differences at the wavelengths 3.7, 10.8, and 12 μm.

Atmospheric correction is performed, either assuming the standard polar atmosphere model (Stamnes *et al.*, 2007) or deriving the aerosol properties (e.g., aerosol optical thickness over snow) using spectral or multi-angular spaceborne observations (Istomina *et al.*, 2011; Mei *et al.*, 2012).

In the case of Lambertian underlying surfaces, the signal as detected on a satellite can be represented as follows:

$$R(\mu_0, \mu, \varphi) = R_{atm}(\mu_0, \mu, \varphi, \tau) + \frac{A_{surf}T}{1 - A_{surf}r} \tag{4.29}$$

where:

T is the total atmospheric transmittance from a satellite to the snowfield and back
A_{surf} is the surface albedo to be found
r is the spherical albedo of atmosphere
$R_{atm}(\mu_0, \mu, \varphi, \tau_{aer})$ is the atmospheric path reflectance.

It follows from this equation:

$$A_{surf} = \frac{R_{mes}(\mu_0, \mu, \varphi) - R_{atm}(\mu_0, \mu, \varphi, \tau)}{(R_{mes}(\mu_0, \mu, \varphi) - R_{atm}(\mu_0, \mu, \varphi, \tau))r + T} \tag{4.30}$$

This equation can be used to find the surface albedo (and also snow grain size) under the assumption of the Lambertian ground. Zege *et al.* (2011) proposed a way to account for the effects of non-Lambertian underlying surfaces.

The vertical snow inhomogeneity can impact the quality of retrievals. Snow has a layered structure (Colbeck, 1991). Therefore, complications occur when several wavelengths are used to retrieve the snow grain size. The penetration depth of solar light in snow is different for different wavelengths; therefore, the assumption that the grain size is the same if multiple wavelengths are used is violated. This introduces a bias in the retrievals for layered snow layers. One way around this problem is to use the two-layered snow model (Li *et al.*, 2001). In particular, it follows:

$$R(\mu_0, \mu, \varphi) = R(\mu_0, \mu, \varphi, \tau) + \frac{At^2 K_0(\mu_0)K_0(\mu)}{1 - Ar} \tag{4.31}$$

Here r, t are the spherical albedo and transmittance of the upper layer, respectively, $R(\mu_0, \mu, \varphi, \tau)$ is the reflection function of the upper layer (for the case of an underlying black surface), A is the spherical albedo of the second layer. The functions $R(\mu_0, \mu, \varphi, \tau)$, $r(\tau)$ and $t(\tau)$ can be calculated either using the radiative transfer equation solution or an approximation, as described above.

The upper/bottom snow effective radius and upper snow ice water path can be retrieved using Equation 4.31. For this, one needs to fit the observations and calculations at several wavelengths.

Clearly, if the penetration depth of radiation is smaller than the thickness of the upper snow layer, then the properties of the second (buried) snow layer (or layers) cannot be determined. The analysis of Jacobians given in Figure 4.7 (Kokhanovsky *et al.*, 2011) suggests that optical remote sensing gives the sub-surface grain size only (1–2 cm from the surface).

The study of snow microstructure at larger depths is hardly possible using optical methods (except when techniques based on optical measurements along vertical snow walls are employed (Nikolaeva and Kokhanovsky, 2012). The same is true for

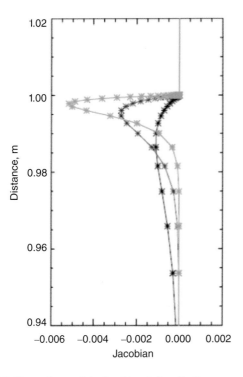 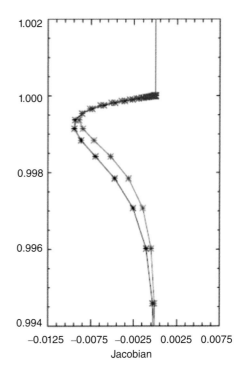

Figure 4.7 Dependence of the Jacobian J_a for effective radius of ice crystals in snow on the distance from the snow bottom at the wavelengths 865 nm (black curve), 1020 nm (red curve), and 1240 nm (green curve) (left panel). Right panel: the same except at the wavelengths 1610 nm (red) and 2190 nm (black). The LOWTRAN aerosol model with the aerosol optical thickness equal to 0.05 was used. The snow geometrical thickness is equal to 1 m and the length of side of fractal particles is equal to 50 μm (first panel), 300 μm (second panel), 750 μm (third panel). The concentration of soot is equal to 10-8. The solar zenith angle is equal to 60° and the observation is at the nadir direction (Khokhanovsky and Rozanov 2012. Reproduced with permission of Springer).

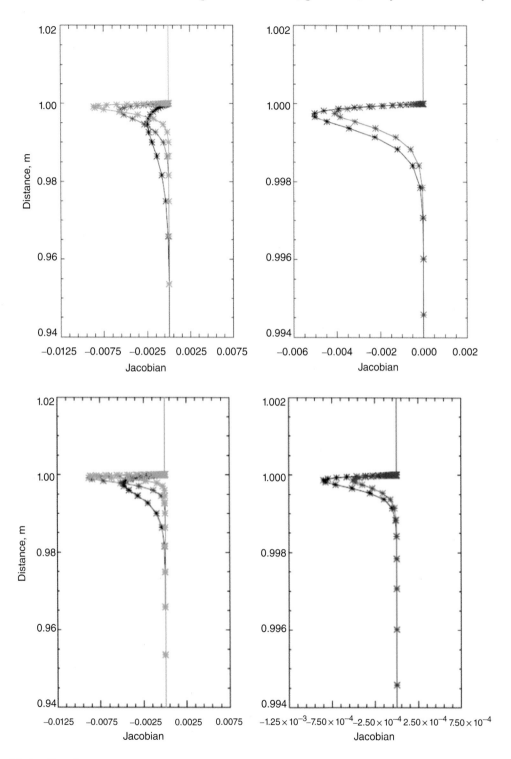

Figure 4.7 (*continued*)

the snow pollution determination. Only pollution in upper snow layer/layers can be studied (up to 10 cm or so – see Figure 4.8).

The snow horizontal inhomogeneity is usually ignored. For retrieval purposes, it is generally assumed that the snow surface is ideally flat and is horizontally homogeneous. However, this is rarely the case, especially if the instrument is located on a satellite and has a poor spatial resolution. Patches of vegetation and forest can be present in the satellite scene (Tedesco and Kokhanovsky, 2007) But, even in the absence of vegetation, the prevailing winds and ice stress can produce horizontal snow inhomogeneities (ridges, sastrugi, see, e.g., the MISR browse imagery presented for the 79° backward viewing camera in Figure 4.9, http://eosweb.larc .nasa.gov/HPDOCS/misr/misr_html/Antarctica_ice_waves.html). This will bias retrievals. In particular, if the solar illumination is perpendicular to sastrugi, then albedo decreases due to the substantial trapping of photons by vertical walls of

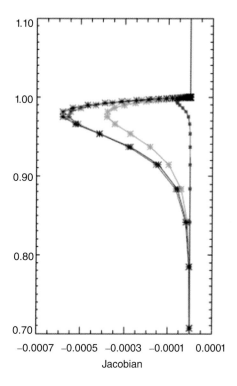
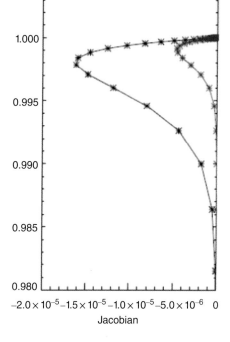

Figure 4.8 Dependence of the Jacobian for soot concentration $J_c(\lambda, z)$ on the distance from the snow bottom at the wavelengths 400, 443, 555, 670 nm (left, larger Jacobians at maximum correspond to a smaller wavelength) and at wavelengths 865 and 1029nm (right, larger Jacobians at maximum correspond to a smaller wavelength). The LOWTRAN aerosol model with the aerosol optical thickness equal to 0.05 was used. The snow geometrical thickness is equal to 1 m and the length of side of fractal particles is equal to 50 μm (first panel), 300 μm (second panel), 750 μm (third panel). The concentration of soot is equal to 10^{-8}. The solar zenith angle is equal to 60° and the observation is at the nadir direction (Khokanovsky and Rozanov 2012. Reproduced with permission of Springer).

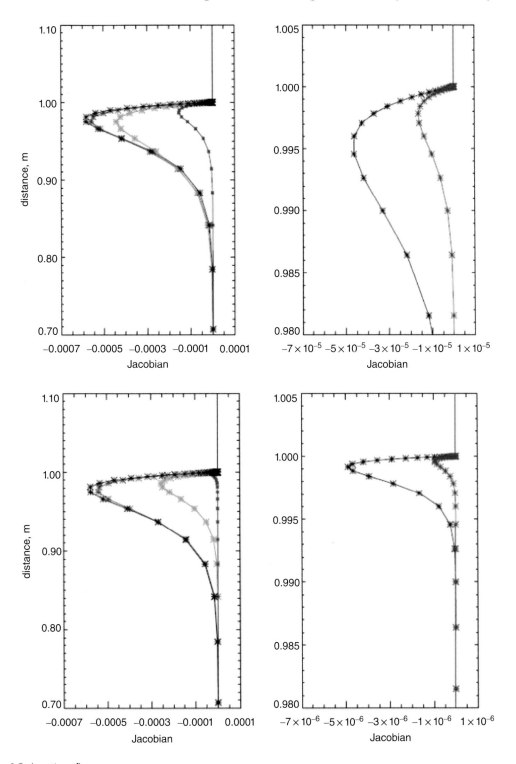

Figure 4.8 (*continued*)

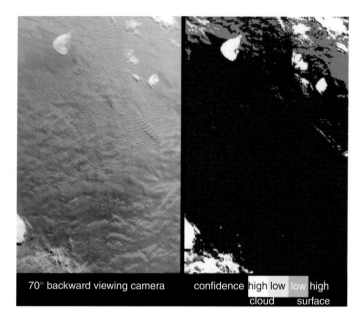

Figure 4.9 The MISR browse image over Antarctica for the 70° backward viewing camera (December 16, 2004; (http://eosweb.larc.nasa.gov /HPDOCS/misr/misr_html /antarctica_ice_waves.html) (NASA).

sastrugi (Kuchiki *et al.*, 2011; Zuravleva and Kokhanovsky, 2011). This decrease in albedo due to purely geometrical effects (up to 30% for the single scattering albedo equal to 0.98; Zuravleva and Kokhanovsky, 2011) can be erroneously interpreted as increase in the grain size (or level of pollution).

4.6 Conclusions

Remote sensing optical data can be used for the determination of subsurface snow properties without disturbing snow. To achieve this aim, the spectral solar light reflectance governed by the snow pollution and snow grain size must be measured in the visible and near-infrared regions. Snow reflectance in the visible region depends on the amount and type of pollutants. Fresh and clean snow is very reflective, and its albedo is close to 1.0, almost irrespective of grain size. On the other hand, snow with larger grains is less reflective in the near infrared, compared to snow with fine grains. This is due to the enhanced light absorption by large ice grains compared with small ones.

This difference in reflectance can be actually measured and provides a means for the determination of the grain effective radius and, therefore, the separation of different types of snow (e.g., snow with very fine ($a^{ef} < 0.1$ mm), fine (0.1–0.25 mm), medium (0.25–0.5mm), coarse (0.5–1.0mm), very coarse (1.0–2.5 mm), and extreme (> 2.5 mm) grains (http://www.crrel.usace.army.mil/library /booksnongovernment/Seasonal_Snow.pdf). The size of snow grains is important for determining snow albedo and the thermal environment within the snowpack, and also for detecting of signs of snowmelt and snowfall. In this chapter, we have summarized some techniques used for estimating snow grain size and albedo and the level of pollution using spaceborne observations. Similar techniques are used for ground- based and airborne optical snow remote sensing.

Acknowledgments

This work was supported by the BMBF Project CLIMSLIP and FP7 Project SIDARUS. It was conducted as part of the GCOM-C/SGLI Snow Project supported by Japan Aerospace Exploration Agency (JAXA). The author is grateful to O.L. Hadley for providing the experimental data (see Figure 4.2) in a tabular form. The author is grateful for a cooperation and discussions on snow grain size retrievals with T. Aoki, G. Heygster, M. Hori, A. Lyapustin, F. Seidel, V.V. Rozanov, M. Tedesco, and E.P. Zege.

References

Aoki, T., Hori, M., Motoyoshi, H., *et al.* (2007). ADEOS-II/GLI snow/ice products – Part II: Validation results using GLI and MODIS data. *Remote Sensing of Environment* **111**, 274–290.

Bourdelles, B. & Fily, M. (1993). Snow grain-size determination from Landsat imagery over Terre Adélie, Antarctica. *Annals of Glaciology* **17**, 86–92.

Chandrasekhar, S. (1960). *Radiative Transfer*. New York: Dover.

Colbeck, S.C. (1991). The layered structure of snow layers. *Reviews of Geophysics* **29**(1), 81–96.

Dozier, J. & Painter, T.H. (2004). Multispectral and hyperspectral remote sensing of alpine snow properties. *Annual Review of Earth and Planetary Sciences* **32**, 465–494.

Fily, M., Bourdelles, B., Dedieu, J.P. & Sergent, C. (1997). Comparison of in situ and Landsat Thematic Mapper derived snow grain characteristics in the Alps. *Remote Sensing of Environment* **59**(3), 452–460.

Green, R.O., Dozier, J., Roberts, D.A. & Painter, T.H. (2002). Spectral snow reflectance models for grain size and liquid water fraction inmelting snow for the solar reflected spectrum. *Annals of Glaciology* **34**, 71–73.

Green, R.O., Painter, T.H., Roberts, D.A. & Dozier, J. (2006). Measuring the expressed abundance of the three phases of water with an imaging spectrometer over melting snow. *Water Resources Research* **42**, W10402. doi:10.1029/2005WR004509.

Hadley, O. & Kirchstetter, W. (2012). Black-carbon reduction of snow albedo. *Nature Climate Change* **2**, 437–440.

Hori, M., Aoki, T., Stamnes, K., Chen, B. & Li, W. (2007). ADEOS-II/GLI snow/ice products – part III: Retrieved results. *Remote Sensing of Environment* **111**, 274–319.

Istomina, L.G., von Hoyningen-Huene, W., Kokhanovsky, A.A., Schultz, E. & Burrows, J.P. (2011). Remote sensing of aerosols over snow using infrared AATSR observations. *Atmospheric Measurement Techniques* **4**, 1133–1145.

Klein, A.G. & Stroeve, J. (2002). Development and validation of a snow albedo algorithm for the MODIS instrument. *Annals of Glaciology* **34**, 45–52.

Kokhanovsky, A.A. (2006). *Cloud Optics*. Dordrecht: Springer.

Kokhanovsky, A.A. & Breon, F.-M. (2012). Validation of an analytical snow BRDF model using PARASOL multi-angular and multispectral observations. *IEEE Geoscience and Remote Sensing Letters* **5**, 928–932.

Kokhanovsky, A.A. & Rozanov, V.V. (2012). The retrieval of snow characteristics from optical measurements. In: Kokhanovsky, A.A. (ed). *Light Scattering Reviews* **6**, 289–331.

Kokhanovsky, A.A. & Zege, E.P. (2004). Scattering optics of snow. *Applied Optics* **43**, 1589–1602.

Kokhanovsky A.A., Rozanov, V.V., Aoki, T., Odermatt, D., Brockmann, C., Krüger, O., Bouvet, M., Drusch, M. & Hori, M. (2011). Sizing snow grains using backscattered solar light. *International Journal of Remote Sensing* **32**, 6975–7008.

Kuchiki, K., Aoki, T., Niwano, M., Motoyoshi, H. & Iwabuchi, H. (2011). Effect of sastrugi on snow bidirectional reflectance and its applications to MODIS data. *Journal of Geophysical Research* **116**. doi: 10.1029/2011JD016070.

Li, W., Stamnes, K., Chen, B. & Xiong, X. (2001). Snow grain size retrieved from near-infrared radiances at multiple wavelengths. *Geophysical Research Letters* **28**, 1699–1702.

Liou, K.N. (2002). *An Introduction to Atmospheric Radiation*. New York, NY, Academic Press.

Lyapustin, A., Tedesco, M., Wang, Y., Aoki, T., Hori, M. & Kokhanovsky, A. (2009). Retrieval of snow grain size over Greenland from MODIS. *Remote Sensing of Environment* **113**, 1976–1987.

Macke, A., Mueller, J. & Raschke, E. (1996). Scattering properties of atmospheric ice crystals. *Journal of the Atmospheric Sciences* **53**, 2813–2825.

Mei, L., Xue, Y., de Leeuw, G., *et al.* (2012). Aerosol optical depth retrieval in the Arctic region using MODIS over snow. *Remote Sensing of Environment* **128**, 234–245.

Mie, G. (1908). Beiträge zur Optik trüber Medien, speziell kolloidaler Metallösungen. *Annals of Physics* **330**, 377–445.

Mishchenko, M.I. (2011). Directional radiometry and radiative transfer: A new paradigm. *Journal of Quantitative Spectroscopy and Radiative Transfer* **112**, 2079–2094.

Nikolaeva, O.V. & Kokhanovsky, A.A. (2012). Theoretical study of solar light reflectances from vertical snow surfaces. *Cryosphere Discussions* 4205–4231.

Nolin, A.W. & Dozier, J. (2000). A hyperspectral method for remotely sensing the grain size of snow. *Remote Sensing of Environment* **74**, 207–216.

Painter, T.H., Dozier, J., Roberts, D.A., Davis, R.E. & Green (2003). Retrieval of subpixels snow-covered area and grain size from imaging spectrometer data. *Remote Sensing of Environment* **85**, 64–77.

Scambos, T.A., Haran, T.M., Fahnestock, M.A., Painter, T.H. & Bohlander, J. (2007). MODIS-based mosaic of Antarctica (MOA) data sets: continent-wide surface morphology and snow grain size. *Remote Sensing of Environment* **111**, 242–257.

Seidel, K. & Martinec, J. (2004). *Remote Sensing in Snow Hydrology*. Berlin, Springer-Praxis.

Stamnes, K., Li, W., Eide, H., Aoki, T., Hori, M. & Storvold, R. (2007). ADEOS-II GLI snow/ice products – Part I. Scientific basis. *Remote Sensing of Environment* **111**, 258–273.

Stroeve, J., Box, J.E., Gao, F., Liang, S., Nolin, A. & Schaaf, C. (2005). Accuracy assessment of the MODIS 16-day albedo product for snow: comparisons with Greenland in situ measurements. *Remote Sensing of Environment* **94**, 46–60.

Tedesco, M. & Kokhanovsky, A.A. (2007). The semi-analytical snow retrieval algorithm and its application to MODIS data. *Remote Sensing of Environment* **111**, 228–241.

Tedesco, M. & Kokhanovsky, A.A. (2010). Errata of the paper: "The semi-analytical snow retrieval algorithm and its application to MODIS data". *Remote Sensing of Environment* **115**, 255.

Tsang, L., Chen, C.-T., Chang, A. T. C., Guo, J. & Ding, K.-H. (2000). Dense media radiative transfer theory based on quasicrystalline approximation with applications to passive microwave remote sensing of snow. *Radio Science* **311**, 731–750.

van de Hulst, H.C. (1980). *Multiple Light Scattering*, v.1, 2. New York, NY, Academic Press.

van de Hulst, H.C. (1981). Light Scattering by Small Particles. New York, NY, Dover.

Warren, S.G. (1982). Optical properties of snow. *Reviews of Geophysics and Space Physics* **20**, 67–89.

Warren, S.G. (2013). Can be black carbon in snow be detected by remote sensing? *Journal of Geophysical Research* (in press).

Warren, S.G. & Brandt, R.E. (2008). Optical constants of ice from ultraviolet to the microwave: A revised compilation. *Journal of Geophysical Research* **113**, D14220. doi: 10.1029/2007JD009744.

Wiebe, H., Heygster, G., Zege, E.P., Aoki, T. & Hori, M. (2013). Snow grain size retrieval SGSP from optical satellite data: validation with ground measurements and detection of snowfall events. *Remote Sensing of Environment* **128**, 11–20.

Zege E.P., Katsev, I.L. & Ivanov, A.P. (1991). *Image Transfer through a Scattering Medium*. Heidelberg: Springer.

Zege, E.P., Katsev, I.L., Malinka, A.V., Prikhach, A.S., Heygster, G. & Wiebe, H. (2011). Algorithm for retrieval of the effective snow grain size and pollution amount from satellite measurements. *Remote Sensing of Environment* **115**, 2674–2685.

Zuravleva, T.B. & Kokhanovsky, A.A. (2011). Influence of surface roughness on the reflective properties of snow. *Journal of Quantitative Spectroscopy and Radiative Transfer* **112**, 1353–1368.

Acronyms

AVIRIS	Airborne Visible/Infrared Imaging Spectrometer
SZA	Solar zenith angle
VZA	Viewing zenith angle
EGS	Effective grain size
LUT	Look up table
MERIS	MEdium Resolution Imaging Spectrometer
GLI	Global Imager
MODIS	Moderate-resolution Imaging Spectroradiometer
MISR	Multi-angle Imaging SpectroRadiometer
JAXA	Japan Aerospace Exploration Agency

Websites cited

http://snow.engin.umich.edu/

http://eosweb.larc.nasa.gov/HPDOCS/misr/misr_html/Antarctica_ice_waves.html

http://www.crrel.usace.army.mil/library/booksnongovernment/Seasonal_Snow.pdf

5 Remote sensing of snow depth and snow water equivalent

Marco Tedesco[1], Chris Derksen[2], Jeffrey S. Deems[3] and James L. Foster[4]

[1] The City College of New York, City University of New York, New York, USA
[2] Environment Canada, Toronto, Canada
[3] CIRES National Snow and Ice Data Center, and CIRES/NOAA Western Water Assessment, University of Colorado, Boulder, USA
[4] NASA Goddard Space Flight Center, Greenbelt, USA

Summary

Snow is a major source for freshwater production and a key player in water resources management. It modulates the Earth's energy budget through its high albedo, thermal insulating properties and its impact on temperatures at the land-atmosphere interface. Knowledge of snow depth and snow water equivalent (SWE, the amount of water stored within the snowpack) spatial distribution and temporal evolution is, therefore, critical for hydrological, climatological and water management applications.

In this chapter, we present remote sensing techniques for estimating snow depth and SWE from both spaceborne and airborne measurements. First, we introduce photogrammetry and LiDAR-based approaches, followed by techniques based on gamma radiation. We then focus our attention on the use of spaceborne gravimetry for estimating the water storage at global scale. Lastly, we describe and report techniques that utilize both passive and active microwave data.

5.1 Introduction

Estimating snow depth and snow water equivalent (SWE, the amount of water stored within the snowpack) is crucial for hydrological, climatological and water management applications. Snow represents a major source of municipal, industrial, and irrigation water for many regions in the world, while estimates of snow mass can also be used to support the management of resources for hydropower production and flood forecasting. The presence of snow on the ground also affects the Earth's energy budget due to its high albedo, thermal insulating properties and its impact on temperatures at the land-atmosphere interface.

Several remote sensing methods and techniques have been utilized to estimate snow depth and SWE at different spatial and temporal scales. In many cases, snow depth is first estimated from remote sensing observations, and then SWE is obtained by multiplying the estimated snow depth by snow density. In the following, we first describe photogrammetric approaches – gamma radiation and LiDAR-based approaches. We then put special emphasis on microwave-based approaches, in view of the large use of these techniques to estimate snow depth and SWE at the global scale.

5.2   Photogrammetry

Photogrammetry, (defined as the practice of determining the geometric properties of objects from photographic images) can be used for snow depth retrieval from airborne measurements, though it is not very common. Three-dimensional coordinates (the position on the plane and the height of the snowpack, in this case) are determined from two or more photographic images taken from different positions through the identification of common points. Rays are constructed from the camera location to the point on the object and are then used to determine the three-dimensional location of the points through triangulation.

Cline (1993) showed that sufficient snow feature exists in the Alpine test area selected in the study to permit automated pattern matching in many locations, allowing the generation of a digital elevation model (DEM) of the snow surface, which can be used to map changes in snow depth. However, saturation, due to the relative angles between the sun, the snow and the camera, can remove textual information for pattern matching, thereby introducing uncertainty into the generation of the surface elevation.

5.3   LiDAR

A rapidly evolving technique for snow depth mapping uses surface elevation models derived from LiDAR (Light Detection And Ranging). LiDAR is an active ranging system which, used in combination with Global Positioning and Inertial Measurement Systems, can provide high-resolution, high-accuracy surface elevation maps. This technique has the additional advantage of relying on a direct time-of-flight measurement, not on an inversion of attenuated backscatter. An airborne LiDAR system is composed of a laser scanning system, a global positioning system (GPS), and an inertial measuring unit (IMU). The emitted laser pulse is reflected off multiple features (vegetation and bare ground or snowpack) back to the platform, where its time-of-flight (t) is recorded and the range to target (R) computed as:

$$R = c\frac{t}{2} \tag{5.1}$$

where c is the speed of light.

Examples of airborne LiDAR system parameters and geometries are shown in Figure 5.1 (from Deems *et al.*, 2013). Here, R is the range measurement, θ is the scan angle, h is the platform height, γ is the beam divergence, AL is the laser spot footprint and SW is the swath width. GPS and inertial navigation systems are on the platform and time-synchronized with the laser-scanning system.

The laser pulse widens from its initial radius at an angle determined by the beam divergence and, thus, the laser footprint is larger at longer ranges. This beam spreading allows portions of each pulse to be reflected by multiple targets within the footprint and, commonly, to produce ground returns even in forested areas. Most systems record more than one range per pulse. The majority of airborne systems in use are discrete-return recording systems, where multiple return signals are logged by an analog detector as the return energy exceeds a predetermined threshold or a constant fraction of local return energy peaks. These systems are limited in the number of returns logged per pulse (commonly up to 4–6 total), and are subject to increased error when target complexity distorts the shape of the return energy waveform. Systems that record the full return energy waveform are becoming more common. These systems digitally sample the return energy at a high temporal resolution, allowing detection of an effectively infinite number of returns through waveform processing conducted in real-time or as a post-processing step.

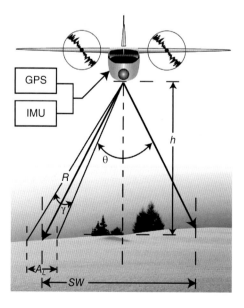

Figure 5.1 Airborne LiDAR system geometry (Deems *et al.*, 2013. Reproduced with permission of the International Glaciological Society).

Airborne LiDAR surveys are commonly planned to achieve a specific ground point spacing or point density (analogous to resolution) at a nominal flight altitude above ground level (AGL). Higher point density (lower point spacing) allows more accurate surface modeling, especially in sloped or vegetation-covered terrain. Laser system parameters, including pulse rate (or pulse repetition frequency, PRF), scan rate, scan angle, and flight parameters such as ground speed, swath overlap, and flight line layout, can all be adjusted to provide an optimal point sampling pattern for the terrain of interest.

The recent generation of commercial airborne sensors employs techniques allowing multiple laser pulses in the air (known as Multiple-Pulse-in-Air, MPiA, or Multiple-Time-Around, MTA). Analogous to phase ambiguity in active radar systems, at high ranges and high PRF, a laser pulse is emitted prior to receipt of the return from the previous pulse, resulting in ambiguous range attribution. One established technique for resolving the range ambiguity is to apply a random timing offset, or "jitter", to the pulse emission timing, and then assign return signals to emitted pulses, such that the variance among nearby pulses is minimized (see Rieger *et al.* (2011) for a full discussion of this technique). The MPiA capability allows surveys to be conducted at higher flight elevations (AGL) while maintaining high pulse rates, broadening the swath covered per flight line at minimal point spacing penalty, which can substantially reduce flight time and cost, while expanding the survey area.

Snow depth from LiDAR data can be obtained from two co-registered LiDAR images, one each for snow-free and snow-covered dates, followed by differencing the snow surface and bare-ground elevations (e.g., Hopkinson *et al.*, 2004; Deems *et al.*, 2006; Mott *et al.*, 2011; Deems *et al.*, 2013). Differencing can be conducted using raw point cloud elevations, interpolated surfaces, or a combination of the two, depending on the required final product type and resolution (see Deems *et al.*, 2013).

LiDAR errors can be separated into horizontal (geolocation) and vertical (altitude) errors, with horizontal errors being generally one order of magnitude greater than vertical errors. Glennie *et al.* (2007) provide an excellent analysis of error sources in LiDAR systems. Error sources include GPS (position), IMU (attitude), timing clock (range), boresight (co-registration between laser and IMU system), lever arm (IMU and scanner offset), and scan angle (scanner angular resolution). GPS error affects both horizontal and vertical accuracy, and is commonly estimated to contribute on the order of 10 cm to the overall error budget (Hodgson and Bresnahan, 2004).

Horizontal errors are dominated by IMU and boresight errors which, being angular errors, are proportional to the measurement range. Vertical errors at short ranges are predominantly due to timing clock accuracy, directly affecting the range measurement accuracy. As range (flight elevation) increases, the angular IMU and scan system errors exert a greater influence on vertical error. A careful boresight calibration, reliable GPS ground control and post-processing, and use of the most precise IMU available would constitute a best-practices approach to controlling error propagation in LiDAR surveys for snow depth mapping.

Figure 5.2 (a) 1 m resolution gridded LiDAR snow depths over a coincident orthophotograph. Yellow dots indicate the locations of manual depth surveys. (b) Snow volume calculated from ground-based LiDAR (Deems *et al.*, 2013). The LiDAR data used to produce Figure 5.2a concerning snow depth estimates are available at http://nsidc.org/data/nsidc-0157.html (Miller, 2004) and were collected within the Small Regional Study Area (SRSA) as part of the Cold Land Processes Field Experiment (CLPX) in Northern Colorado on 8–9 April and 18–19 September, 2003. The data set consists of color infrared orthophotography, LiDAR elevation returns (raw/combined, filtered to bare ground/snow, and filtered to top of vegetation), elevation contours (0.5 m) and snow depth contours (0.1 m). Reprinted from the *Journal of Glaciology* (Deems *et al.*, 2013. Reproduced with permission of the International Glaciological Society).

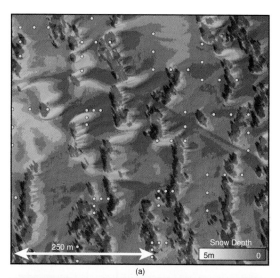

(a)

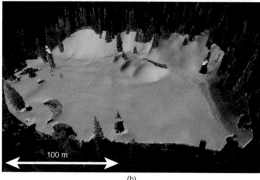

(b)

As an example of applications, Figure 5.2 shows 1 m resolution gridded LiDAR snow depths displayed over a coincident orthophotograph (Figure 5.2a), and snow volume calculated from ground-based LiDAR.

For further reading, we refer the reader to Glennie *et al.* (2013) for general LiDAR overview and emerging technologies, and to Deems *et al.* (2013) for a review of LiDAR mapping procedures and error sources, potential errors unique to snow surface remote sensing in the near-infrared and visible wavelengths, and recommendations for projects using LiDAR for snow-depth mapping.

5.4 Gamma radiation

SWE can be inferred at local spatial scales from gamma radiation airborne measurements. In this case, the natural terrestrial gamma radiation emitted from the potassium, uranium, and thorium radioisotopes in the upper layer of soil is sensed from a an aircraft flying ≈ 150 m above the ground (Carroll and Carroll, 1989; Peck *et al.*, 1971). Water mass in the snow cover attenuates, or blocks, the terrestrial radiation signal. Consequently, the difference between airborne radiation measurements made over bare ground and snow-covered ground can be used to calculate a mean areal SWE value with a root mean square error of less than half an inch. The National Operational Hydrologic Remote Sensing Center (NOHRSC) maintains an Airborne Gamma Radiation Snow Survey Program to make airborne SWE measurements, with over 1900 flight lines, covering portions of 29 states and seven Canadian provinces. Airborne gamma surveys are generally conducted along selected flight lines; a flight when no snow is on the ground is conducted before snow accumulates, and one or multiple flights can be conducted over the same lines during winter. The technique provides no information on snow depth, only SWE. Airborne SWE (in g/cm^2) is calculated using the following relationship:

$$SWE = \frac{1}{A} \left(\ln\left(\frac{C_0}{C}\right) - \ln\frac{(100 + 1.11M)}{(100 + 1.11M_0)} \right) \tag{5.2}$$

where:

C and C_0 are the uncollided terrestrial gamma count rates over snow and bare ground, respectively

M and M_0 represent the percent soil moisture over snow and bare ground, respectively

A is the radiation attenuation coefficient in water ($cm^2\ g^{-1}$).

The inverse radiation attenuation coefficients used in Equation 5.1 for the potassium, thorium, and total count windows are 14.34, 18.85, and 17.73, respectively. SWE values are calculated for each of the three radioisotope photopeaks. A weighted SWE is calculated by multiplying each of the three independent SWE values by a weighting coefficient (which sums to unity) and summing the results. The potassium, thorium, and total count weighting factors are 0.346, 0.518, 0.136, respectively.

Background radiation and soil moisture values (C_0 and M_0) are collected once under no-snow cover conditions and are used to calibrate flight lines. Flight lines are typically ~ 16 km long, with a ≈ 300 m wide measurement swatch covering an area of $\approx 5 - 8$ square kilometers. Radiation data collected over each flight line are an integrated value over this area. Consequently, gamma airborne SWE estimated values represent a mean areal value. As an example, Figure 5.3 shows interpolated SWE from airborne Gamma data for the period February 10 – 14, 2010, produced by NOHRSC.

Because gamma radiation is attenuated by water in all phases, changes in soil moisture between the fall and winter flights must be accounted for. Also, in forested environments, the vegetation biomass can attenuate gamma radiation, hence increasing the error on estimated SWE. Offenbacher and Colbeck (1998)

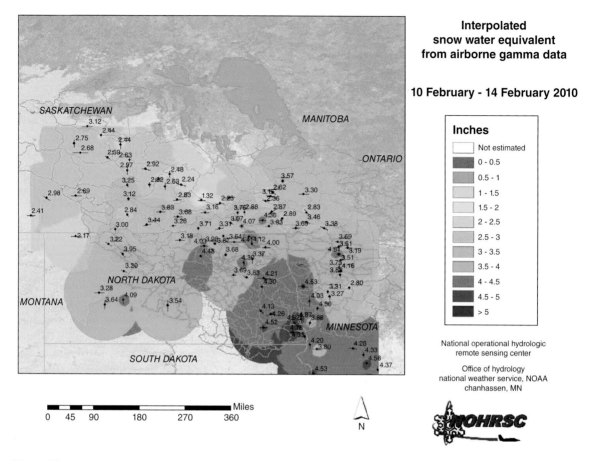

Figure 5.3 Interpolated SWE from airborne Gamma data for the period February 10–14, 2010 (NOHRSC, NOAA).

investigated several aspects of the use of natural gamma-ray emissions to determine the mass of snow covering the ground, describing the interactions of gamma rays with water mass, different techniques and the sources of error.

5.5 Gravity data

The estimation of SWE at large spatial scales is possible, using gravity data collected by the Gravity Recovery and Climate Experiment (GRACE) satellites. GRACE measurements can be used to estimate changes in terrestrial water storage and therefore, in principle, it is possible to derive SWE from GRACE data. It is important, however, to note that the changes measured by GRACE are also due to other factors besides SWE and it is, therefore, crucial to separate the different aspects of mass change contributing to the measured quantities. Because of this, several applications have been developed in which GRACE data is analyzed in conjunction with model outputs (e.g., land surface models), or GRACE data is assimilated into land surface models (e.g., Forman *et al.*, 2012). Frappart *et al.* (2006) showed that GRACE retrievals correlate well with the high-latitude zones of significant snow accumulation.

Regional means computed for four large boreal basins (Yenisey, Ob, MacKenzie and Yukon) show a good agreement at the seasonal scale between the snow mass solutions and model predictions. Niu *et al.* (2007) also showed that SWE can be derived from GRACE terrestrial water storage change in regions where the ground is not covered by snow in a summer month, if accurate changes in below-ground water storage can be provided by a land surface model. As an example, Figure 5.4a shows a map of SWE (in mm) for March of 2004, obtained using a GRACE terrestrial water storage (TWS) derived from a gravity field release by the Center for Space Research (CSR, University of Texas at Austin) minus unfiltered model outputs of below ground water storage change (Niu *et al.*, 2007). To demonstrate the feasibility of retrieving SWE from GRACE, Niu *et al.* (2007) compare the SWE fields in Figure 5.4a with ground-based US Air Force (USAF) Environmental Technical Application Center (ETAC) SWE climatology (Figure 5.4b) obtained from snow depth multiplying snow density (300 kg m³), highlighting the similar nature of the spatial patterns obtained with the two data sets.

Frappart *et al.* (2011) computed land water and snow mass anomalies versus time from GRACE data using the RL04 GeoForschungZentrum (GFZ) release. These estimates are then used to characterize the hydrology of the Arctic drainage system and are compared to snow water equivalent and snow depth climatologies, and to snowfall for validation purposes. GRACE data has also been assimilated into land models to improve SWE estimates. Su *et al.* (2010) establish a multi-sensor snow data assimilation system over North America via incorporating both GRACE

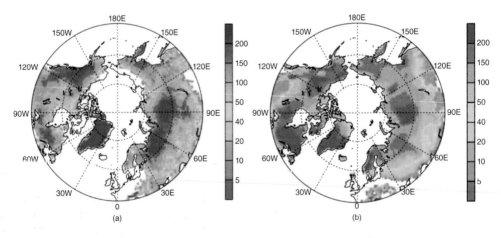

Figure 5.4 (a) Example of GRACE-based SWE (in mm) for March of 2004. (b) Ground-based USAF ETAC SWE climatology. (Niu *et al.*, 2007. Reproduced with permission of John Wiley & Sons Ltd).

terrestrial water storage and MODIS snow cover fraction information into the Community Land Model. Results show that this multi-sensor approach can provide significant improvements over a MODIS-only approach.

5.6 Passive microwave data

Approaches based on passive microwave data collected by spaceborne sensors have been proposed since the 1980s (e.g., Hall *et al.*, 1978, 1982, 1984, 1987), and operational algorithms have been used to produce global maps of snow depth and SWE on a daily basis at spatial resolutions on the order of tens of kilometers (Figure 5.5).

Generally, the microwave bands available on spaceborne sensors that are used for the retrieval are K-band ($\approx$19 GHz) and Ka-band ($\approx$37 GHz). Some sensors, as in the case of the AMSR-E, also collect data in the X-band ($\approx$10 GHz), which potentially can be used to address issues related to deep snow conditions. For retrieval purposes, the microwave signal recorded by passive microwave sensors over a snow covered terrain is approximated as the signal naturally emitted by the underlying soil attenuated by a homogeneous snow layer. The contribution from the dry snow layer itself is assumed to be negligible, because of the low emissivity of dry snow.

The attenuation of the microwave signal for a dry snow covered terrain increases with grain size and frequency (Mätzler, 1987), and it is due to the volumetric

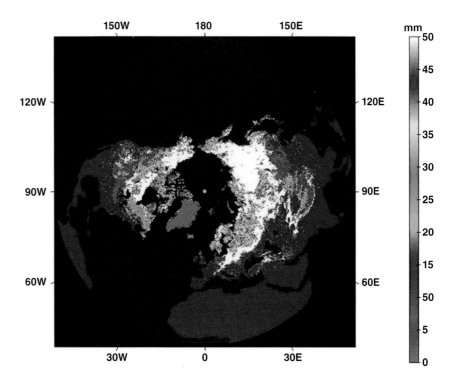

Figure 5.5 Five-day composite AMSR-E SWE product over the northern hemisphere for the period March 22–26, 2006. (NASA/NSIDC).

scattering occurring in the snowpack. Because of the dense packing of the ice particles, the independent scattering assumption is not valid, and theoretical efforts have been made to understand the interaction between snow particles and electromagnetic waves in the microwave region (e.g., Tsang and Kong, 2001; Tsang *et al.* 2000). Moreover, snow crystals tend to cluster in a snowpack after metamorphosis, and studies have been carried out to account for the clustering of non-spherical snow grains in the snowpack (e.g., "stickyness" parameter: Ding *et al.*, 2010; Zurk *et al.*, 1995).

Figure 5.6 shows the temporal trend of air temperature (upper plot), snow depth (second plot from the top) recorded by the World Meteorological Organization station of Njurba (WMO # 246390, Lat. 63.28 N, Lon. 118.33 E), QuikSCAT backscatter coefficients at vertical and horizontal polarizations (third plot from the top), brightness temperatures recorded by the Special Sensor Microwave Imager (SSM/I), with vertical polarization (second plot from the bottom) and horizontal polarization (bottom plot) between January 1st, 1999 and December 31st, 2003 (Tedesco and Miller, 2007). From the figure, we observe that the low frequencies (e.g., $\approx$19 GHz) show a relatively weaker sensitivity to snow depth on the ground during dry snow conditions than the higher frequencies (e.g., $\approx$37 GHz). This is a consequence of the extinction introduced by the volumetric scattering. The effect of wet snow on microwave brightness temperatures is discussed in Chapter 6 of this book.

Figure 5.7 shows modeled emissivity values at the frequencies of 10 GHz (Figure 5.7a) and 37 GHz (Figure 5.7b) as a function of the observation angle, assuming three different snow depth values of 0.5 m, 1 m and 1.5 m and a grain size of 0.5 mm. The remaining model parameters used in the simulations are reported in the inset of Figure 5.7b. The model used is based on the Dense Medium Radiative Transfer Theory (DMRT; e.g., Tsang *et al.*, 2000), which accounts for multiple volumetric scattering.

At 10 GHz, the signal is mostly dominated by the emissivity of the soil underlying the snowpack, and the sensitivity to snow depth is negligible. This is not the case at 37 GHz, where the emissivity of the snowpack reduces from $\approx$0.9 at the nadir ($\approx$244 K, assuming the snow temperature to be 271 K) in the case of the 0.5 m deep snowpack, to $\approx$0.8($\approx$215 K) in the case of the 1.5 m deep snowpack. Note that the decrease of emissivity with depth is not linear, and that a "saturation" depth value exists, above which the emissivity (and, consequently, the brightness temperature) will not decrease any more, though snow depth is still increasing. This is a consequence of the fact that when snow depth exceeds such a "saturation" value, the signal from the soil will be masked by the presence of the overlying snowpack. The saturation depth is inversely proportional to the frequency, and depends on snow parameters (e.g., grain size, density), being usually of the order of 60–80 cm at 37 GHz (the highest frequency used for snow depth retrieval), but it can be as low as 30–40 cm, or even lower, in the case of large grains.

The maximum snow depth that can be estimated from passive microwave observations cannot be greater than the saturation depth. For example, in Figure 5.6, the brightness temperature at 85 GHz is mostly affected by the top layer of the

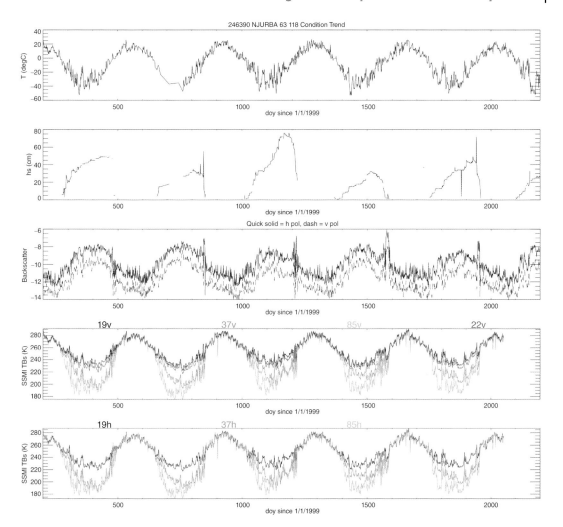

Figure 5.6 Temporal trend of air temperature (°C) and snow depth (cm) recorded at the WMO statin of Njurba (WMO id # 246390), together with QuikSCAT backscatter coefficient (third plot from the top), SSM/I brightness temperatures with vertical polarization plotted in the fourth plot from the top and horizontal polarization in the bottom plot (Tedesco & Miller 2007. Reproduced with permission of Elsevier).

snowpack and, therefore, cannot be used for snow depth retrieval purposes (unlike the 37 GHz channel, which is sensitive to snow depth variability).

There can be several sources of error in the retrieval of snow depth or SWE from spaceborne passive microwave data. Grain size evolution is a significant source of uncertainty for snow depth retrieval from passive microwave observations (e.g., Tedesco and Narvekar, 2010; Tedesco et al., 2010). After snow deposition, snow crystals metamorphose in response to vapor gradients within the snowpack, or as a result of melting and refreezing cycles.

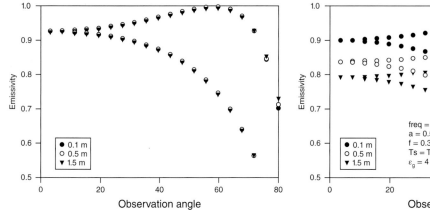

Figure 5.7 Modeled emissivity at (a) 10 GHz and (b) 37 GHz of a snowpack with different depths of 0.1, 0.5 and 1.5 m. Snow temperature is assumed to be 271 K, ground temperature 271 K. The snow particle radius is 0.5 mm and snow density is 0.3 g/cm³. Black (white) symbols refer to vertical (horizontal) polarization. Image and data courtesy of Tedesco (2003).

Although several models (either empirical or physically based) have been developed to predict the seasonal growth of snow crystals (e.g., Navarre, 1974; Marbouty, 1980; Jordan, 1991; Brun et al., 1992), it is not straightforward to select a general model that will account for regional to global scale conditions. Retrieval errors are also associated with the presence of multiple features within the same pixel, and to the heterogeneity of the scene observed by the sensor (e.g., mixed pixel effect). For example, the presence of forest attenuates the radiation emitted by the underlying snowpack, affecting the retrieval accuracy of the algorithm. The presence of water bodies within the area under study can also affect the retrieval of snow depth or SWE, because of the strong brightness temperature gradient between liquid and frozen water (Rees et al., 2006; Derksen et al., 2006, 2009).

Topography is also a source of uncertainty, and estimating snow storage in the mountainous region is extremely challenging. This is due either to the snow depth spatial variability or to the different observation angles within the sensor footprint because of the complex topography. Lastly, despite the fact that early studies have neglected atmospheric effects, recent work shows that they should be accounted to reduce retrieval errors further (e.g., Wang and Tedesco, 2007).

Empirical techniques based on the inverse relationship between brightness temperature and snow depth were the first to be proposed in the literature (e.g., Chang et al., 1978, 1979, 1986, 1989, 1980, 1982, 1985, 1987a, 1987b; Grody and Basist, 1996; Goodison and Walker, 1995; Tait, 1998). In particular, because of the different sensitivity of brightness temperature to snow depth and SWE at different frequencies (e.g., K- and Ka-band, Figure 5.6), algorithms based on the direct relationship between snow depth and brightness temperature spectral difference were proposed (Chang et al., 1987a):

$$SD = R_c(Tb_{19} - Tb_{37}) \qquad (5.3)$$

Such algorithms are computationally inexpensive (hence ideal for large scale applications) but they are limited by their simplicity. The regression coefficient R_c was initially treated as a constant in both space and time and was set to 1.59 cm/K in the case of snow depth and 4.8 mm/K in the case of SWE (assuming a fixed snow density of 0.3 g/cm^3). This value was obtained from ground radiometric observations conducted in conjunctions with snow properties measurements during the 1980s (Foster, Pers. Comm.). In forested-covered areas, Chang and Chiu (1991) proposed the formula:

$$\text{SWE} = a(\text{Tb}_{19H} - \text{Tb}_{37H}) + f(\text{Tb}'_{19H} - \text{Tb}'_{37H}) \qquad (5.4)$$

where:

f is the forest cover fraction within the satellite footprint (ranging from 0 to 1)
Tb'_{19H} and Tb'_{37H} are the brightness temperature measured for a snow-free forest-covered pixel.

The expression for R_c was, therefore, later modified by Foster $et\ al.$ (1997) to account for the presence of forest, through a revised formula for the retrieval coefficient given by:

$$R_c = 1.59/(1 - f) \qquad (5.5)$$

An alternative linear regression algorithm was proposed by Hallikainen (1984) in which SWE is computed as:

$$\text{SWE} = a\Delta\text{Tb} + b \qquad (5.6)$$

with Tb, begin:

$$\Delta\text{Tb} = (\text{Tb}_{19V} - \text{Tb}_{37V})_{\text{winter}} - (\text{Tb}_{19V} - \text{Tb}_{37V})_{\text{autumn}} \qquad (5.7)$$

The so-called Spectral and Polarization Difference (SPD) algorithm was also proposed by Aschbacher (1989) for the retrieval of SWE when no information on the land cover categories in the area of interest is available. The algorithm is based on a combination of channels to generate the SPD:

$$\text{SPD} = [\text{Tb}_{19V} - \text{Tb}_{37V}] + [\text{Tb}_{19V} - \text{Tb}_{19H}] \qquad (5.8)$$

that is then used to estimate SD or SWE following:

$$\text{SD} = A_0 * \text{SPD} - A_1 \qquad (5.9)$$
$$\text{SWE} = B_0 * \text{SPD} - B_1 \qquad (5.10)$$

The coefficients appearing in the above equations were computed to be $A_0 = 0.68$, $A_1 = -0.67$ and $B_0 = 2.20$, $B_1 = -7.11$ for all data, and $A_0 = 0.72$, $A_1 = -1.24$ and $B_0 = 2.02$, $B_1 = -7.42$ if the maximum daily temperature is lower than 0°C.

In the mid-2000s, studies addressing the spatio-temporal variability of the retrieval coefficients were published in the literature. Foster *et al.* (2005) proposed an algorithm in which R_c depends on monthly climatology and on snow classes defined following Sturm *et al.* (1995). Grippa *et al.* (2004) estimated seasonal SWE in Siberia, also computing R_c in a dynamic fashion, and Kelly *et al.* (2003) proposed a dynamic algorithm based on the original baseline algorithm by Chang *et al.* (1987a), which accounts for the evolution of grain size and density through an exponential growth model for grain size and through the analysis of the outputs of an electromagnetic model.

Empirical approaches have the advantage of being computationally inexpensive and are, therefore, well suited for large-scale applications. On the other hand, they suffer from the local nature of the coefficients relating snow depth and SWE to the measured brightness temperatures, as well as from their spatio-temporal static nature. Empirical algorithms are typically derived from field campaigns that capture only a narrow range of snowpack conditions; hence, algorithm uncertainty can increase dramatically when these conditions change due to unique meteorological forcing.

Theoretical and semi-empirical approaches making use of electromagnetic models have also been proposed in the literature (e.g., Tedesco *et al.*, 2004; Tedesco and Kim, 2006; Pulliainen *et al.*, 1999; Tsang *et al.*, 1992). In this case, snow depth is generally estimated by fitting the snow parameters in the electromagnetic model that minimize the difference between measured and modeled brightness temperatures. For example, in the Helsinki University of Technology (HUT) snow emission model-based iterative inversion algorithm (e.g., Pulliainen *et al.*, 1999), it is assumed that the random error in brightness temperature predictions and the fluctuation of snow grain size are normally distributed, and the conditional probability for SWE and grain size D_0. The cost function to be minimized for yielding the maximum likelihood estimate for SWE and grain size D_0 is defined as:

$$C(SWE, D_0) = \frac{1}{2\sigma^2} \{[Y1(SWE, D_0) - X1]^2 + [Y2(SWE, D_0) - X2]^2\}$$
$$+ \frac{1}{2\lambda^2}(D_0 - \langle D_0 \rangle)^2 \qquad (5.11)$$

where:

$X1$ and $X2$ are the observations (e.g., measured brightness temperatures at two frequencies)
$Y1$ and $Y2$ are the modeled brightness temperatures
$\langle D_0 \rangle$ is the expected value (*a priori* information) of snow grain size (diameter)
σ is the standard deviation of the channel differences
λ is the standard deviation of snow grain size.

Other approaches also making use of electromagnetic models rely on more elaborated numerical methods to identify those parameters that minimize the difference between measured and simulated brightness temperatures. For example, Tedesco *et al.* (2004) proposed an approach in which artificial neural networks

(ANNs) are trained with the outputs of an electromagnetic model, and snow parameters are obtained still from the minimization of measured and simulated brightness temperatures.

Although approaches based on electromagnetic models are generally expected to provide better results than empirical methods, large retrieval errors can still exist, due, among other things, to the *ill-posed* nature of the problem. An ill-posed problem is characterized by the following conditions:

- that it is not possible to affirm that the solution always exists;
- that if the solution exists then it is not unique; and
- that different sets of snow parameters may correspond to the same set of brightness temperatures and a small change in the problem may lead to a big change in the solution.

The ill-posed nature of the problem in the case of snow depth/SWE retrieval from microwave data consists of the fact that there are several combinations of snow parameters that can correspond to a given set of measured brightness temperatures. For example, a relatively thin snowpack with large snow grains can have similar brightness temperature values to a deep snowpack with relatively small grains. Though the brightness temperatures in the two examples can be different if computed by means of an electromagnetic model, such difference is small compared to the magnitude of the error associated to the data and to the convergence criteria associated to the minimization techniques.

Without *a priori* knowledge of the snowpack conditions (e.g., thin with large grains vs. thick with small grains), the solution computed from the minimization of the measured and modeled brightness temperatures might not be stable around a minimum, and small perturbations in the input (e.g., the measured minus modeled brightness temperature) could result in large changes in the output (e.g., snow estimated parameters).

To mitigate this aspect, recent studies have been assimilating ancillary information into remote sensing retrieval schemes. Takala *et al.* (2011), for example, proposed an algorithm assimilating synoptic weather station data on snow depth with satellite passive microwave radiometer data, to produce a 30-year-long time series of seasonal SWE for the northern hemisphere, pointing out the benefits of the developed assimilation approach with respect to a stand-alone passive microwave data algorithm.

Operational daily estimates of snow depth and SWE from spaceborne microwave observations are produced and distributed by several agencies or groups. Microwave brightness temperatures measured by AMSR-E were used to generate the NASA SWE product for land surfaces (Tedesco *et al.*, updated 2012). The product was terminated following the failure of the instrument, due to a problem with the rotation of its antenna in October 2011. The AMSR-E SWE product suite is still distributed for the period 2002–2011 and consists of daily, pentad (five-day) maximum and monthly average SWE estimates that, together, comprise the only NASA satellite-based SWE product available to the scientific community. Further information on the current SWE AMSR-E operational algorithm can be found at http://nsidc.org/data/docs/daac/ae_swe_ease-grids.gd.html.

The operational Microwave Integrated Retrieval System (MIRS) at NOAA produces microwave products from various satellites with different instrumental configurations. The goal of the operational MIRS is to produce advanced near-real-time surface and precipitation products in all-weather and over all-surface conditions, using brightness temperatures from available microwave instruments (http://www.ospo.noaa.gov/products/atmosphere/mirs/index.html). In particular, SWE is made available to users with different types and formats, through a multi-year stratified phase approach.

The European Space Agency (ESA) GlobSnow project aims at creating a global database of SWE for the Northern Hemisphere building on the results reported in Pulliainen (2006). An assimilation technique, using a Bayesian approach, weighs the spaceborne data and a reference snow depth field, interpolated from discrete synoptic observations with their estimated statistical accuracy. The results indicate that, over northern Eurasia and Finland, the employment of spaceborne data using the assimilation technique improves the snow depth and SWE retrieval accuracy when compared with the use of values obtained from the interpolation from synoptic observations (Pulliainen, 2006).

In general, this approach provides improved results (reduced RMSE; higher correlation) when evaluated with independent reference data by comparison to standalone passive microwave retrieval algorithms. However, it must be pointed out that this kind of approach has the advantage of using discrete synoptic observations, where the other approaches discussed rely on the sole use of remotely sensed data. More information about the GlobSnow project can be found at www.globsnow.info.

5.7   Active microwave data

The total backscattered power from a snow-covered terrain is due to the contribution of different processes that can be divided into two categories: surface scattering and volume scattering. The incident wave is partially reflected and partially transmitted at the air-snow interface. The transmitted portion of the radiation is attenuated by scattering by ice particles, by scattering at the boundaries with air and ground interfaces and by scattering at the boundaries of different snow layers. The backscattering from snow-covered terrain σ_{tot}^0 can, therefore, be expressed as:

$$\sigma_{tot}^0 = \sigma_{air-snow}^0 + \sigma_{vol}^0 + \sigma_{snow-ground}^0 \times f(\theta_i, d, \tau_{as}, \tau_{sg}) \tag{5.12}$$

where:

σ_i is the incidence angle
σ_t is the transmitted angle
d is the snow depth
τ_{as} and τ_{sg} are, respectively, the Fresnel power transmission for the interface air-snow and snow-ground.

The first term in the equation describes the backscattering form air-snow boundary, the second describes the volume scattering, and the last term describes the backscattering from snow-ground boundary attenuated by a factor depending on the snowpack properties and incidence angle. The surface scattering depends on the permittivities of adjacent layers on the incidence angle and surface parameters. The latter ones can be described statistically by means of the standard deviation of the surface height variation, the surface correlation length and the correlation function.

The contribution of the surface scattering by the air-snow interface in the case of dry snow is, generally, very low. This mainly depends on the fact that the permittivity contrast between air and dry snow permittivity is low and, at low frequencies, the sensitivity to the surface roughness is weak. Figure 5.6 shows the sensitivity of backscattering coefficients measured by the QuikSCAT instrument (13.4 GHz) to snow depth. Many of the spaceborne active microwave sensors have been collecting data in the C- and X-bands. These bands are not optimal for snow depth retrieval, in view of the low sensitivity of the measured backscattering coefficients to snow depth. Nevertheless, the higher sensitivity of Ku-band active data to snow depth collected by sensors such as QuikSCAT has allowed the development of algorithms for accumulation estimates over ice sheets (see Chapter 8) or snow depth over land.

For example, Hallikainen *et al.* (2003) examined the use of data from QuikSCAT and SSM/I for monitoring key snow parameters in Finland, indicating that a Ku-band scatterometer with a fixed incidence angle can provide reasonable accuracy for retrieval of dry SWE and that retrieval accuracy is better using the combined active/passive, rather than only the SSM/I data. Similarly, Tedesco and Miller (2007) analyzed the trend of QuikSCAT backscatter coefficients in conjunction with K- and Ka-band brightness temperatures measured by SSM/I with respect to snow depth values at different locations in the northern hemisphere during the period 1999–2004. The authors quantify the dynamic range of spaceborne Ku-band scatterometer data over snow-covered areas at very large spatial scale, versus that of passive microwave brightness temperatures. Potential improvement on snow depth retrieval related to the combined use of active and passive data, with respect to the sole use of passive data, is indicated.

The potential application of active Ku-band data to SWE retrieval has also been confirmed using airborne data. Yueh *et al.* (2009) studied Ku-band polarimetric scatterometer (POLSCAT) data acquired from five sets of aircraft flights in the winter months of 2006–2008 during the second Cold Land Processes Experiment (CLPX-II) in Colorado. The authors observe about 0.15–0.5 dB increases in backscatter for every 1 cm of SWE accumulation for areas with short vegetation. Moreover, the comparison between POLSCAT and QuikSCAT data points out the effects of mixed terrain covers in the coarse-resolution QuikSCAT data. Studies to support the CoreH20 mission (one of the three Earth Explorer Core mission candidates that, however, was not selected to move to the next phase in May 2013), have been focusing on the combination of two synthetic aperture radars at 9.6 and 17.2 GHz to retrieve snow depth and other snow parameters (see Chapter 15; https://earth.esa.int/web/guest/content?p_r_p_564233524_assetIdentifier= coreh20).

Beside those approaches based on the relationships between backscattering coefficients and snow depth/SWE, techniques using synthetic aperture radar (SAR) have also been reported in the literature. Bernier and Fortin (1998) assessed the potential of C-band synthetic aperture radar (SAR) data to determine SWE. The authors used airborne data collected during three winters over a watershed in the Appalachian Mountains in Southern Quebec, Canada, together with extensive simultaneous ground measurements. To estimate the SWE of a given snowpack, they developed a model linking the scattering coefficient to the physical parameters of the snow cover and the underlying soil, based on the ratio of the scattering coefficient of a field covered by snow to the scattering coefficient of a field without snow. Results indicated that volume scattering from a dry snow cover with SWE <20 cm is undetectable, and pointed to a unique relationship between the backscattering power ratio and the thermal resistance of the snowpack. Since linear relationships between SWE and R have been observed, the authors suggest that it should be possible to estimate the SWE of shallow dry snow cover with C-band SAR data, using few ground truthing data in an open area when the soil is frozen.

Bernier *et al.* (1999) developed a relationship between the backscattering ratios of a winter image and a reference (snow-free) image to estimate the snowpack thermal resistance using the Radarsat SAR. Shi and Dozier (2000a, 2000b) obtained density estimates from fully polarimetric measurements and estimated snow depth and particle size using SIR-C/X-SAR imagery using a physically-based first order backscattering model.

Interferometric approaches have also been proposed in the literature to estimate SWE. Guneriussen *et al.* (2001) first described the theoretical relation between interferometric phase and changes in SWE, with results from experiments using European Remote sensing Satellite 1 (ERS-1) and ERS-2 tandem data. The basic assumption in this study is that the main scattering contribution from a dry snow cover is from the snow-ground interface. However, because the radar wave will be refracted in the snow, small changes in the snow properties between two SAR images will change the interferometric phase.

Building on this, Engen *et al.* (2004) used interferometric synthetic aperture radar (InSAR) data for deriving SWE of dry snow-covered ground from ERS-1 data. Rott *et al.* (2003) investigated the feasibility for retrieving SWE from repeat-pass SAR interferometry, using theoretical modeling and analysis of repeat pass SAR data, indicating that the C-band coherence often deteriorates rapidly in case of snowfall, whereas coherence is much better preserved at L-band.

5.8 Conclusions

Estimates of snow depth and SWE at different spatial and temporal scales can benefit hydrological, climatological and other societal applications, such as water management. Snow depth can be estimated by means of photogrammetric techniques, from LiDAR and RADAR measurements and GPS data. SWE can also directly be retrieved from gamma radiation, or through the combination of gravity measurements and model outputs. Spaceborne passive microwave measurements have been used to generate daily, global estimates of snow depth and SWE. The sensitivity of microwave data to sub-surface processes, together with the relatively weak sensitivity to atmospheric parameters, the existence of a long-term data set (more than 30 years) and the large spatial coverage, are the major drivers of the wide use of passive microwave observations for snow depth/SWE estimates.

Obviously, any of the approaches described above has advantages and disadvantages. The coarse spatial resolution of passive microwave data limits its use in relatively small basins, where techniques based on optical data might be more suitable. At the same time, optical data are limited by the presence of cloud coverage and by other factors such as illumination and spatial coverage.

Over the past decades, many remote sensing scientists have been focusing on the development of theoretical tools that have allowed the scientific community to study the interaction between snow particles and electromagnetic radiation, and to develop and refine retrieval tools from both airborne and spaceborne platforms. Although much progress has been made over the past years, major challenges still exist. Besides intrinsic issues (e.g., wet snow opacity to microwaves, saturation depth, etc.), the refinement of remote sensing approaches for snow depth and SWE retrieval can be performed, for example, by improving the algorithm's capability to account for the spatio-temporal dynamic nature of the driving factors (e.g., grain size, density). In many approaches, snow depth is first estimated from remote sensing observations, then SWE is derived by multiplying snow depth estimates by snow density, which is also unknown. Therefore, the improvement of SWE retrieval does not only depend on improving snow depth estimates, but also on improving snow density values used to extract SWE from snow depth.

Moreover, the *ill-posed* nature of the retrieval problem in the case of passive microwave data poses great challenges on those algorithms that are solely based on remote sensing data. In this regard, over the past years there has been an increasing interest in remote sensing algorithms making use of ancillary information, with the best results obtained when using observations of snow parameters to derive the initial conditions.

One of the aspects that has allowed these new algorithms to be developed is the availability of increased computational resources compared with even just a few years ago. It is helpful to remember here that the first remote sensing based algorithms were developed in the late 1970s, when computers were starting to be developed, electromagnetic models accounting for the multiple scattering mechanisms were being conceived and the generation of modeled snow depth fields at global scale was a far possibility. Today, the outcome of the efforts of many remote sensing experts, in concert with the birth and evolution of land system models, the increasing computational power of computers and the long-term time series of passive microwave observations, offer the possibility of producing improved estimates.

Nevertheless, if it is true that maximizing the benefit from the combination of multiple techniques and tools is crucial for the reduction of errors and uncertainty, it is also true that the intrinsic improvement of each of the components of a retrieval system represents a crucial step for future developments. Therefore, focusing on how to assess and refine current remote sensing approaches that rely only on remote sensing observations is still a fundamental step toward the overall improvement of estimates of snow parameters.

References

Aschbacher J. (1989). Land surface studies and atmospheric effects by satellite microwave radiometry. PhD thesis dissertation, University of Innsbruck, Austria.

Bernier, M. & Fortin, J-P. (1998). The potential of times series of C-Band SAR data to monitor dry and shallow snow cover, *Geoscience and Remote Sensing, IEEE Transactions on Geoscience and Remote Sensing* **36**(1), 226–243.

Bernier, M., Fortin, J.-P., Gauthier, Y., Gauthier, R., Roy, R. & Vincent, P. (1999). Determination of snow water equivalent using RADARSAT SAR in eastern Canada, *Hydrological Processes* **13**, 3041–3051.

Brun, E., P. David, Sudul M. & Brunot G. (1992). A numerical model to simulate snowcover stratigraphy for operational avalanche forecasting, *Journal Of Glaciology* **38**, 13–22.

Carroll, S.S. & Carroll, T.R. (1989). Effect of uneven snow cover on airborne snow water equivalent estimates obtained by measuring terrestrial gamma radiation. *Water Resources Research* **25**(7), 1505–1510.

Chang, A.T.C. & Chiu, L.S. (1991). *Satellite estimation of snow water equivalent: classification of physiographic regimes*. Proc. Of the International Geoscience and Remote Sensing Symposium, Espoo, Finland.

Chang, A.T.C., Hall, D.K., Foster, J.L., Rango, A. & Schmugge, T.J. (1978). *Studies of Snowpack Properties by Passive Microwave Radiometry*. NASA Tech. Memo. 79761.

Chang, A.T.C., Hall, D.K., Foster, J.L., Rango, A. & Shiue, J. (1979). *Passive Microwave Sensing of Snow Characteristics Over Land, Satellite Hydrology*. Proc. of the 5th Annual William T. Pecora Memorial Symp., Sioux Falls, South Dakota.

Chang, A.T.C., Foster, J.L. & Hall, D.K. (1986a). Microwave Snow Signatures (1.5 mm to 3 cm) Over Alaska. *Cold Regions Science and Technology* **13**, 153–160.

Chang, A.T.C., Foster, J.L. & Hall, D.K. (1986b). Nimbus-7 SMMR Derived Global Snow Cover Parameters. *Annals of Glaciology* **9**, 39–44.

Chang, A.T.C, Foster, J.L. & Hall, D.K. (1987a). Nimubus-7 SMMR derived global snow cover parameters. *Annals of Glaciology* **9**, 39–44.

Chang, A.T.C., Foster, J.L., Gloersen, P., *et al.* (1987b). Estimating Snow Parameters in the Colorado River Basin by Microwave Radiometry. In: *Large Scale Effects of Seasonal Snow Cover*, pp. 403–413. Vancouver, British Columbia, Canada.

Chang, A.T.C., Foster, J.L. & Hall, D.K. (1989). *Effect of Vegetation Cover on Microwave Snow Water Equivalent Estimates*. Proc. on the Int. Symp. on Remote Sens. of Water Resources, pp. 137–145. The Netherlands Society for Remote Sensing.

Chang, A.T.C., Foster, J.L., Hall, D.K. & Rango, A. (1980). *Monitoring Snowpack Properties by Passive Microwave Sensors On-board Aircraft and Satellites*. Proc. of Workshop on Microwave Remote Sensing of Snowpack Properties, Ft. Collins, CO.

Chang, A.T.C., Foster, J.L., Hall, D.K., Rango, A. & Hartline, B. (1982). Snow Water Equivalent Determination by Microwave Radiometry. *Cold Regions Science and Technology* **5**, 259–267.

Chang, A.T.C., Foster, J.L., Owe, M., Hall, D.K. & Rango, A. (1985). Passive and Active Microwave Studies of Snowpack Properties, Results of March, 1981 Aircraft Mission. *Nordic Hydrology* **16**(2), 57–66

Cline, D. W. (1993). *Measuring Alpine snow depths by digital photogrammetry, Part 1, Conjugate point identification*. 50th Eastern snow Conference and 63rd Western snow conference, Quebec City, Quebec, Canada.

Deems JS, Fassnacht SR and Elder K (2006). Fractal distribution of snow depth from LiDAR data. *Journal of Hydrometeorology* **7**(2), 285–297 (doi: 10.1175/JHM487.1).

Deems, J.S., Painter T.H., Finnegan D.H. (2013). LiDAR measurements of snow depth: a review. *Journal of Glaciology* **59**(215), 467–479. doi:10.3189/2013JoG12J154.

Derksen, C., Strapp, W. & Walker, A. (2006). *Passive microwave brightness temperature scaling over snow covered boreal forest and tundra*. Proceedings, IGARSS 2006.

Derksen, C., Silis, A., Sturm, M., *et al.* (2009). Northwest Territories and Nunavut Snow Characteristics from a Subarctic Traverse: Implications for Passive Microwave Remote Sensing, *Journal of Hydrometeorology* **10**(2), 448–463.

Ding, K H., Xu, X. & Tsang, L. (2010). Electromagnetic Scattering by Bicontinuous Random Microstructures with Discrete Permittivities. *IEEE Transactions on Geoscience and Remote Sensing* **48**(8), 3139–3149

Engen, G., Guneriussen, T., Overrein, Y. (2004). Delta-K interferometric SAR technique for snow water equivalent (SWE) retrieval, *IEEE Geoscience and Remote Sensing Letters* **2**, 57–61.

Forman, B.A., Reichle, R.H. & Rodell, M. (2012). Assimilation of terrestrial water storage from GRACE in a snow-dominated basin, *Water Resources Research* **48**, W01507. doi:10.1029/2011WR011239.

Foster, J., Chang, A.T.C. & Hall, D. (1997). Comparison of snow mass estimates from a prototype passive microwave snow algorithm, a revised algorithm and a snow depth climatology. *Remote Sensing of Environment* **62**(2), 132–142.

Foster, J.L., Sun, C.J., Walker, J.P., *et al.* (2005). Quantifying the uncertainty in passive microwave snow water equivalent observations. *Remote Sensing of Environment* **94**(2), 187–203.

Frappart, F., Ramillien, G., Biancamaria, S., Mognard, N.M. & Cazenave, A. (2006). Evolution of high-latitude snow mass derived from the GRACE gravimetry mission (2002–2004). *Geophysical Research Letters* **33**, L02501. doi:10.1029/2005GL024778.

Frappart F., Ramillien G. & Famiglietti J.S. (2011). Water balance of the Arctic drainage system using GRACE gravimetry products. *International Journal of Remote Sensing* **32**(2), 431–453. doi : 10.1080/01431160903474954.

Glennie, C. (2007). Rigorous 3D error analysis of kinematic scanning LIDAR systems. *Journal of Applied Geodesy* **1**, 147–157. doi:10.1515/JAG.2007.

Glennie, C.L., W.E. Carter, R.L. Shrestha & W.E. Dietrich (2013). Geodetic imaging with airborne LiDAR: the Earth's surface revealed. *Reports on progress in physics. Physical Society (Great Britain)* **76**(8), 086801. doi:10.1088/0034-4885/76/8/086801.

Goodison, B.E. & Walker, A.E. (1995). Canadian development and use of snow cover information from passive microwave satellite data, In: Choudhury, B.J. Kerr, Y.H. Njoku, E.G. & Pampaloni, P. (eds). *Passive microwave remote sensing of land-atmosphere interactions.* Zeist: VSP BP.

Grody, N.C. & Basist, A.N. (1996). Global identification of snowcover using SSM/I measurements. *IEEE Transactions on Geoscience and Remote Sensing* **34**(1), 237–249.

Guneriussen, T., Hogda, K.A., Johnsen, H. & Laukknes, I. (2001). InSAR for estimation of changes in snow water equivalent of dry snow. *IEEE Transactions on Geoscience and Remote Sensing* **39**(10), 2101–2108.

Hall, D.K., Chang, A., Foster, J.L., Rango, A. & Schmugge, T. (1978). *Passive Microwave Studies of Snowpack Properties.* NASA Tech. Memo. 78089.

Hall, D.K., Foster, J.L. & Chang, A.T.C. (1982). Measurement and Modeling of Microwave Emission from Forested Snowfields in Michigan. *Nordic Hydrology* **13**, 129–138.

Hall, D.K., Foster, J.L. & Chang, A.T.C. (1984). Nimbus 7 Polarization Responses to Snow Depth in the Mid-Western US. *Nordic Hydrology* **15**(5), 1–8.

Hall, D.K., Chang, A.T.C. & Foster, J.L. (1987). Distribution of Snow Extent and Depth in Alaska as Determined from Nimbus-7 SMMR Maps (1982–1983), In: *Large Scale Effects of Seasonal Snow Cover, Vancouver, British Columbia, Canada,* pp. 343–352.

Hallikainen, M.T. (1984). Retrieval of snow water equivalent from Nimbus-7 SMMR data: effect of land cover categories and weather conditions. *IEEE Journal of Ocean and Engineering* **9**(5), 372–376.

Hallikainen, M., Halme, P., Lahtinen, P. & Pulliainen, J. (2003). Combined active and passive microwave remote sensing of snow in Finland. Proc. Of IEEE geoscience and remote sensing symposium, IGARSS 2003, Toulouse, France.

Hodgson, M.E. & Bresnahan, P. (2004). Accuracy of airborne LiDAR derived elevation: empirical assessment and error budget. *Photogrammetric Engineering and Remote Sensing* **70**(3), 331–339.

Hopkinson C, Sitar M, Chasmer L and Treltz P (2004). Mapping snowpack depth beneath forest canopies using airborne LiDAR. *Photogrammetric Engineering and Remote Sensing* **70**(3), 323–330.

Jordan, R. (1991). *A one-dimensional temperature model for a snow cover.* CRREL Special Report 91–16, 64 pp.

Kelly, R.E.J., Chang, A.T.C., Tsang, L. & Foster, J.L. (2003). Development of a prototype AMSR-E global snow area and snow depth algorithm. *IEEE Transactions on Geoscience and Remote Sensing* **41**(2), 230–242.

Larson, K.M., Gutmann, E.D., Zavorotny, V.U., Braun, J.J., Williams, M.W. & Nievinski, F.G. (2009). Can we measure snow depth with GPS receivers? *Geophysical Research Letters* **36**, L17502. doi:10.1029/2009GL039430.

Marbouty, D. (1980). An experimental study of temperature-gradient metamorphism, *Journal of Glaciology* **26**, 303–312.

Mätzler, C (1987). Applications of the interaction of microwaves with the natural snow cover. *Remote Sensing Reviews* **2**, 259–387.

Miller, S. 2004. *CLPX-Airborne: Infrared Orthophotography and LiDAR Topographic Mapping.* Boulder, Colorado, USA: NASA DAAC at the National Snow and Ice Data Center.

Mott, R., Schirmer, M. & Lehning, M. (2011). Scaling properties of wind and snow depth distribution in an Alpine catchment. *Journal of Geophysical Research* **116**(D6), D06106. doi: 10.1029/ 2010JD014886.

Navarre, J.P. (1974). Modele undimensionnel d'evolution de la neige depose. *La Meteroroloie* 109–120.

Niu, G.-Y., Seo, K.-W., Yang, Z.-L., *et al.* (2007). Retrieving snow mass from GRACE terrestrial water storage change with a land surface model. Geophysical Research Letters **34**, L15704. doi:10.1029/2007GL030413.

Offenbacher, E.L. & Colbeck, S.C. (1998). *Remote Sensing of Snow Covers Using the Gamma-Ray Technique.* CRREL Report.

Peck, E.L., Bissell, V.C., Jones, E.B. & Burge, D.L. (1971). Evaluation of Snow Water Equivalent by Airborne Measurement of Passive Terrestrial Gamma Radiation. *Water Resources Research* **7**(5), 1151–1159. doi:10.1029/WR007i005p01151.

Pulliainen, J. (2006). Mapping of snow water equivalent and snow depth in boreal and sub-arctic zones by assimilating space-borne microwave radiometer data and ground-based observations *Remote Sensing of Environment* **101**(2), 257–269.

Pulliainen, J.T., Grandell, J. & Hallikainen, M. (1999). HUT snow emission model and its applicability to snow water equivalent retrieval. *IEEE Transactions on Geoscience and Remote Sensing* **37**(3), 1378–1390.

Rees, A., Derksen, C., English, M., Walker, A. & Duguay, C. (2006). Uncertainty in snow mass retrievals from satellite passive microwave data in lake-rich high-latitude environments. *Hydrological Processes* **20**(4), 1019–1022.

Rieger, P. & Ullrich, A. (2011). Resolving range ambiguities in high-repetition rate airborne LiDAR applications. *Proceedings of SPIE* **8186**(9), 81860A–8.

Rott, H., Nagler, T. & Scheiber, R. (2003). *Snow mass retrieval by means of SAR interferometry*. In: 3rd FRINGE Workshop (European Space Agency ESA), available at http://earth.esa.int/workshops/fringe03/participants/17/paper_Fringe03_Rott_etal.pdf.

Shi, J. & Dozier, J. (2000a). Estimation of snow water equivalence using SIR–C/X SAR, Part I: inferring snow density and subsurface properties. *IEEE Transactions on Geoscience and Remote Sensing* **38**(6), 2465–2474.

Shi, J. & Dozier, J. (2000b). Estimation of snow water equivalence using SIR–C/X SAR, Part II: Inferring snow depth and particle size. *IEEE Transactions on Geoscience and Remote Sensing* **38**(6), 2475–2488.

Sturm, M., Holmgren, J. & Liston, G.E. (1995). A seasonal snow cover classification system for local to global applications. *Journal of Climate* **8**(5), 1261–1283.

Su H., Yang Z.-L., Dickinson R.E., Wilson C.R., Niu G.-Y. (2010). Multisensor snow data assimilation at the continental scale: The value of Gravity Recovery and Climate Experiment terrestrial water storage information. *Journal of Geophysical Research: Atmospheres (1984–2012)* **115**, D10, 2156–2202; 10.1029/2009JD013035.

Tait, A. (1998). Estimation of snow water equivalent using passive microwave radiation data. *Remote Sensing of Environment* **64**(3), 286–291.

Takala,M., Luojus, K., Pulliainen, J., Derksen, C., Lemmetyinen, J., Kärnä, J.-P., Koskinen, J. & Bojkov, B. (2011). Estimating northern hemisphere snow water equivalent for climate research through assimilation of space-borne radiometer data and ground-based measurements. *Remote Sensing of Environment* **115**(12), 3517–3529. doi:10.1016/j.rse.2011.08.014.

Tedesco, M. (2003). *Microwave remote sensing of snow*. PhD Thesis, Institute of Applied Physics Carrara, Firenze, Italy.

Tedesco, M. & Kim, E.J. (2006). A study on the retrieval of dry snow parameters from radiometric data using a dense medium model and genetic algorithms. *IEEE Transactions on Geoscience and Remote Sensing* **44**(8), 2143–2151.

Tedesco, M. & Miller, J. (2007). Observations and statistical analysis of combined active–passive microwave space-borne data and snow depth at large spatial scales. *Remote Sensing of Environment* **111**(2–3), 382–397.

Tedesco, M. & Narvekar, P. (2010). Assessment of the NASA AMSR-E SWE Product. *IEEE Journal of Selected Topics in Applied Earth Observations and Remote Sensing* **3**(1), 141–159.

Tedesco, M., Pulliainen, J., Pampaloni, P. & Hallikainen, M. (2004). Artificial neural network based techniques for the retrieval of SWE and snow depth from SSM/I data. *Remote Sensing of Environment* **90**(1), 76–85.

Tedesco, M., Reichle, R., Loew, A., Markus, T. & Foster, J.L. (2010). Dynamic Approaches for Snow Depth Retrieval From Spaceborne Microwave Brightness Temperature. *IEEE Transactions on Geoscience and Remote Sensing* **48**(4), 1955–1967.

Tedesco, M., Kelly, R.E.J., Foster, J.L. & Chang, A.T.C (2011 updated). AMSR-E/Aqua Daily L3 Global Snow Water Equivalent EASE-Grids V002. National Snow and Ice Data Center Digital Media, Boulder Colorado, USA.

Tsang, L. & Kong, J. (2001). *Scattering of electromagnetic waves: Advanced topics.* Wiley-Interscience.

Tsang, L., Chen, Z., Oh, S., Marks, R.J. & Chang, A.T.C. (1992). Inversion of snow parameters from passive microwave remote sensing measurements by a neural network trained with a multiple scattering model. *IEEE Transactions on Geoscience and Remote Sensing* **30**(5), 1015–1024.

Tsang, L., Chi-Te Chen, A., Guo, J. & Ding, K. (2000). Dense media radiative transfer theory based on quasicrystalline approximation with applications to passive microwave remote sensing of snow. *Radio Science* **35**(3).

Wang, J. & Tedesco, M. (2007). Identification of atmospheric influences on the estimation of snow water equivalent from AMSR-E measurements. *Remote Sensing of Environment* **111**(2–3), 398–408.

Yueh, S.H., Dinardo, S.J., Akgiray, A., West, R., Cline, D.W. & Elder, K. (2009). Airborne Ku-Band Polarimetric Radar Remote Sensing of Terrestrial Snow Cover. *IEEE Transactions on Geoscience and Remote Sensing* **47**(10), 3347.

Zurk, L.M., Tsang, L., Ding, K.H. & Winebrenner, D.P. (1995). Monte Carlo Simulations of the Extinction Rate of Densely Packed Spheres with Clustered and Non Clustered Geometries. *Journal of the Optical Society of America* **12**(8), 1771–1781.

Acronyms

AGL	Altitude above ground level
ANNs	Artificial neural networks
CLPX-II	Second Cold Land Processes Experiment
CSR	Center for Space Research
DEM	Digital elevation model
DMRT	Dense Medium Radiative Transfer Theory
ERS-1 and ERS-2	European Remote sensing Satellite 1 and 2
ESA	The European Space Agency
ETAC	Environmental Technical Application Center
GFZ	GeoForschungZentrum
GPS	Global positioning system
GRACE	Gravity Recovery and Climate Experiment
HUT	Helsinki University of Technology
IMU	Inertial measuring unit
InSAR	Interferometric synthetic aperture radar
LiDAR	Light Detection And Ranging
MIRS	Microwave Integrated Retrieval System
MpiA	Multiple-Pulse-in-Air
MTA	Multiple-Time-Around
NOHRSC	National Operational Hydrologic Remote Sensing Center
POLSCAT	Polarimetric scatterometer

PRF	Pulse repetition frequency
SAR	Synthetic aperture radar
SPD	Spectral and Polarization Difference
SWE	Snow water equivalent
TWS	Terrestrial water storage
USAF	US Air Force
WMO	World Meteorological Organization

Websites cited

http://nsidc.org/data/docs/daac/ae_swe_ease-grids.gd.html

http://www.ospo.noaa.gov/Products/atmosphere/mirs/index.html

www.globsnow.info

https://earth.esa.int/web/guest/content?p_r_p_564233524_assetIdentifier=
   coreh20

https://earth.esa.int/web/guest/content?p_r_p_564233524_assetIdentifier=
   coreh20

6 Remote sensing of melting snow and ice

Marco Tedesco[1], Thomas Mote[2], Konrad Steffen[3], Dorothy K. Hall[4] and Waleed Abdalati[5]

[1] The City College of New York, City University of New York, USA
[2] University of Georgia, Athens, Georgia, USA
[3] Swiss Federal Institute for Forest, Snow and Landscape Research WSL, Lausanne, Switzerland
[4] NASA/Goddard Space Flight Center, Greenbelt, USA
[5] University of Colorado, Boulder, USA

Summary

Studying the spatio-temporal evolution of melting snow and ice is important, among other things, for climatological and hydrological applications. Over the large ice sheets of Greenland and Antarctica, melting is one of the drivers for mass losses, either through direct runoff or through the impact on ice dynamics. Over land, melting and refreezing cycles contribute to increase snow grain size through constructive metamorphism, with the consequence of affecting albedo and the surface energy balance through the amount of solar radiation absorbed by the snowpack. Moreover, knowledge of the dates of melt onset of seasonal snow cover is fundamental for water resources management and flood forecasting.

This chapter describes remote sensing tools and techniques for the detection and spatio-temporal analysis of melting snow and ice. In the first section, general considerations concerning techniques for melt detection, using either optical or microwave data, are reported, together with a brief description of the electromagnetic properties of both dry and wet snow. The successive section shows techniques and results of remote sensing of melting snow over land, using either optical or microwave techniques. This is followed by two sections focusing on remote sensing of snow and ice on either the Greenland and Antarctica ice sheets.

6.1 Introduction

Monitoring melting snow and ice at large spatial scales is crucial for climatological and hydrological applications. For example, melting and refreezing cycles are responsible for increased snow grain size through constructive metamorphism,

Remote Sensing of the Cryosphere, First Edition. Edited by M. Tedesco.
© 2015 John Wiley & Sons, Ltd. Published 2015 by John Wiley & Sons, Ltd.
Companion Website: www.wiley.com/go/tedesco/cryosphere

thereby affecting albedo and the surface energy balance through the amount of solar radiation absorbed by the snowpack. Knowledge of the dates of melt onset of seasonal snow cover is fundamental for water resources management and flood forecasting. In Antarctica and Greenland, melt processes are the main drivers for mass losses, either through direct runoff or through the impact on ice sheet dynamics. In Greenland, most of the melting occurs over the ice sheet whereas, in Antarctica, melting dominates on ice shelves, with a potential impact on the rate of grounded ice flow.

Melting is the outcome of a suite of intimately connected processes, which are dependent on the components of the surface energy balance. Ground measurements of energy balance components that are necessary to estimate or to model melting are sparse, are difficult to collect and are not cost-effective. Usually, near-surface air temperature ($\approx 2-3$ m above ground level) is used as a proxy for melt detection from ground measurements (either using hourly or daily averaged values), assuming that melting occurs when the recorded near-surface temperature exceeds the melting point (e.g., $0°C$). However, melting can occur even when the near-surface air temperature is below $0°C$, as a consequence of radiative forcing.

Remote sensing is a powerful tool for monitoring melting at large spatial scales and at multiple spatial and temporal resolutions. Through spaceborne observations, it is indeed possible to detect melting over areas where it is not possible by means of ground observations and with a temporal resolution up to a few hours. Because of the long heritage of spaceborne data, especially those collected in the microwave spectrum (1979 to date), it is possible to study melting trends for a period longer than 30 years. This information can be, in turn, used to relate melting features with climate drivers (e.g., Tedesco and Monaghan, 2009; Tedesco *et al.*, 2009) or can be assimilated into hydrological tools and combined with complementary products to provide improved estimates and the forecast of runoff (e.g. Yan *et al.*, 2009), a crucial parameter for hydrological applications and water resources management.

6.2 General considerations on optical/thermal and microwave sensors and techniques for remote sensing of melting

6.2.1 *Optical and thermal sensors*

Wet snow can be mapped from remote sensing measurements using optical/thermal and microwave data. Approaches using optical and/or thermal data use the capabilities of some sensors to estimate surface temperature and complement this information with albedo data. Spaceborne land surface temperature (LST) or ice surface temperature (IST) can be used to replace or augment air temperature records obtained from a sparse network of *in situ* observations, and can also be used to monitor surface melting over large areas of ice sheets and ice caps (e.g., Hall *et al.*, 2013). Typically, a threshold on LST below the freezing point is selected to indicate surface melt.

Initial work with LST estimated by the Moderate Resolution Imaging Spectroradiometer (MODIS) by Hall *et al.* (2006, 2008a) used a threshold of 0°C on LST to detect melt. When the LST product was compared to *in situ* measurements using multiple thermal-infrared radiometers and data collected on the ground, the error was found to be < 1°C (Wan *et al.*, 2002), leading Hall *et al.* (2009) to use a threshold of LST $\geq$ – 1°C. However, when compared with *in situ* (thermochron) surface temperatures collected at Summit (Greenland), the LST swath product was on average $\approx$3°C higher in the winter (Koenig and Hall, 2010). We point out that, in the case of remotely sensed LST, the quantity used for detecting melting is the surface (i.e., skin) temperature, rather than the temperature a few meters above the surface. In the case of LST, work is ongoing to understand biases and uncertainty on the melting patterns detected with this approach, through the comparison of LST-based maps with other maps obtained from other sensors, (as, for example, microwave radiometers and scatterometers).

6.2.2 *Microwave sensors*

Many remote sensing techniques developed to detect wet snow make use of microwave sensors. There are several reasons for this. Compared to optical and near-infrared instruments, microwave sensors provide a direct measurement of the changes in the electromagnetic properties of the surface from dry to wet conditions. These electromagnetic properties will be discussed later in this chapter. While visible and thermal-infrared sensors cannot see through clouds, microwave measurements are largely insensitive to weather conditions. Moreover, unlike optical sensors, microwave sensors do not require solar illumination and can, therefore, map melting at a temporal resolution from a few hours to days, compared with optical/thermal sensors, for which temporal resolution is variable and depends on cloud presence within the scene.

Another important aspect of microwave remote sensing is that the emitted or backscattered radiation can originate from below the surface, while the emitted radiation recorded by thermal sensors is confined to the surface. Consequently, microwave sensors can detect melt, even when the surface is frozen and the near-surface temperature is below freezing but sub-surface melting is occurring because of radiative forcing.

Compared to optical/thermal data, microwave sensors have a much greater Ground Instantaneous Field of View (GIFOV), on the order of tens of kilometers (depending on the frequency), which makes the data suitable for large spatial scale studies. Also, the large swath of microwave sensors, together with their small sensitivity to the atmosphere, allows one to obtain information over most of the Earth on a daily basis. Coverage improves at the poles, where the same sensor can collect data over the same area multiple times a day (up to every few hours). Depending on the applications, such spatial and temporal resolutions may or may not be sufficient. For example, in the case of flash flood forecasting at a basin scale, both high temporal and spatial resolutions would be desirable while, in the case of climatological studies over an ice sheet, coarser spatio-temporal resolution would suffice, given a sufficiently long time series.

6.2.3 Electromagnetic properties of dry and wet snow

In the following, we report a brief overview of the electromagnetic properties (e.g., brightness temperature, backscattering) of dry snow and the changes associated with the presence of liquid water.

In dry snow, volumetric scattering is dominant over absorption. The appearance of liquid water within the snowpack changes this, with volumetric scattering being reduced, while absorption increases. As an example, Figure 6.1a shows a simplistic schematic diagram of the different contributions to the recorded microwave brightness temperature in the case of dry and wet snow conditions. Brightness temperature recorded from dry snow is lower than that recorded from wet snow, because the presence of dry snow on soil attenuates the microwave radiation emitted by the soil. When liquid water forms in the snow, the wet snow layer absorbs the radiation from the bottom snow layer and soil, and most of the recorded signal is generated by the top wet layer, which can be approximated as a black body.

As a first approximation, wet snow permittivity can be expressed as:

$$\varepsilon = \varepsilon_{ds} + \Delta\varepsilon \tag{6.1}$$

where ε_{ds} is the permittivity of the dry snow, and the second term takes into account the presence of liquid water.

Expressions for can be found in the literature using different approaches (see, for example, Frolov and Macharet, 1999; Ambach *et al.*, 1974; Sihvola 1999). Wet snow can also be modeled as ice particles surrounded by a thin film of water with the effective dielectric constant of ice spheres with a water coating (Jin, 1993). We refer the reader to Chapter 2 for more details on the permittivity of dry and wet snow and other components of the cryosphere.

As the liquid water content (LWC) within the snowpack increases, so does the absorption, as a consequence of the increase of the imaginary part of snow permittivity. In the case of passive microwave sensors, this has the consequence of suddenly and considerably increasing the recorded microwave brightness temperature (with respect to the dry snow case, Figure 6.1b). This continues until a threshold value for the LWC is reached, above which an increase in the LWC is not followed by a further increase in the brightness temperature. This threshold value on LWC depends on the initial snow conditions (e.g., dry snow grain size, density, etc.) but it is generally relatively small (around or less than a few percent

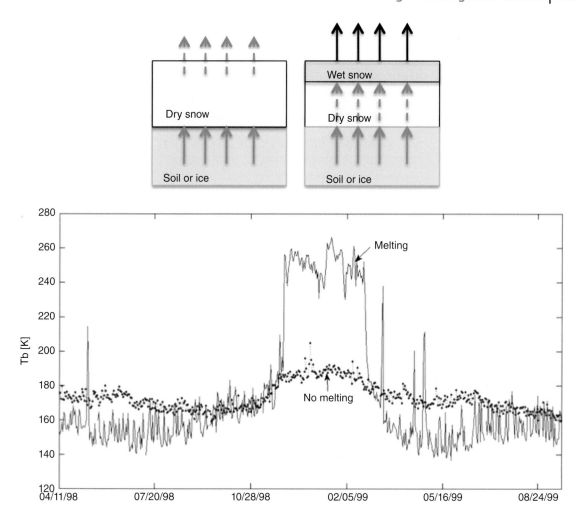

Figure 6.1 (a) The presence of dry snow on soil attenuates the microwave radiation naturally emitted by the soil, therefore reducing the measured brightness temperature with respect to the bare soil case (left). Higher brightness temperature values are recorded in the case of the presence of wet snow (right), which absorbs the radiation from the bottom layer and emits a signal stronger than that of the dry snow covering soil or ice. (b) Annual time series of 19.35 GHz, horizontal polarization, SSM/I brightness temperature for two pixels over Antarctica. The continuous line refers to data measured over an area where melting occurs during summer, while the dashed line and black dots refer to an area where no melting is occurring. (Tedesco, 2009. Reproduced by permission of Elsevier).

of liquid water by volume). Because of the saturation effect on LWC, approaches based on microwave remote sensing cannot provide independent, quantitative estimates of the LWC: the estimated parameters are mainly constrained to melt extent and duration. When snow refreezes, volumetric scattering becomes dominant again, and brightness temperature decreases as a consequence of the loss of LWC (Figure 6.1b).

In the case of active microwave remote sensing, the measured backscattering signal decreases as LWC increases (the opposite of what happens in the case of passive microwave data), because surface scattering becomes dominant over volumetric scattering, reducing the measured backscattering (e.g., Ulaby *et al.*, 1981; Cagnati *et al.*, 2004; Tedesco *et al.*, 2006; Figure 6.2).

6.3   Remote sensing of melting over land

Using active microwave data, Koskinen *et al.* (1997), Nagler *et al.* (2000) and Nagler and Rott (2000) proposed and applied a wet snow mapping algorithm based on change detection and repeat pass Synthetic Aperture Radar (SAR) images. Wet snow detection is then used to produce snow maps in drainage basins in the Austrian Alps with steep topography, using RADARSAT and European Remote Sensing Satellite (ERS) SAR images, and results are compared with those obtained from LANDSAT-5 Thematic Mapper (Nagler *et al.*, 2000; Nagler and Rott, 2000). Results reported in these studies indicate that the agreement between the SAR- and LANDSAT-based approaches are satisfactory, with differences being observed mainly over areas with broken snow cover. The technique was further expanded in Nagler *et al.* (2008), where a data assimilation scheme for remote sensing snow cover products and meteorological data for short-term runoff forecasting is proposed. Evaluation of the runoff forecasts reveals good agreement with measurements, confirming the potential usefulness of the assimilation scheme for operational use.

Hillard *et al.* (2003) evaluated the backscatter response of the NSCAT scatterometer to freeze and thaw cycles over the upper Mississippi River basin of the north central USA and the Boreal Ecosystem Atmosphere Study (BOREAS) region in central Canada. Estimates of snowmelt conditions from NSCAT measurements are compared with outputs of the Variable Infiltration Capacity (VIC) hydrology model (e.g., Liang *et al.*, 1994). The authors compare images of the NASA Scatterometer (NSCAT) backscattering with daily and hourly-modeled snow surface wetness and temperature, and show that the outputs of the VIC model agreed with the backscatter for snow surface wetness on some days but not on others, with the discrepancies due to differences in overpass times, vegetation on the ground and their freeze-thaw state, and liquid moisture content.

Lampkin and Yool (2004) defined their near-surface moisture index (NSMI) that models relative moisture using visible and thermal data. Field measurements of surface wetness, surface/near surface grain size, average pack temperature, and surface temperature for late February and March, are obtained from data collected by the Advanced Spaceborne Thermal Emission and Reflection Radiometer (ASTER) and validated against field observations in Fraser, Colorado.

Further work has involved the inversion of backscattering models to yield estimates of the relative changes of snow wetness from SAR (Koskinen *et al.*, 2010).

An approach to estimate timing of snowmelt and freeze-up using SSM/I data was proposed by Ramage and Isacks (2002, 2003) and was further applied in

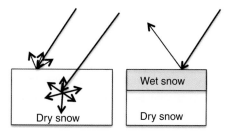

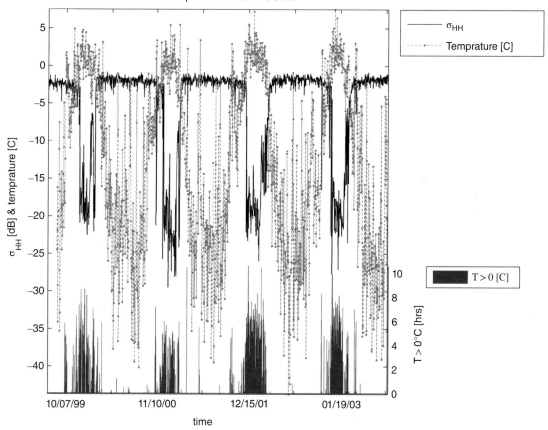

Figure 6.2 (a) The presence of wet snow reduced the radiation scattered back to the sensor, hence reducing the backscattering signal because of the surface scattering and increased absorption. (b) Time series of QuikSCAT backscattering coefficient over an area on the Larsen Ice Shelf, Antarctica, and surface air temperature measured by an automatic weather station within the pixel under observation (Data courtesy of Steiner and Tedesco).

Ramage *et al.* (2006, 2007), Apgar *et al.* (2007), Kopczynski *et al.* (2008) and Yan *et al.* (2009) over land and extended over the Greenland ice sheet in (Tedesco, 2007, discussed later in this chapter). The algorithm makes use of time series of brightness temperatures acquired during both day and night time at ≈37 GHz (vertically polarized). Both melt onset dates and melt presence are identified when the average daily brightness temperature and the brightness temperature diurnal amplitude variations (DAV), defined as the running difference between the early-morning and late-afternoon brightness temperature observations, exceed threshold values.

The idea behind this algorithm is that melt onset is characterized by melting during daytime and refreezing at night. The approach has also been applied to freeze-thaw processes over all surfaces, including snow cover (Kim *et al.* 2011), and a similar approach has been used for detecting melting of snow cover over sea ice (Drobot and Anderson, 2001; Markus *et al.*, 2009). In the DAV approach, histograms of brightness temperatures measured during both dry and wet snow conditions are assumed to be modeled by means of a bimodal distribution, with the left (right) normal distribution containing values representative of dry (wet) snow conditions. Wet snow is assumed to occur when the measured brightness temperature belongs to the right normal distribution (e.g., brightness tempera-ture is greater than a threshold value Tc) and the DAV is greater than a threshold value DAVc. Both Tc and DAVc are spatially and temporally fixed (e.g., Ramage and Isacks 2002; Tedesco, 2007) and are set through the comparison of spaceborne and ground-based observations.

A novel methodology, in which the threshold values are computed in a dynamic fashion (both spatial and temporal), was introduced by Tedesco *et al.* (2009). Here, the threshold value DAVc is computed for each pixel as the mean of the DAV values for the months of January and February, plus a fixed value of 10 K. The threshold value on the brightness temperature is computed from the parameters of the bimodal distribution as follows: for each pixel and year, the five param-eters describing the bimodal distribution (e.g., mean and standard deviation of the two normal distributions and the probability that a value belongs to one or the other normal distribution) are computed through a fitting procedure, mini-mizing the mean square error between the values of the brightness temperature histogram (January through August) and those of the bimodal distribution. The optimal threshold value is then computed by minimizing the probability of erro-neously classifying a dry pixel as a wet pixel and vice versa (Tedesco *et al.*, 2009).

The algorithm was applied to study pan-Arctic terrestrial snowmelt trends for the period 1979–2008. The results indicated statistically significant negative trends for melt onset and end dates, as well as for the length of the melt season, with melting starting (finishing) ≈0.5 days/year (≈ 1 day/year) earlier over the past 30 years and the length of the melting season shortening by ≈0.6 days/year. These results are consistent with other published results (e.g., Derksen and Brown, 2012).

As mentioned above, one major advantage of passive microwave data is the length of the data record, starting in 1979. This allows one to study melting trends in the context of climate variability. Tedesco *et al.* (2009) studied the correlation

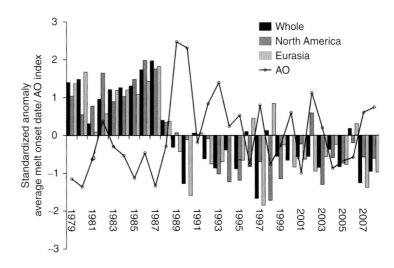

Figure 6.3 Standardized anomaly (1979–2008) of spatially averaged melt onset for the pan-Arctic region (black bars), for North America (NA, 50% gray) and Eurasia (EU, 25% gray). The January through March averaged AO indices are reported as a black line with circles (Tedesco *et al.*, 2009; reproduced by permission of John Wiley & Sons).

between the melting trends and the Arctic Oscillation (AO), and found that the AO index variability can explain up to 50% of the melt onset variability over Eurasia and only 10% of that over North America. This is consistent with spatial patterns of surface temperature changes related to the AO (Figure 6.3).

In another study, Takala *et al.* (2009) used four different algorithms to determine the snowmelt date from both SMMR and SSM/I data for a nearly 30-year period over Eurasia. The authors considered different methodologies, from thresholding channel differences to artificial neural networks (ANNs) and time series analysis. Results were compared with ground-based observations of snowmelt status and an algorithm was selected indicating, consistent with Tedesco *et al.* (2009), a statistically significant trend of an earlier snow clearance for the whole Eurasia.

Wang *et al.* (2011) combined active and passive microwave satellite data to obtain an integrated pan-Arctic melt onset data set from multiple algorithms developed for the northern high-latitude land surface, ice caps, large lakes, and sea ice. The authors found that tree fraction and latitude could explain more than 60% of the variance in melt onset date, with the former exerting a stronger influence on the melt onset date than the latter, though elevation was also found to be an important factor. Interestingly, the authors found that melt onset progresses fastest over land areas of uniform cover or elevation, slowing down in mountainous areas, on ice caps, and over forest and tundra regions.

6.4   Remote sensing of melting over Greenland

6.4.1   *Thermal infrared sensors*

Several approaches have used thermal infrared satellite data, particularly from MODIS, to determine melt from surface (i.e. skin) temperatures of the ice sheets.

Hall *et al.* (2006) used the 5 km resolution standard MODIS LST product (MOD11C1) to examine melt on the Greenland ice sheet from years 2000–2005. They attempted to avoid the cloud contamination problem by including cloud contaminated areas as melting if melt occurred the previous and subsequent days, following a cloud-covered period. Hall *et al.* (2008b) also compared LSTs derived from thermal infrared sensors on both MODIS and ASTER to *in situ* observations from the Greenland Climate Network's (GC-Net) automatic weather stations. Melt was also identified using the Enhanced Thematic Mapper Plus (ETM+) on LANDSAT 7.

Hall *et al.* (2009) used MODIS LST and albedo products to detect melting over the Greenland ice sheet. Results were compared with those obtained from microwave data measured by the Seawind QuikSCAT scatterometer. Hall *et al.* (2012) demonstrated a new clear-sky surface temperature of the Greenland ice sheet using a new ice-surface temperature (IST) algorithm, which can be used to assess surface melt (Figure 6.4). Note that, the selection of a threshold on the LST for melt detection and the potential impact of radiative forcing on measured LST can represent a source of uncertainty that needs to be further addressed.

6.4.2   *Microwave sensors*

Wismann (2000) used radar cross section obtained by the C-band scatterometers aboard the first and second European Remote Sensing Satellite (ERS-1 and ERS-2) measurements over Greenland to detect and monitor snow melt for the period of August 1991 to December 1999. A threshold of 3dB was chosen to separate dry and wet snow conditions, indicating that more or less conservative choices (e.g., 2 or 4 dB) do not impact the general trends. Nghiem *et al.* (2001) proposed the use of active microwave data collected by the SeaWinds Ku-band (13.4 GHz) scatterometer on the QuikSCAT satellite to map snowmelt regions on the Greenland ice sheet. This approach is similar to the one previously applied to other areas (Nghiem and Tsai, 2001). The authors were capable of identifying multiple melting and refreezing events, and related melting patterns to topography.

Mote *et al.* (1993) proposed the use of a simple 19 GHz horizontally polarized brightness temperature difference of 31 K above the winter mean to identify areas with melt. Mote and Anderson (1995) proposed a melt detection approach based on a microwave-emission model, which was used to simulate 37 GHz brightness temperatures associated with melt conditions for locations across the Greenland ice sheet associated with 1% liquid water content. The algorithm used the simulated values as threshold and compared them to daily, gridded Scanning Multi-channel Microwave Radiometer (SMMR) and Special Sensor microwave Imager (SSM/I) passive-microwave data.

The method was validated by comparing volumetric water content simulated using a snowpack energy and mass balance model and observed meteorological data versus swath (Mote and Rowe, 1996) and gridded SSM/I data (Mernild *et al.*, 2011). The approach was then applied by Mote (2000, 2003) to estimate runoff rates, mass balance, and surface elevation changes for the Greenland ice sheet,

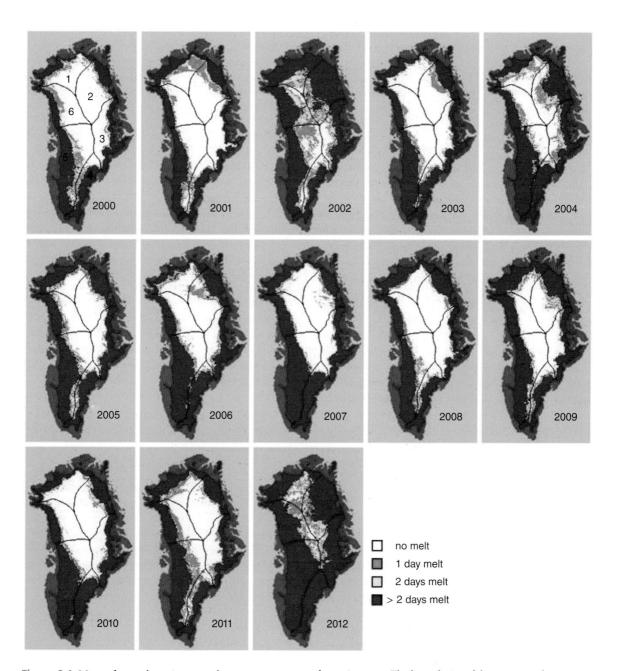

Figure 6.4 Maps of annual maximum melt extent constructed from MODIS IST data of the Greenland ice sheet for the study period (March 2000 through August 2012). The non-ice-covered land surrounding the ice sheet is shown in green. The boundaries of the six major drainage basins of the Greenland ice sheet (after Zwally *et al.*, 2005) are superimposed on the maps (Data and image courtesy of NASA).

to study the large increase in melting occurring in 2007 (Mote, 2007), and to examine the influence of low frequency variations in atmospheric circulation (i.e., teleconnections), such as the North Atlantic Oscillation (NAO), on Greenland surface melt (Mote (1998); Battacharya *et al.* (2009)) and, finally, to contribute to large, interdisciplinary studies monitoring climate change in the Arctic (e.g., Box *et al.*, 2011).

Abdalati and Steffen (1995) defined the so-called cross-polarized gradient ratio (XPGR, defined as the normalized difference between the 19 GHz horizontally-polarized and 37 GHz vertically polarized brightness temperatures) to develop a threshold-based methodology for detecting wet snow over the Greenland ice sheet from microwave brightness temperatures. Using *in situ* observations, the authors established a threshold value of −0.025 on XPGR, above which snow is classified as wet. The algorithm was used to map wet snow for the years 1988 through 1991, and to study the interannual variability of melt extent over the entire Greenland.

The approach was later refined and extended to SMMR (Abdalati and Steffen, 1997). *In situ* measurements suggested a threshold value on XPGR of −0.0158 for SSM/I and −0.0265 for SMMR. With the updated XPGR algorithm, the authors performed a spatio-temporal analysis of melting for the period 1979–1994, identifying a notable increasing trend in melt area between the years 1979 and 1991, which came to an abrupt halt in 1992 after the eruption of Mt. Pinatubo. An analysis of near-surface (air) temperature at six coastal stations was also reported, with the results suggesting that a +1°C temperature increase would correspond to an increase in melt area of 73 000 km^2.

Ashcraft and Long (2005) used the sensitivity of the horizontal to vertical polarization ratio of brightness temperature to surface wetness to map melting over Greenland. In particular, the authors indicated that the 19/37 GHz frequency ratio is sensitive to a frozen surface layer over wet snow, being associated with the freeze stage of the melt cycle. Results from remote sensing observations were supported by model results and *in situ* measurements.

Ashcraft and Long (2006) compared the outputs of six different melt detection algorithms, using either active or passive microwave spaceborne observations among themselves, and with the outputs of a simple physical model, discussing the merits and limitations of the various methods. For example, one of the algorithms considered in their study made use of a combination of winter and wet snow brightness temperatures for the threshold value as:

$$Tc = T_{winter} * a + T_{wet\_snow} * (1-a)$$

where:

T_{winter} and T_{wet_snow} are, respectively, the winter (June through August) and the brightness temperature of a wet snowpack (set to 273 K)

a is a mixing coefficient set to the value of 0.47, derived assuming a wet layer of 4.7 cm and a LWC = 1%.

Tedesco (2007) proposed a technique for monitoring snowmelt over the Greenland ice sheet based on DAV values, mentioned in the previous section. In the study, surface (air) temperature values, either recorded by ground-based stations or derived from model, were used for calibrating and validating the technique and the results compared with those obtained using backscattering coefficients recorded by QuikSCAT during an extreme melting event occurring in June 2002. The algorithm was then applied to updated satellite measurements for the period 1979 to date (and annually updated), to study anomalies of the number of melting days for the period 1979–2012 (Tedesco et al., 2011c) and identify melting records (e.g., Tedesco et al., 2008, 2011a, 2011b).

Surface melting in 2012 set two new records according to passive microwave observations (e.g., Tedesco, 2007, 2009), during the satellite era (1979 to date), the first concerning maximum melt extent and the second concerning the seasonal melt intensity (measured through the so-called Melting Index, MI, obtained as the number of melting days times the area subject to melting). Melt extent over the Greenland ice sheet reached record values for the days July 11–12 (e.g., Nghiem et al. 2012), reaching up to ≈97% of the ice sheet on a single day. This was confirmed by the analysis of different methods (e.g., those based on Mote and Anderson, 1995; Tedesco, 2007, 2009) using spaceborne passive microwave observations. This was more than three times the 1981–2010 averaged value of ≈25% for the same period.

In 2012, melting started about two weeks earlier than the average at low elevations, and was sustainably higher than the values of the previous record year (2010) for most of June, July and mid August (Figure 6.5a). Melting lasted up to 140 days for some areas in the southwest at low elevations (Figure 6.5b). The 2012 anomaly for the number of melting days (e.g., number of melting days in 2012 minus the 1980–2010 mean) exceeded +25–30 days in the south and +40–50 days in the northwest region. This means that areas in northwest Greenland between 1400 and 2000 m asl had nearly two months more melt than during the 1981–2010 normal. The standardized melting index (SMI) for 2012, obtained by subtracting MI of its mean (1979–2012) and dividing by its standard deviation (Figure 6.5c), computed a modified version of ≈+2.4, against a previous record in 2010 of ≈+1.3 (in this case, the zero value indicates the mean values). More details about the 2012 records over Greenland can be found in Tedesco et al. (2013).

The combination of active and passive microwave data, together with optical/thermal data allowed detection of an extreme melting event that occurred in 2012, when nearly the entire ice sheet of Greenland experienced some degree of melting at its surface. Figure 6.6 shows the extent of surface melt over Greenland's ice sheet on July 8 (left) and July 12 (right). On July 8, about 40% of the ice sheet had undergone thawing at or near the surface. Melting had dramatically accelerated, and an estimated 97% of the ice sheet was undergoing melting on July 12.

All of the methods based on microwave measurements described so far make use of a threshold value on brightness temperature or backscattering to detect melting. Nevertheless, other approaches using the intrinsic properties of the spaceborne signal have been also proposed in the literature (Joshi et al., 2001; Liu et al., 2005; Wang & Yu, 2011). Such approaches are based on the magnitude of

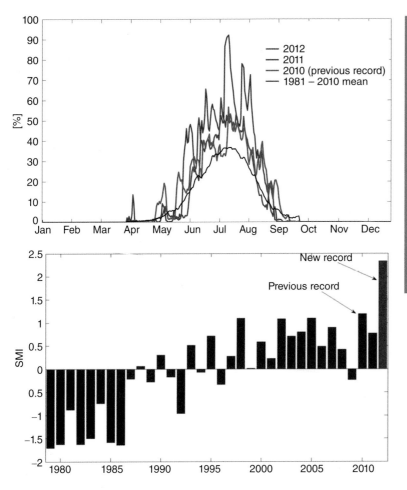
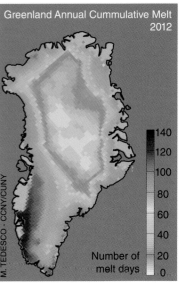

Figure 6.5 (a) Melt extent (as a percentage of the Greenland ice sheet) time series derived from spaceborne passive microwave observations using the algorithm in Tedesco (2009) in 2012 (red), 2011 (blue), 2010 (green, being the previous record) and for the 1981–2010 mean (black). (b) Anomaly of the number of melting days for 2012 (1979–2011 baseline). (c) Standardized melting index for the period 1979–2012 over the entire Greenland ice sheet. (Data courtesy of and images, M. Tedesco).

the relative change within a backscattering or brightness temperature time-series. The overall hypothesis is that large-amplitude signal changes can be associated with the appearance or disappearance of wet snow and can be studied by means of differential operators, such as the derivative-of-Gaussian (Joshi *et al.*, 2001) or wavelet-based detection methods.

For example, Joshi *et al.* (2001) computed the melt extent, duration and melt season on the Greenland ice sheet by means of an edge detection technique applied to passive microwave data from the SSM/I and SMMR instruments for the period 1979 to 1997. This was accomplished by means of a derivative-of-Gaussian

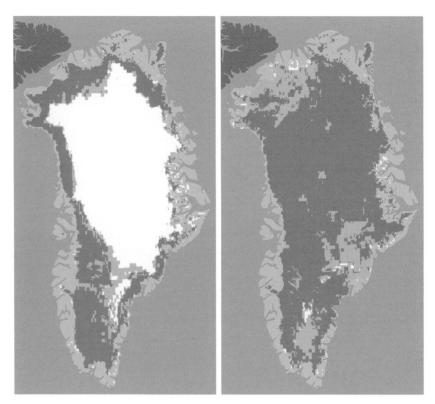

Figure 6.6 Extent of surface melt over the Greenland ice sheet on (a) July 8 and (b) July 12. The areas classified as "probable melt" (light pink) correspond to those sites where at least one satellite detected surface melting. The areas classified as "melt" (dark pink) correspond to sites where two or three satellites detected surface melting (NASA).

edge detector to identify edges corresponding to the onset and end of melt. These approaches will be further discussed in the following section, concerning melting over Antarctica.

6.5   Remote sensing of melting over Antarctica

Zwally and Fiegles (1994) estimated the extent and duration of surface melting on the Antarctic ice shelves and margins of the ice sheet from satellite passive microwave data for the period 1978–1987. Melting was assumed to occur when the measured brightness temperature exceeded the annual mean by 30 K, similar to the approach proposed in Mote *et al.* (1993). The authors indicated that most Antarctic surface melting occurs during December and January and correlated their results with regional air temperatures and to katabatic-wind effects.

Building on Zwally and Fiegles (1994), Torinesi *et al.* (2003) proposed to use a threshold on brightness temperature computed as:

$$Tc = T_{winter} + \Delta T$$

where $\Delta T = 3\sigma$ and σ is the winter brightness temperature standard deviation.

The authors concentrated on the period 1980–99 and defined the cumulated product of the surface area affected by melting and the duration of the melting event, called cumulative melting surface (CMS). A description of these approaches and a comparative discussion can be found in Tedesco (2009).

Because the threshold value is generally set directly on brightness temperatures, the minimum value of LWC to which the algorithm is sensitive is different for different approaches. Indeed, the amount of liquid water content in two pixels having a wet snow brightness temperature value of, for example, 240 K will be different, depending on the dry snow brightness temperature value for those pixels. For the pixel with a higher (lower) dry snow brightness temperature value, a lower (higher) amount of liquid water content will be necessary to raise the brightness temperature value to 240 K.

Building on this concept, Tedesco (2009) proposed a physically-driven approach considering a fixed minimum liquid water content value, rather than a fixed brightness temperature value. The threshold value on brightness temperature was computed as a function of the dry snow initial brightness temperature, using the outputs of a multi-layer electromagnetic model. The use of a dynamic algorithm, with a threshold based on a fixed LWC simulated by a multi-layer electromagnetic model, is conceptually similar to that of Mote and Anderson (1995). Results reported in Kuipers Munneke *et al.* (2012) indicated no statistically significant trend in either continent-wide or regional meltwater volume for the 31-year period 1979–2010, consistent with previous studies (e.g., Tedesco and Monaghan, 2009).

Barrand *et al.* (2012) and Trusel *et al.* (2012) made use of the enhanced spatial resolution QuikSCAT data set to study melting at high spatial resolution over the Antarctic peninsula and the whole continent, using a threshold value above which melting is assumed to be occurring.

Similar to what happened in the case of Greenland, techniques complementary to those based on threshold values have been proposed for Antarctica. For example, Liu *et al.* (2005) proposed a method for estimating the onset date, end date, duration and spatial extent of snowmelt, using a wavelet transform of daily brightness temperature observations. The authors identified an optimal edge strength threshold to separate actual snowmelt edges from weak edges caused by noisy perturbations and other non-melt processes using variance analysis and bimodal Gaussian curve fitting. As a last step, the authors used a neighborhood operator to detect and correct possible errors in the melt computations that were purely based on temporal analysis of individual brightness temperature curves, relying on a spatial autocorrelation analysis. The method was applied to SSM/I data collected in 2001–2002 over the Antarctic ice sheet and the results evaluated through visual interpretation of brightness temperature time series and examination of historical near-surface air temperature records.

Steiner and Tedesco (2013) used enhanced resolution (2.25 km) scatterometer data collected by SeaWinds on the QuikSCAT satellite between 1999 and

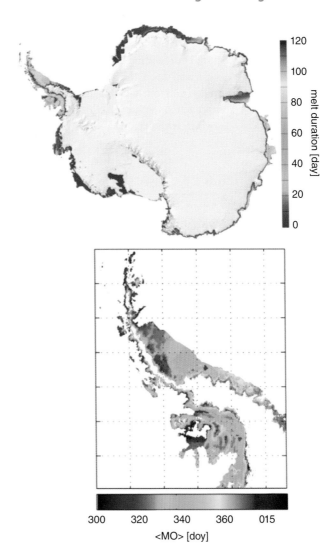

Figure 6.7 (a) Maps of mean (2000–2009) seasonal melt duration estimated from the SeaWinds sensor on QuikSCAT, using 3 dB below winter mean threshold approach. (b) Maps of mean melt onset day of the year over the Antarctic peninsula. Data courtesy of Steiner and Tedesco.

2009 in conjunction with continuous wavelets to identify those changes in the backscattering coefficient time series that propagate through the low (high) order. This was helpful to identify persistent versus sporadic melting, the latter identified using a fixed threshold algorithm.

As an example, Figure 6.7 shows maps of mean (2000–2009) seasonal melt duration estimated from the SeaWinds sensor on QuikSCAT, using both threshold- and wavelet-based approaches, together with maps of melt onset day over the Antarctic Peninsula obtained with the two approaches. Algorithms that do not rely on threshold values, such as that proposed in Steiner and Tedesco (2012), do not require a definition of persisting melting based on the consecutive number of melting days ,as done in previous studies, but the persistency of melting is derived from the intrinsic properties of the backscattering time series. Also, unlike previously published algorithms making use of wavelets, the proposed approach does not require *a priori* knowledge of the threshold values on the wavelet, and it can be applied without pre-processing.

6.6 Conclusions

Monitoring melting at the hemispheric scale is crucial for studying the evolution of surface mass balance of the cryosphere, for monitoring and managing water resources and for studying the impact of recently observed warming surface temperature on snow and ice. Optical, thermal infrared and microwave data can be used to detect the presence of melting and to map wet snow over large areas with varying spatio-temporal resolution. In the case of optical/infrared data, melting is detected from the analysis of albedo changes or by assuming that wet snow is present above a selected threshold value on the estimated skin temperature. In the case of microwave measurements, melting is detected because of the physical change in the electromagnetic properties of the snow or ice in the microwave region induced by the presence of liquid water.

Each of the techniques and algorithms discussed above has strengths and limitations. Thermal infrared data can generate maps at a relatively high spatial resolution (from a few meters to hundreds of meters), but temporal resolution is limited by the presence of clouds. On the other hand, microwave measurements do not require solar illumination, are not considerably affected by the atmosphere and provide data over most of the Earth, at least on a daily basis. Moreover, the availability of long-term spaceborne microwave measurements allows one to study the evolution of melting and other parameters for more than three decades. Results from remote sensing algorithms can be combined with modeling or assimilation tools to generate a robust and more exhaustive set of tools and techniques, in which the weaknesses of each tool is mitigated by the strength of another tool.

The combination of the abovementioned remote sensing techniques also holds great potential, allowing studies of melt processes at multiple spatial and temporal resolutions and the detection of extreme events. One example is the extreme melting that occurred over Greenland in July 2012, when ≈97% of the Greenland ice sheet was subject to melting (e.g., Nghiem *et al.*, 2012), setting a new record for the satellite era. The extreme melting was detected by both active and passive sensors and was confirmed by the analysis of MODIS surface temperature and ground observations. The availability of multiple sensors and techniques provided more confidence on the out-of-the-ordinary event and allowed to monitor such even on a near real-time basis.

References

Abdalati, W. & Steffen, K. (1995). Passive microwave-derived snow melt regions on the Greenland ice sheet. *Geophysical Research Letters* **22**(7), 787–790. American Geophysical Union.

Abdalati, W. & Steffen, K. (1997). Snowmelt on the Greenland ice sheet as Derived from Passive Microwave Satellite Data. *Journal of Climate* **10**(2), 165–175.

Ambach, W. & Denoth, A. (1974). Studies on the dielectric properties of snow, *Zeitschrift fur Gletscherkunde und Glazialgeologie* **8**, 1974, 113–123.

Apgar, J.D., Ramage, J.M., McKenney, R.A., & Maltais, P. (2007). AMSR-E algorithm for snowmelt onset detection in sub-arctic heterogeneous terrain. *Hydrological Processes* **21**(12), 1587–1596.

Ashcraft, I.S., & Long, D.G. (2005). Differentiation between melt and freeze stages of the melt cycle using SSM/I channel ratios. *IEEE Transactions on Geoscience and Remote Sensing* **43**(6), 1317–1323.

Ashcraft, I.S., & Long, D.G. (2006). Comparison of methods for melt detection over Greenland using active and passive microwave measurements. *International Journal of Remote Sensing* **27**(12), 2469–2488.

Barrand N.E., D.G. Vaughan, N. Steiner, M. Tedesco, P. Kuipers Munneke, M.R. van den Broeke & J.S. Hosking (2012). Trends in Antarctic Peninsula surface melting conditions from observations and regional climate modeling. Accepted in *Journal of Geophysical Research*.

Bhattacharya, I., Jezek, K., Wang, L., & Liu, H. (2009). Surface melt area variability of the Greenland ice sheet: 1979–2008. *Geophysical Research Letters* **36**, L20502.

Box, J.E., Ahlstrøm, A., Cappelen, J., Fettweis, X., Decker, D., Mote, T., van As, D., van de Wal, R., Vinther, B., Wahr, J. (2011). Greenland in "State of the Climate in 2010". *Bulletin of the American Meteorological Society* **92**, S156–S160.

Cagnati, A., Crepaz, A., Macelloni, G., Pampaloni, P., Ranzi, R., Tedesco, M., Tomirotti, M., *et al.* (2004). Study of the snow melt-freeze cycle using multi-sensor data and snow modelling. *Journal of Glaciology* **50**(170), 419–426.

Derksen, C. & R. Brown (2012), Spring snow cover extent reductions in the 2008–2012 period exceeding climate model projections. *Geophysical Research Letters* **39**, L19504, doi:10.1029/2012GL053387.

Drobot, S., & Anderson, M. (2001). An Improved Method for Determining Snowmelt Onset Dates over Arctic Sea Ice Using Scanning Multichannel Microwave Radiometer and Special Sensor Microwave/Imager Data. *Journal of Geophysical Research* **106**(D20), 24,033–24,049.

Frolov A.D., Macheret Y.Y. (1999), On dielectric properties of dry and wet snow. *Hydrological Processes* **13**, 1755–1760.

Hall, D.K., Williams Jr., R.S., Casey, K.A., DiGirolamo, N.E., & Wan, Z. (2006). Satellite-derived, melt season surface temperature of the Greenland ice sheet (2000–2005) and its relationship to mass balance. *Geophysical Research Letters*, **33**, L11501, doi:10.1029/2006GL026444.

Hall, D.K., Williams Jr., R.S., Luthcke, S.B., & Digirolamo, N.E. (2008a). Greenland ice sheet surface temperature, melt and mass loss: 2000–06. *Journal of Glaciology* **54**, 81–93.

Hall, D.K., Box, J.E., Casey, K.A., Hook, S.J., Shuman, C.A., & Steffen, K. (2008b). Comparison of satellite-derived and in-situ observations of ice and snow surface temperatures over Greenland. *Remote Sensing of Environment* **112**, 3739–3749.

Hall, D.K., Nghiem, S.V., Schaaf, C.B. & DiGirolamo, N.E. (2009). Evaluation of surface and near surface melt characteristics on the Greenland ice sheet using MODIS and QuikSCAT data. *Journal of Geophysical Research*, **114**, F04006, doi:10.1029/2009JF001287.

Hall, D.K., Comiso, J.C., Digirolamo, N.E., Shuman, C.A., Key, J.R., & Koenig, L.S. (2012). A satellite-derived climate-quality data record of the surface temperature of the Greenland ice sheet. *Journal of Climate*, in press.

Hall, D.K., J.C. Comiso, N.E. DiGirolamo, C.A. Shuman, J.E. Box and L.S. Koenig, (2013) Variability in the surface temperature and melt extent of the Greenland ice sheet from MODIS. *Geophysical Research Letters* **40**, 2114–2120, doi: 10.1002/grl.50240.

Hillard, U., Sridhar, V., Lettenmaier, D.P., & McDonald, K.C. (2003). Assessing snowmelt dynamics with NASA scatterometer (NSCAT) data and a hydrologic process model. *Remote Sensing of Environment* **86**(1), 52–69.

Jin, Y.Q. (1993). *Electromagnetic scattering modelling for quantitative remote sensing*. World Scientific, Singapore.

Joshi, M., Merry, C.J., Jezek, K.C., & Bolzan, J.F. (2001). An edge detection technique to estimate melt duration, season and melt extent on the Greenland ice sheet using passive microwave data. *Geophysical Research Letters* **28**(18), 3497–3500.

Kim, Y., Kimball, J.S., McDonald, K.C., & Glassy, J. (2011). Developing a Global Data Record of daily landscape freeze/thaw status using satellite microwave remote sensing. *IEEE Transactions on Geoscience and Remote Sensing* **49**(3), 949–960, doi: 10.1109/TGRS.2010.2070515.

Koenig, LS, & Hall, DK (2010). Comparison of satellite, thermochron and air temperatures at Summit, Greenland, during the winter of 2008/09. *Journal of Glaciology* **56**(198), 735–741.

Kopczynski, S., Ramage, J., Lawson, D., Goetz, S., Evenson, E., Denner, J., & Larson, G. (2008). Passive microwave (SSM/I) satellite predictions of valley glacier hydrology, Matanuska Glacier, Alaska. *Geophysical Research Letters* **35**(16), doi: 10.1029/2008GL034615.

Koskinen, J.T., Pulliainen, J.T., & Hallikainen, M.T. (1997). The use of ERS-1 SAR data in snow melt monitoring. *IEEE Transactions on Geoscience and Remote Sensing* **35**(3), 601–610.

Koskinen, J.T., Pulliainen, J.T., & Luojus, K.P., & Takala, M. (2010). Monitoring of Snow-Cover Properties During the Spring Melting Period in Forested Areas. *IEEE Transactions on Geoscience and Remote Sensing* **35**(3), 601–610.

Kuipers Munneke, P., G. Picard, M.R. van den Broeke, J. T. M. Lenaerts & E. van Meijgaard (2012), Insignificant change in Antarctic snowmelt volume since 1979, *Geophysical Research Letters* **39**, L01501, doi:10.1029/2011GL050207.

Lampkin, D.J., & Yool, S.R. (2004). Monitoring mountain snowpack evolution using near-surface optical and thermal properties. *Hydrological Processes* **18**(18), 3527–3542, doi:10.1002/hyp.5797

Liang, X., Lettenmaier, D.P., Wood, E.F., & Burges, S.J. (1994). A Simple hydrologically Based Model of Land Surface Water and Energy Fluxes for GSMs. *Journal of Geophysical Research* **99**(D7), 14415–14428.

Liu, H., Wang, L., & Jezek, K.C. (2005). Wavelet-transform based edge detection approach to derivation of snowmelt onset, end and duration from satellite passive microwave measurements. *International Journal of Remote Sensing* **26**(21), 4639–4660.

Markus, T., Stroeve, J., & Miller, J. (2009). Recent changes in Arctic sea ice melt onset, freezeup, and melt season length. *Journal Of Geophysical Research* **114**, C12024, doi:10.1029/2009JC005436.

Mernild, S., Mote, T.L., & Liston, G.E. (2011). Greenland Ice Sheet surface melt extent and trends, 1960–2010, *Journal of Glaciology* **57**, 621–628, doi: 10.3189/002214311797409712.

Mote, T.L. (1998). Mid-tropospheric circulation and surface melt on the Greenland ice sheet Part I: Atmospheric teleconnections. *International Journal of Climatology* **18**, 111–130.

Mote, T.L. (2000). Ablation estimates for the Greenland ice sheet from passive microwave measurements. *Professional Geographer* **52**, 322–331.

Mote, T.L. (2003). Estimation of runoff rates, mass balance, and elevation changes on the Greenland ice sheet from passive microwave observations. *Journal of Geophysical Research* **108**(D2), 1–7.

Mote, T.L. (2007). Greenland surface melt trends 1973 – 2007: Evidence of a large increase in 2007. *Geophysical Research Letters* **34**(22), L22507.

Mote, T.L. & Anderson, M.R. (1995). Variations in snowpack melt on the Greenland ice sheet based on passive-microwave measurements. *Journal Of Glaciology* **41**(137), 51–60.

Mote, T.L., & Rowe, C.M. (1996). A comparison of microwave radiometric data and modeled snowpack conditions for Dye-2, Greenland. *Meteorology and Atmospheric Physics* **59**, 245–256.

Mote, T.L., Anderson, M.R., Kuivinen, K.C., & Rowe, C.M. (1993). Spatial and temporal variations of surface melt on the Greenland ice sheet. *Annals of Glaciology* **17**, 233–238.

Nagler, T., & Rott, H. (2000). Retrieval of Wet Snow by Means of Multitemporal SAR Data, IEEE Transactions on Geoscience and Remote Sensing. *IEEE Transactions on Geoscience and Remote Sensing* **32**(2), 754–765.

Nagler, T., Rott, H., & Glendinning, G. (2000). Snowmelt runoff modelling by means of RADARSAT and ERS SAR. *Canadian Journal of Remote Sensing* **26**(6), 512–520.

Nagler, T., Rott, H., Malcher, P., & Muller, F. (2008). Assimilation of meteorological and remote sensing data for snowmelt runoff forecasting. *Remote Sensing of Environment* **112**(4), 1408–1420, doi:10.1016/j.rse.2007.07.006

Nghiem, S.V., & Tsai, W.Y. (2001). Global snow cover monitoring with spaceborne K(u)-band scatterometer. *IEEE Transactions on Geoscience and Remote Sensing* **39**(10), 2118–2134.

Nghiem, S.V., Steffen, K., Kwok, R., & Tsai, W.Y. (2001). Detection of snowmelt regions on the Greenland ice sheet using diurnal backscatter change. *Journal of Glaciology* **47**(159), 539–547.

Nghiem, S.V., D.K. Hall, T.L. Mote, M. Tedesco, M.R. Albert, K. Keegan, C.A. Shuman, N.E. DiGirolamo & G. Neumann (2012), The extreme melt across the Greenland ice sheet in 2012. *Geophysical Research Letters* **39**, L20502, doi:10.1029/2012GL053611.

Ramage, J.M., & Isacks, B.L. (2002). Determination of melt-onset and refreeze timing on southeast Alaskan icefields using SSM/I diurnal amplitude variations. *Annals of Glaciology* **34**, 391–398.

Ramage, J.M., & Isacks, B.L. (2003). Interannual variations of snowmelt and refreeze timing on southeast-Alaskan icefields, USA. *Journal of Glaciology* **49**(164), 102–116.

Ramage, J.M., McKenney, R.A., Thorson, B., Maltais, P., & Kopczynski, S.E. (2006). Relationship between passive microwave-derived snowmelt and surface-measured discharge, Wheaton River, Yukon Territory, Canada. *Hydrological Processes* **20**(4), 689–704.

Ramage, J.M., Apgar, J.D., Mckenney, R.A., & Hanna, W. (2007). Spatial variability of snowmelt timing from AMSR-E and SSM / I passive microwave sensors, Pelly River, Yukon Territory, Canada. *Hydrological Processes* **21**(12), 1548–1560, doi:10.1002/hyp

Sihvola A. (1999). *Electromagnetic mixing formulas and applications*. IEE Electromagnetic Waves Series 47, IET, UK.

Steiner, N. & Tedesco, M. (2013) A wavelet melt detection algorithm applied to enhanced resolution scatterometer data over Antarctica (2000–2009). *The Cryosphere Discussions* **7**(3), 2635–2678, doi:10.5194/tcd-7-2635-2013.

Takala, M., Pulliainen, J., Metsamaki, S.J., & Koskinen, J.T. (2009). Detection of Snowmelt Using Spaceborne Microwave Radiometer Data in Eurasia From 1979 to 2007. *IEEE Transactions on Geoscience and Remote Sensing* **47**(9), 2996–3007.

Tedesco, M. (2007). Snowmelt detection over the Greenland ice sheet from SSM/I brightness temperature daily variations. *Geophysical Research Letters* **34**(2), doi: 10.1029/2006GL028466.

Tedesco, M. (2009). Assessment and development of snowmelt retrieval algorithms over Antarctica from K-band spaceborne brightness temperature (1979–2008). *Remote Sensing of Environment* **113**(5), 979–997.

Tedesco, M., & Monaghan, A.J. (2009). An updated Antarctic melt record through 2009 and its linkages to high-latitude and tropical climate variability. *Geophysical Research Letters* **36**, L18502.

Tedesco, M., Kim, E.J., England, A.W., De Roo, R.D., & Hardy, J.P. (2006). Brightness temperatures of snow melting/refreezing cycles: Observations and modeling using a multilayer dense medium theory-based model. *IEEE Transactions on Geoscience and Remote Sensing* **44**(12), 3563–3573.

Tedesco, M., Serreze, M. & Fettweis, X. (2008). Diagnosing the extreme surface melt event over southwestern Greenland in 2007. *The Cryosphere Discussions* **2**, 383–397.

Tedesco, M., Brodzik, M., Armstrong, R., Savoie, M., & Ramage, J. (2009). Pan arctic terrestrial snowmelt trends (1979–2008) from spaceborne passive microwave data and correlation with the Arctic Oscillation. *Geophysical Research Letters* **36**(21), L21402. doi:10.1029/2009GL039672

Tedesco, M., Fettweis, X., Van Den Broeke, M.R., Van De Wal, R. S. W., Smeets, C. J. P. P., Van De Berg, W.J., Serreze, M.C., *et al.* (2011a). The role of albedo and accumulation in the 2010 melting record in Greenland. *Environmental Research Letters* **6**(1), 014005, doi: 10.1088/1748-9326/6/1/014005.

Tedesco M., Fettweis, X., van den Broeke, M.R., van de Wal, R.S.W., Smeets, C.J.P.P., van de Berg, W.J., Serreze, M.C. & Box, J.E. (2011b). Record summer melt in Greenland in 2010. *Eos, Transactions, American Geophysical Union* **92**(15), 126.

Tedesco, M., Box, J.E., Cappellen, J., Mote, T., van de Wal, R.S.W. & Wahr, J. (2011c). [Greenland Ice Sheet] The Arctic [in State of the Climate in 2011]. *Bulletin of the American Meteorological Society* Special Supplement S148 – S151.

Tedesco M., Fettweis, X., Mote, T., Wahr, J., Alexander, P., Box, J. & Wouters, B. (2012). Evidence and analysis of 2012 Greenland records from spaceborne observations, a regional climate model and reanalysis data. *The Cryosphere Discussions* **6**, 4939–4976. www.the-cryosphere-discuss.net/6/4939/2012/ doi:10.5194/tcd-6-4939-2012

Torinesi O., Fily, M. & Genthon, C. (2003). Interannual variability and trend of the Antarctic summer melting period from 20 years of spaceborne microwave data. *Journal of Climate* **16**, 1047–1060.

Trusel, L.D., Frey, K.E. & Das, S.B. (2012). Antarctic surface melting dynamics: Enhanced perspectives from radar scatterometer data. *Journal of Geophysical Research* **117**, F02023, doi:10.1029/2011JF002126.

Ulaby, F.T., Moore, R.K. & Fung, A.K. (1981). Microwave Remote Sensing: Active and Passive, Vol. I – Microwave Remote Sensing Fundamentals and Radiometry. Addison-Wesley, Advanced Book Program, Reading, Massachusetts, 456 pages.

Wan, Z., Zhang, Y., & Li, Z.L. (2002). Validation of the land surface temperature products retrieved from Terra Moderate Resolution Imaging Spectroradiometer data. *Remote Sensing of Environment* **83**, 163–180.

Wang, L., & Yu, J. (2011). Spatiotemporal Segmentation of Spaceborne Passive Microwave Data for Change Detection. *IEEE Geoscience and Remote Sensing Letters* **8**, 909–913.

Wang, L., Wolken, G.J., Sharp, M.J., Howell, S.E.L., Derksen, C., Brown, R.D., Markus, T., *et al.* (2011). Integrated pan-Arctic melt onset detection from satellite active and passive microwave measurements, 2000–2009. *Journal of Geophysical Research* **116**(D22), 2000–2009, doi:10.1029/2011JD016256

Wismann, V. (2000). Monitoring of seasonal snowmelt on Greenland with ERS scatterometerdata. *IEEE Transactions on Geoscience and Remote Sensing* **38**(4 Part 2), 1821–1826.

Yan, F., Ramage, J., & McKenney, R. (2009). Modeling of high-latitude spring freshet from AMSR-E passive microwave observations. *Water Resources Research* **45**(11), W11408.

Zwally, H.J., & Fiegles, S. (1994). Extent and Duration of Antarctic Surface Melting. *Journal of Glaciology* **40**(136), 463–476.

Zwally, H.J., Giovinetto, M.B., Li, J., Cornejo, H.G., Beckley, M.A., Brenner, A.C., Sabba, J.L. & Li, D. (2005). Mass changes of the Greenland and Antarctic ice sheets and shelves and contributions to sea-level rise: 1992–2002. *Journal of Glaciology* **51**, 509–527.

Acronyms

ANN	Artificial Neural Networks
ASTER	Advanced Spaceborne Thermal Emission and Reflection Radiometer
BOREAS	Boreal Ecosystem Atmosphere Study
CMS	Cumulative Melting Surface
DAV	Diurnal Amplitude Variations
ERS	European Remote Sensing Satellite
ETM+	Enhanced Thematic Mapper Plus
GC-NET	Greenland Climate Network
GIFOV	Ground Instantaneous Field of View
IST	Ice Surface Temperature
LST	Land Surface Temperature
LWC	Liquid Water Content
MI	Melting index
MODIS	Moderate Resolution Imaging Spectroradiometer
NAO	North Atlantic Oscillation
NSCAT	NASA Scatterometer
NSMI	Near-Surface Moisture Index
SAR	Synthetic Aperture Radar
SMMR	Scanning Multi-channel Microwave Radiometer
SSM/I	Special Sensor microwave Imager
VIC	Variable Infiltration Capacity

7

Remote sensing of glaciers

Bruce H. Raup[1], Liss M. Andreassen[2], Tobias Bolch[3] and
Suzanne Bevan[4]

[1] NSIDC, University of Colorado, Boulder, USA
[2] Norwegian Water Resources and Energy Directorate, Oslo, Norway
[3] University of Zurich, Zurich, Switzerland
[4] Swansea University, UK

Summary

Many elements of the cryosphere respond to changes in climate, but mountain glaciers are
particularly good indicators of climate change, because they respond more quickly than
most other ice bodies on Earth. Changes in glaciers are easily noticed by specialists and
non-specialists alike, in ways that other climate indicators, such as ocean temperature or
statistics of atmospheric circulation indices, are not. Remote sensing methods are capable
of measuring many parameters of mountain glaciers and the changes they exhibit, leading
to greater insight into processes affecting changes in glaciers and, hence, climate.

Field-based measurements are indispensable, as they yield high-precision data and give
key insights into processes. However, due to expense and difficult logistics, such measurements are limited to a small number of sites. Remote sensing can cover large numbers of
glaciers per image, and some long-term data collections (e.g., Landsat) are available for
free. Algorithms and computational resources are now capable of producing maps of glacier
boundaries at useful accuracy over large regions in a short time. New sensors will be coming
online soon that will continue and extend this capability.

Mass balance is an important parameter indicating the health of a glacier. Mass balance
can be estimated from satellite data by the "geodetic method" of measuring volume changes,
where the change in volume is estimated by subtracting two digital terrain models of the
glacier surface. A more recent approach to detect mass changes in land ice is through measurement of the gravitational field using the Gravity Recovery and Climate Experiment
(GRACE) satellite system, which measures changes in mass below the orbit track.

Advances have been made recently in remote sensing of glaciers on a number of fronts,
including more complete and more accurate glacier inventories, improved glacier mapping
techniques, and new insights from gravimetric satellites. Through international cooperative
efforts such as the Global Land Ice Measurements from Space (GLIMS) initiative and the
Global Terrestrial Network for Glaciers (GTN-G), satellite remote sensing of glaciers has led
to the ability to produce glacier outlines quickly over large regions, leading to the production

Remote Sensing of the Cryosphere, First Edition. Edited by M. Tedesco.
© 2015 John Wiley & Sons, Ltd. Published 2015 by John Wiley & Sons, Ltd.
Companion Website: www.wiley.com/go/tedesco/cryosphere

of nearly complete global glacier inventories. These remote sensing products are being used to better understand climate, hydrological systems, and water resources, as our environment continues to change.

7.1 Introduction

Many elements of the cryosphere respond to changes in climate, but mountain glaciers are particularly good indicators of climate change because they respond more quickly than most other ice bodies on earth (Lemke *et al.*, 2007; Kääb *et al.*, 2007). Changes in glaciers are easily noticed by specialists and non-specialists alike in ways that other climate indicators, such as ocean temperature or statistics of atmospheric circulation indices, are not.

The most important parameter indicating the health of a glacier is its mass balance – the difference between accumulation, in the form of snow, avalanches and wind-blown snow, and ablation, which occurs principally through melt for land-terminating glaciers and melt plus iceberg calving for glaciers terminating in lakes or the ocean. The elevation at which accumulation and ablation occur in equal amounts over the course of a year is known as the equilibrium line altitude, or ELA.

End-of-season mass balance can be measured at a point on a glacier, and this quantity is generally positive (mass gain) in the accumulation area above the ELA, and negative (mass loss) in the ablation area below the ELA. Integrating such point measurements over the entire surface of the glacier yields the mass balance for the whole glacier for that year. The quantity is usually expressed as a water-equivalent change in thickness, in meters. Changes in volume or mass, measured with remote sensing methods, can be related directly to *in situ* measurements of mass balance.

There are long records of field-based measurements of mass balance (e.g., Zemp *et al.*, 2009), and these are valuable due to their detail and record length. However, field measurements on glaciers are laborious and expensive, and thus remote sensing techniques are necessary to extend our understanding to larger spatial scales and more glacier systems.

Glacier volume or mass, and their changes, can be investigated through remote sensing observations of glacier area or length, changes in elevation inferred from digital elevation models of the glacier surface estimated at two different times, or the end-of-summer snow line, which approximates the ELA (Braithwaite, 1984; Rabatel *et al.*, 2005). The connection between glacier geometry, as seen in 2-D imagery and mass balance, is complex, so some sort of modeling must be done to relate the two. Simple glacier flow models, together with long records of glacier length, have been used to estimate changes in regional climate (Oerlemans, 2005).

To estimate total glacier volume over large areas from satellite imagery, scaling relationships between glacier length or area and thickness or volume have been used (Bahr *et al.*, 1997; Raper and Braithwaite, 2005; Radić and Hock, 2010). However, scaling approaches have large uncertainty and are suitable for large samples of glaciers only; the method can estimate the average volume, for example, for a system of glaciers, but not an individual one. The uncertainties of the scaling approaches can be reduced by taking into account empirical relationships between the glacier shear stress and the topography. Recent modeling advances, based on these parameters, allow the calculation of an approximate glacier bed topography and, hence, the glacier thickness and volume (Farinotti *et al.*, 2009; Paul and Linsbauer, 2012).

Glacier monitoring has been an internationally coordinated activity since 1894 (Haeberli, 1998), yet only recently are we beginning to have a nearly complete and accurate map of glaciers on Earth. Following decades of field-based measurements, many national programs are using remote sensing techniques to map glaciers in their respective countries (Paul *et al.*, 2002; Casassa *et al.*, 2002; Bajracharya and Shrestha, 2011; Lambrecht and Kuhn, 2007; Andreassen *et al.*, 2008; Bolch *et al.*, 2010a). International initiatives such as Global Land Ice Measurements from Space (GLIMS; Raup *et al.*, 2007) are bringing these results together into a single database (http://glims.org) and are driving more analysis in order to complete the data set. Remote sensing data are now used to measure many parameters of glaciers, including location, areal extent and changes in area, mass and volume changes, ice velocity, distribution of supraglacial features such as rock debris, ponds, and lakes, steepness, distribution of glacier area over elevation, and others. The following sections give a brief overview of how satellite data can be used to derive many of these measurements from space.

7.2 Fundamentals

Optical satellite imagers (instruments that produce imagery) are much like modern digital cameras, in that they sense and record the intensity of electromagnetic radiation (radiance) in several different parts of the spectrum (called bands or channels), which can be combined or analyzed in many different ways to derive information about the materials on the ground (see also Chapter 2). Materials have unique reflectance spectra which depend on their molecular make-up. Figure 7.1 shows high resolution spectra of several materials typically found in images of glacierized terrain.

For example, the local maximum in the green part of the visible spectrum for deciduous vegetation is due to the chlorophyll content of the leaves. The reflectance is higher still in the near-infrared part of the spectrum. By contrast, fresh snow has high reflectance across the visible parts of the spectrum, then drops to much lower values in the infrared.

The shape of the reflectance curve of a material is known as its spectral signature. Automatic methods for discriminating materials on the basis of their different spectral signatures, specifically adapted to the task of mapping glaciers, rely on differences in spectral signature. For clean glacier ice, automatic methods can be used to discriminate glaciers from surrounding materials. Many glaciers have rock and morainal debris covering their surfaces, making it nearly impossible to

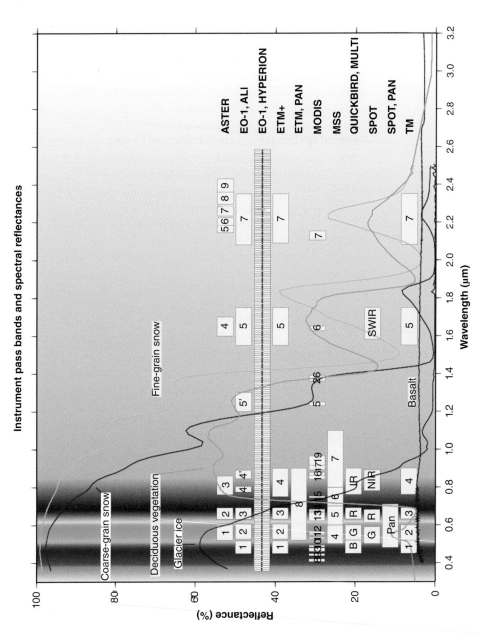

Figure 7.1 Reflectance spectra of several typical materials encountered in satellite imagery of glacierized terrain, and pass bands of several sensors in the visible and near-infrared (VNIR), and short-wave infrared (SWIR) parts of the spectrum.

discriminate between glacier debris cover and the rock, moraines, or forefield near the glacier. Other information must usually be brought to bear, which could be a digital terrain model or thermal information. Simple methods can be augmented with more sophisticated techniques, such as texture analysis or human interpretation, to compensate for the fact that debris-covered glacier ice is spectrally similar to surrounding materials.

Another challenge is determining locations of ice flow divides, where different parts of a contiguous ice mass flow in diverging directions. Ice flow divides generally mark a boundary between separate glaciers, but they are difficult to locate with precision and, typically, elevation data must be used in addition to satellite imagery to estimate ice flow directions.

Optical satellite sensors are characterized by several variables pertaining to their capabilities. Ground instantaneous field of view (GIFOV), which is closely related to spatial resolution, is the size of the area on the ground represented by one pixel in the image. Typically, the size of pixels and their spacing on the ground are essentially the same. Spectral resolution describes the number and spacing of pass bands along the electromagnetic spectrum. This determines the degree to which different materials on the ground can be discriminated.

The instruments discussed here are either multispectral instruments, which have 4–15 bands in the visible, near infrared, and thermal infrared, or hyperspectral instruments, which have hundreds of closely spaced bands. Additionally, some instruments record an integrated signal of radiance over the whole visible spectrum, and the resulting images are called panchromatic images. These are similar to black-and-white photographic images, which are also broadband in their response to light.

Another property is radiometric resolution, which is determined by the number of bits that are used to digitize the radiance signal produced by the light sensors. For example, an 8-bit system quantizes the radiance signal into 256 levels (i.e., 2^8, 0–255) or digital numbers (DN), whereas a 12-bit system produces 2^{12} (4096) levels. An instrument with low radiometric resolution can suffer from saturation of the DN values over bright targets (medium-bright to bright pixels set to the highest DN level), or only a few DN values being useful over dark targets. For example, variably bright snow in an 8-bit image may all be associated with the same DN value (255), while dark regions of the image that are in the shadow of a mountain may be represented by only a few DN values. Mapping the huge variations in brightness of a scene to only 255 levels would be similar to representing shoe sizes with only "small," "medium," and "large," rather than having a numeric scale with 30 levels. A 12-bit system, having 4096 levels, can represent a large range of brightness levels in a scene with high precision.

In addition to typical reflectance spectra, Figure 7.1 also shows the pass bands, or channels, for several optical satellite instruments that have been and are being used in glacier mapping. The primary workhorses for glacier mapping are multispectral imagers.

After their acquisition, raw images from a digital satellite sensor must be geometrically corrected so that the pixels represent a regular grid on the ground and are in a common map projection (e.g., Universal Transverse Mercator, UTM). This is generally done by the data provider. An additional step, important for most glacier mapping efforts, is to orthorectify the imagery. In simple terms, orthorectification adjusts the locations of image pixels to correct for distortions caused by terrain. For example, the peak of a tall mountain on the left side of an image will appear closer to the left edge of the image than it should, due to parallax. Orthorectification corrects for this effect, and produces an image where it appears that the viewer is looking straight down (vertically) on all parts of the image (Schowengerdt, 2007).

7.3　Satellite instruments for glacier research

A variety of satellite data sources can be used for studying glaciers, from declassified Corona photographs (panchromatic images) through LANDSAT, Advanced Spaceborne Thermal Emission and Reflection Radiometer (ASTER), Satellite Pour l'Observation de la Terre (SPOT), to instruments recently or nearly online, such as the high spatial resolution Quickbird or Worldview satellites (Table 7.1).

The first spaceborne sensors to acquire imagery systematically were flying on the US Corona program strategic reconnaissance satellites, in operation between 1960 and 1972. The most useful characteristics of the Corona panchromatic images are their stereo capability and a spatial resolution of approximately 2–8 meters (Dashora *et al.*, 2007). Corona or other reconnaissance images such as Hexagon (see Table 7.1) have recently been used to map earlier extents of glaciers with high precision (Bhambri *et al.*, 2011; Narama *et al.*, 2010; Bolch *et al.*, 2010b), extending the observational record by a decade into the past.

A comprehensive and continuous survey of the Earth's surface began with the launch of the first LANDSAT satellite (originally called Earth Resources Technology Satellite, or ERTS-1) in 1972. The satellite carried the multispectral scanner (MSS), recording in four wavelength bands in the visible and near infrared spectrum with a GIFOV of about 80 m × 80 m. The first sensor to acquire data in the short wave infrared, which makes it suitable for automated glacier mapping, was the Thematic Mapper (TM) aboard the LANDSAT 4 satellite, launched in 1982.

The LANDSAT instruments, from the MSS through the TM instruments on LANDSAT 4 and 5, to the Enhanced Thematic Mapper Plus (ETM+) instrument on LANDSAT 7, have been particularly useful because they collectively cover a long time span and have a wide swath (185 km), allowing the coverage of large areas in one image. In addition, LANDSAT images are now available at no cost as orthorectified products (http://landsat.gsfc.nasa.gov/). The LANDSAT bands were selected to be able to measure radiance in parts of the spectrum with absorption features associated with a number of common natural materials, including bands in the visible and near-infrared (useful for distinguishing vegetation from rocks and from water), and bands in the thermal infrared (useful for measuring thermal inertia, temperature and temperature ranges). Other multispectral imagers, such as the French SPOT satellites, the Indian IRS (Indian Remote Sensing) satellites and the Japanese instrument ASTER on board the Terra satellite, have similar capabilities (e.g., Figure 7.2). Recent developments include ultra high resolution imagery

Table 7.1 Summary of characteristics for optical instruments suitable for mapping glaciers.

Platform	Instrument	Spectral bands	GIFOV (spatial resolution) [m]	Radiometric resolution [bits]	Swath [km]	Stereo	Time of operation
Corona	KH4, KH4-A, KH-4B	PAN	2–8	n/a	≈15 × 220	Yes	1960–1972
Hexagon	KH-9	PAN	6–10	n/a		Yes	1971–1986
LANDSAT 1-4	MSS	VNIR	≈79	8	183 × 172	No	1972–1993
Resurs-F1, Cosmos	KFA-1000	PAN, multi-spectral	5	n/a	75 × 75	Yes	1974–1999
Cosmos	KVR-1000	PAN	2	n/a	40 × 160	No	1981–2005
LANDSAT 4, 5	TM	VNIR, SWIR, TIR	30, 120	8	183 × 172	No	Since 1982
Cosmos	TK-350	PAN	10	n/a	200 × 300	No	Since 1984
SPOT 1	HRV	PAN, VNIR, SWIR	10, 20	8	60 × 60	No	1986–2002
Resurs-F2	MK-4	6, col, pan	6-12		144 × 144	Yes	1988–1995
SPOT2, 3	HRV	PAN, VNIR, SWIR	10, 20	8	60 × 60	Cross-track	Since 1990
IRS	LISS-III	VNIR, SWIR	5,6, 23, 70, 148	7	70 × 70, 140 × 140, 800 × 800	No	Since 1996
SPOT 4	HRVIR	PAN,VNIR,SWIR	10, 20, 1000	8	60 × 60, 2200 × 2200	Cross-track	Since 1998
LANDSAT 7	ETM+	PAN, VNIR, SWIR, TIR	15, 30, 60	8	183 × 172	No	Since 1999
Terra	ASTER	VNIR, SWIR, TIR	15, 30, 90	8	60 × 60	Along-track	Since 2000
Ikonos		PAN, VNIR	1, 4	11	11 × 14		Since 2000
Quickbird		PAN, VNIR	0.65, 2.7	11			Since 2001
SPOT 5	HRS/HRG	PAN, VNIR, SWIR	2,5, 5, 10, 1000	8	60 × 60, 1000 × 1000	Cross-track	Since 2002
IRS-P6	LISS-IV	VNIR, SWIR	5.8, 23.5, 56–70	7	24; 70, 140, 740	No	Since 2003
Cartosat-1 (IRS-P5)		PAN	2,5	12	30 × 30	Along-track	Since 2005
ALOS	PRISM	PAN	2.5	8	35 × 35, 70 × 70	Along-track	Since 2006
ALOS	AVNIR-2	VNIR	10	8	70 × 70	No	Since 2006
Worldview-2		PAN, VNIR	0.5, 1.8	11	16 × 16	Yes	Since 2009
LDCM (LANDSAT 8)	OLI, TIRS	PAN, VNIR, SWIR, TIR	15, 30, 100	12	185 × 180	No	Since 2013
Sentinel-2	MSI	PAN, VNIR, SWIR, TIR	10, 20, 60	12	290 × 290	No	2013 (sched.)

PAN = pan-chromatic. VNIR = visible and near infrared. SWIR = short-wave infrared.

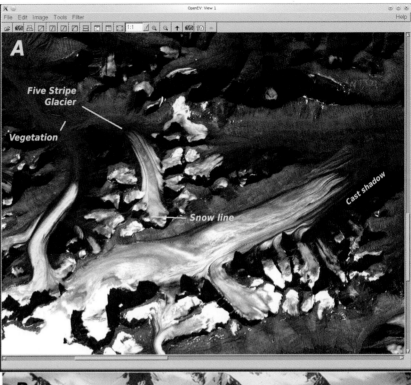

Figure 7.2 (a) Example of an ASTER image, acquired 09-16-2010, of glaciers in the Chugach Range of southern Alaska. Bands 3N, 2, and 1 are displayed as red, green, and blue. (b) Oblique aerial photograph, taken in 1977, of Five Stripe Glacier, visible in the ASTER image at arrow. The snow line is clearly visible in (b) as the transition between bare glacier ice (gray) and snow (white), but the snow line in the ASTER image is almost above the highest elevation of Five Stripe Glacier.

with a resolution of one meter and below (e.g., Ikonos, Quickbird) and fast repeat cycles (e.g., Rapid-Eye).

The instruments described above are *passive*, in that they sense solar radiation scattered from the earth's surface (another class of instruments, "passive microwave sensors", measure microwave radiation emitted and reflected by surfaces). By contrast, radar instruments are *active* systems, where a beam is emitted from the sensor and the magnitude and phase of the backscattered energy is recorded. The United States SEASAT mission (1978) is regarded as the first satellite mission to carry a radar system, and was designed to measure surface properties and topography of the oceans (Evans *et al.*, 2005). Continuous radar missions that are useful for glaciological applications started only in the 1990s, with the European ERS-1, the Russian ALMAZ-1, and the Japanese, J-ERS-1 (Table 7.2).

Most satellite radar applications use synthetic aperture radar (SAR). The resolution of a radar image is inversely proportional to the size of the antenna (see also Chapter 9 for more on SAR and radar systems). The SAR technique synthesizes an antenna longer than is physically possible by using the constant velocity of the radar platform and knowledge of the changing geometry between the instrument and a given location on the ground. This technique increases the spatial resolution of radar systems by orders of magnitude.

Table 7.2 Important radar systems and their characteristics.

Platform	Instrument	Resolution (m)	Band	Pol.	Look angle (°)	Swath width (km)	Operation
SEASAT	SAR	25	L	HH	20	100	Jun–Sep 1978
ALMAZ-1	SAR	12–20	S	HH	25–60	40	1991
ERS-1	SAR	30	C	VV	23	100	1991–2000
J-ERS-1	SAR	18	L	HH	35	75	1992
Space Shuttle	SIR-C	15–25	L	Quad	20–55	30–100	1994
			C	Quad			
			X	VV			
ERS-2	SAR	30	C	VV	23	100	since 1995
Radarsat-1	SAR	10–100	C	HH	20–50	10–500	since 1995
SRTM	SAR	30	C	VV + HH	17–65	225	Feb. 2000
		30	X	VV	55	50	
ENVISAT	ASAR	30	C	Quad	15–45	100	2002–2012
		150				450	
ALOS	PALSAR	7–100	L	Quad	8–60	20–350	since 2006
Radarsat-2	SAR	3–100	C	Quad	10–60	20–500	since 2007
TerraSAR-X	SAR	3.3–17.6	X	Quad	20–60	10–100	since 2007
TanDEM-X	SAR	3.3–17.6	X	Quad	20–60	10–100	since 2010
Sentinel-x	SAR	5–40	C	Quad	20–45	80–400	2013 (sched.)

SAR = Synthetic Aperture Radar; ASAR = Advanced SAR; PALSAR = Phased-Array L-band SAR.

Other systems to measure surface elevation are laser and radar altimeters. Laser altimetry systems such as Geoscience Laser Altimetry System (GLAS), aboard the NASA Ice, Cloud, and land Elevation Satellite (ICESat), or radar altimetry systems such as the ENVISAT RA2 or CryoSat-2, transmit a pulse (laser light or radar beam) to the surface, and measure the time to receive the reflected response, thereby determining the elevation of the surface. ICESat GLAS was designed to study primarily the Greenland and Antarctic ice sheets, but has also been used to study mountain glacier elevation changes from space (Kääb, 2008; Moholdt *et al.*, 2010). The footprint of the transmitted spot for GLAS is approximately 70 m but, as the spacing of laser pulses on the ground is large compared to the size of most glaciers, this instrument can only be used to study a limited number of larger mountain glaciers or ice caps.

New satellites suited to studying mountain glaciers are planned to be launched in the near future. Particularly promising is the European Sentinel-2 satellite, scheduled to be launched in 2014, with a higher temporal and spatial resolution than the LANDSAT TM and ETM+ sensors. The LANDSAT Data Continuity Mission (LCDM) is valuable as well (Table 7.1).

7.4 Methods

7.4.1 *Image classification for glacier mapping*

Image classification is the process of assigning material classes (vegetation, rock, snow, water, etc.) to each pixel in an image. This first step is essential before creating a map of features (e.g., glaciers, lakes) in the imagery, although mapping features (e.g., glaciers) from a classified image (a map of materials) is not trivial, as many features involve a mixture of materials. For example, a glacier is dominantly ice, but it will often contain medial moraines or rock debris cover at the terminus, and the debris cover may even support vegetation. The "glacier" will therefore contain ice, rock, and vegetation classes in the classified image. Many different image classification and feature identification methods have been evaluated for glacier mapping, including manual digitization, supervised classification (maximum likelihood, minimum distance, spectral angle mapper, etc.), unsupervised classification, fuzzy classification techniques, and band arithmetic and threshold techniques (Sidjak and Wheate, 1999; Paul *et al.*, 2002; Albert, 2002; Racoviteanu *et al.*, 2009; Gjermundsen *et al.*, 2011).

Supervised classification begins with the operator selecting representative sets of pixels that are representative of identifiable materials in the image, such as vegetation, snow, glacier ice, and so on. The differences between the spectral signatures of these "training sets" (or end-members) are then used to assign the rest of the pixels in the image to one of those categories. Linear unmixing, a fuzzy classification technique, begins the same way but assigns fractional end-member content to each pixel. Unsupervised techniques such as ISODATA operate similarly, but attempt to identify the end-members automatically by employing cluster detection algorithms. Details on image classification methods can be found in several books (e.g., Schowengerdt, 2007; Richards, 1986).

There is a trade-off between the time required to create glacier outlines from a method and the accuracy of the resulting outlines (Albert, 2002). Manual digitization, done by a glaciologist experienced in studying glaciers in remote sensing imagery, can achieve the highest accuracy in most types of terrain. Errors of interpretation can occur in some situations involving complicated patterns of debris cover on the glacier or seasonal snow attached to the glacier, but humans tend to do better in such situations than current computer algorithms. Mapping glaciers by hand, however, is the most time-consuming, and hence costly, of the available methods, and is therefore impractical for any but small regional studies.

Several studies (Albert, 2002; Paul *et al.*, 2002; Racoviteanu *et al.*, 2009) have concluded that the best trade-off between processing time and accuracy is obtained by band arithmetic techniques, such as computing band ratios or using normalized differences, such as the normalized difference snow index (NDSI; see also Chapter 3). In the band ratio technique, the ratio of bands in the near-infrared (NIR) or red parts of the spectrum and the short-wave infrared (SWIR) is computed, resulting in a "ratio image" that has high values over snow and ice, and low values over rock and vegetation. A threshold is applied to produce a binary raster mask indicating regions of snow/ice and non-snow/ice. Glacier outlines are then derived by converting the raster mask to a vector representation.

Both band ratios NIR/SWIR and red/SWIR have been shown to be well suited for glacier mapping, as they both take advantage of the large difference in spectral signature between snow and other materials in those spectral bands (Figure 7.1). Some studies have shown preference for Red/SWIR, as this discriminates better between ice and snow in shadow cast by adjacent terrain (Paul and Kääb, 2005). Andreassen *et al.* (2008) used the spectral characteristics of snow and ice in LANDSAT bands TM3 (Red) and TM5 (SWIR) for automatic classification of glaciers in Jotunheimen in southern Norway (Figure 7.3). The subset of the full scene used in their analysis contains neither clouds nor stark shadow, and there is no significant debris on the ice. The image is therefore well suited to the band ratio method.

The normalized difference snow index (NDSI; see also Chapter 3), another band arithmetic technique that is defined as (VIS-NIR)/(VIS + NIR), can be used similarly to map glacier outlines (e.g., Sidjak and Wheate, 1999; Racoviteanu *et al.*, 2008b). For LANDSAT TM, for example, NDSI is obtained from the ratio (TM2-TM5)/(TM2 + TM5). It was developed to map snow (Hall *et al.*, 1995; Salomonson and Appel, 2004; see Chapter 3), but since glacier ice is composed of densified snow, the spectral characteristics are similar (Figure 7.1). While the NDSI is useful for glacier mapping, the good performance and simplicity of the band ratio method for clean ice has resulted in the community predominantly using it to do classification (Racoviteanu *et al.*, 2009). The band ratio method is easily automated but, as it relies on differences in spectral signatures of the materials in the image, it erroneously excludes parts of the glacier that are covered with rock debris. While newer techniques, discussed below, address these

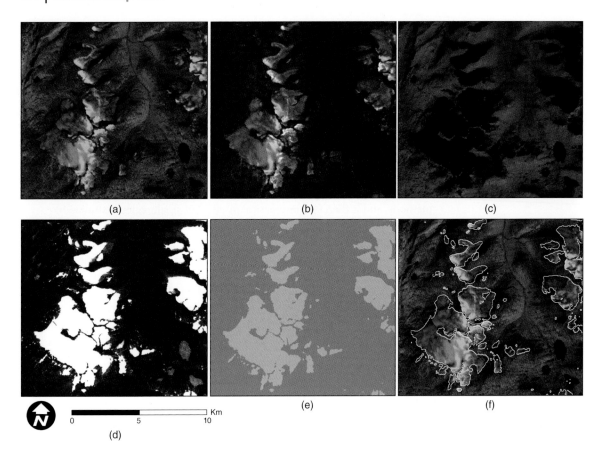

Figure 7.3 Automatic classification of snow and ice for a subset of glaciers in Jotunheimen, southern Norway, using a LANDSAT TM scene from 9 August 2003. (a) Red-green-blue (RGB) composite of TM bands 5, 4 and 3. (b) LANDSAT band TM3 (0.63–0.69 μm). (c) LANDSAT band TM5 (1.55–1.75 μm). (d) Ratio image of TM3/TM5. (e) Thresholded image of TM3/TM5 >2.0 and median filter (3 ∗ 3 kernel). (f) Same as (a), with outlines (in white) derived from raster to vector conversion of (e).

shortcomings, manual editing is generally needed to include debris-covered parts, to exclude seasonal snow and sometimes adjacent lakes, or to include parts of the glacier in heavy shadow.

7.4.2 Mapping debris-covered glaciers

More sophisticated methods than the ones presented above have been developed for automatically classifying debris-covered ice as part of the glacier. Some methods use thermal data, while others use topographic texture, curvature (geomorphometrics; Bishop *et al.*, 2001; Brenning, 2009), velocity, image differencing, or coherence images of SAR data (see next page) for detection of ice motion and to delineate the glacier boundary.

Finding the precise boundary of a glacier can be difficult, even for an experienced glaciologist in the field (Haeberli and Epifani, 1986), but it usually has characteristics that can be used to identify it, including distinctive morphology (such as a lateral concave-up trough or topographic drop at the terminus), or different thermal characteristics from surrounding rocks and tills. Semi-automated methods can also be used to detect glacier boundaries based solely on curvature of the DEM (Raup in Kieffer *et al.*, 2000). This, and other similar approaches, show that even using only the morphology represented in the DEM can often lead to a good first estimate of the location of glacier boundaries.

Paul *et al.* (2004) developed a more automated algorithm which combines topographic information with multispectral imagery into a single algorithm. As a starting point, the algorithm uses a glacier ice region identified using the multispectral band ratio technique described above, and then expands this region, checking to ensure that pixels added to the region classified as "glacier" meet the following criteria:

a) the spectral signature is snow, ice, or rock;
b) the slope angle is low, below a specified threshold;
c) a concave-up area of curvature (a trough) has not been crossed.

Figure 7.4 shows the results of applying this algorithm to LANDSAT 5 (TM) imagery in the Bernese Alps, Switzerland. That study investigated two different DEMs as input – one based on ASTER, and one from the Swiss Federal Office of Topography. This general algorithm shows great promise, as most heavily debris-laden glaciers have similar morphological characteristics (e.g., low slope angles) to the ones investigated by Paul *et al.* (2004). However, the specifics must be modified, depending on the region. Stagnant ice that supports vegetation, varying regional glacier slope angles, icefalls, and heavy thermokarst activity (sub-debris melting, pond growth, and tunnel collapse) can all cause the algorithm to fail.

Another approach is to use the differing thermal properties of the supraglacial debris compared to surrounding rock. When the supraglacial debris is thin enough, the underlying ice cools the debris compared to surrounding rock and tills. For very thin debris, the intermittently exposed ice has a significant influence on the spectral signature of the pixels. Either of these two effects causes the debris-covered parts of the glacier to be distinguishable from the surrounding non-glacier material. Thicker debris violates these conditions, however, so this method appears to work well only for debris cover of less than about 0.5 m (Ranzi *et al.*, 2004; Karimi *et al.*, 2012). Thicker mantles of debris completely obscure the path of radiation between the ice and the sensor, as well as serving as an insulating layer which hides the cold surface of the ice. In addition, the spatial resolution of thermal data is invariably lower than visible and near-infrared imagery from the same platform, making it more difficult to apply such methods on smaller mountain glaciers. Further work is required to make thermally based methods capable of wide operational deployment.

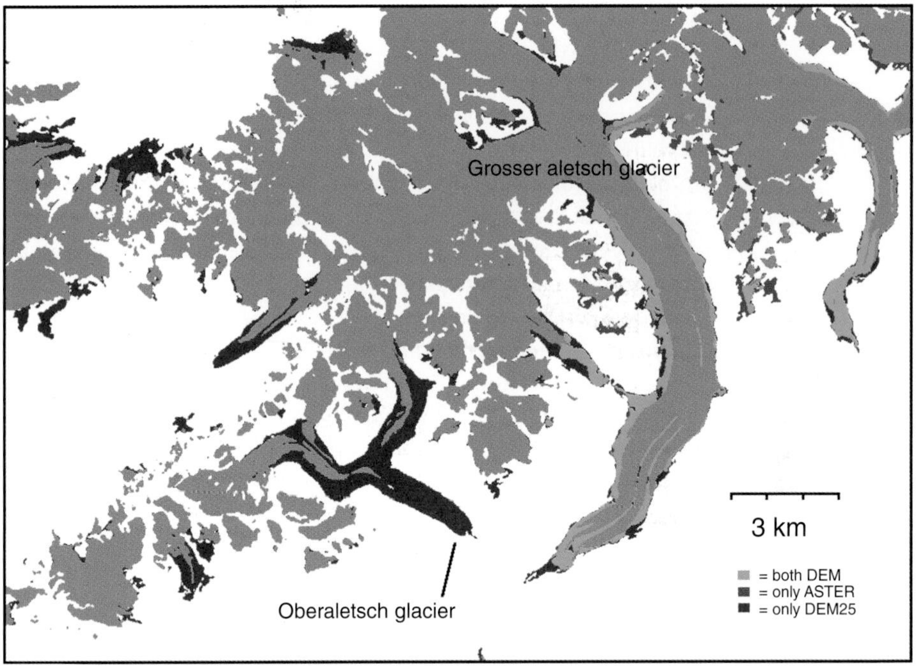

Figure 7.4 Results of applying a debris-covered glacier mapping algorithm to LANDSAT 5 imagery of the Bernese Alps, Switzerland, and two digital elevation models (adapted from Paul *et al.*, 2004).

7.4.3 *Glacier mapping with SAR data*

One main advantage of active radar systems is that they can operate in almost any weather as well as at night, because the beam and the backscattered radiation in the microwave region penetrate through clouds and do not depend on sunlight. In general, SAR data are capable of distinguishing snow and ice from surrounding areas, due to the different backscattering signal from different materials.

A limitation is that spaceborne satellite SAR systems are usually limited to one frequency of the emitted beam, which hampers the correct classification of glaciers (Rott, 1994). This is especially true if glaciers have variable surface roughness, for example due to crevasses, different snow and ice facies (snow, firn, glacier ice, refrozen melt water), and debris cover. The variable nature of the returned signal makes finding the glacier boundary difficult. Hence, SAR data has so far played only a minor role for glacier mapping.

An application where SAR performs better than optical methods in glacier studies is the discrimination of different snow facies. SAR data are often used to map temporal changes of snow and ice areas due to the temporal change of the surface's backscatter characteristics, best observable in the C band (4 to 8 GHz) and X band (8 to 12 GHz) (Nagler and Rott, 2000). The backscattering ratio of a summer and winter image is used to discriminate between ice and snow in summer, while a winter image is best for obtaining the longer term ELA, because the SAR signal integrates over several meters depth in the snow pack and the boundary between ice and firn is therefore averaged over a few years (Floricioiu and Rott, 2001).

Recent efforts have focused on using SAR data for the detection of glaciers in areas with frequent cloud cover, and for debris-covered glaciers where the optical data are limited. One promising approach is the use of SAR coherence images (Atwood *et al.*, 2010), where areas with changing surfaces (e.g., due to glacier motion or down-wasting) decorrelate (the phase of the returned beam changes over time), in contrast to coherent stable areas. In addition, by taking advantage of the phase coherence of the transmitted radar beam, data from multiple passes can be analyzed for differences in the phase of the returned beam, which can be used for mapping topography, detecting subtle changes in the surface (including debris-covered areas), and mapping ice flow velocity. This technique is known as interferometric SAR, or InSAR (see Chapter 9 also for more InSAR applications).

Once glacier outlines are obtained by any method, basic glacier parameters, such as slope, aspect, elevation range, minimum, median, and maximum elevations, etc. can be obtained by examining DEM values within the outline or on the boundary (Paul *et al.*, 2009). Such parameters are generally stored as key parts of glacier inventories, such as the World Glacier Inventory and the GLIMS Glacier Database.

7.4.4 Assessing glacier changes

Time series of satellite imagery, or imagery combined with other data sources, can be used to detect and map changes in glacier extent, thickness, and surface characteristics.

When studying a series of images, the first step is to co-register them using ground control points (GCPs) or tie points. GCPs register the images to known locations on the ground, while tie points are features identifiable in both (or all) images, but may not be precisely located on the ground. This co-registration step is crucial, as many artifacts (false changes) can appear if there are offsets between the images. For change detection involving multi-temporal DEMs, proper co-registration is even more important, since improper co-registration leads to biases in the elevation differences everywhere in the domain, not just at the boundaries of different regions within the domain, as might be the case with classified imagery (Nuth and Kääb, 2011).

Changes in glacier area and length are easy to obtain directly from satellite observations, because they are directly related to changes in the observed two-dimensional (planimetric) information in the imagery. Changes in glacier volume or mass are more important scientifically (due to the connection to climate, and consequences such as sea level change or changes to hydrological resources), but they are more difficult to observe from satellite data.

7.4.5 Area and length changes

A simple yet effective way to detect change is the method of image differencing, where an earlier image is subtracted from a later one taken with identical or similar solar geometry (nearly the same time of day and day of year) (e.g., Sabins, 2007). The subtraction is usually done separately on each band, and results in a grid of numbers that are near zero in areas of no or little change, and which can be positive or negative for changed regions in the image. The grid is typically rescaled to positive numbers for display so that, for a single band, areas of no change appear gray, areas that have become more reflective (e.g., an area newly snow-covered) appear bright, and less reflective parts of the image (e.g., where snow or ice has been removed) look dark. Each difference band can be viewed on its own, or the rescaled difference bands can be displayed as red-green-blue, and the colors then yield information about changes in materials on the ground, since they indicate changes in spectral signature (e.g., by the appearance of vegetation on a moraine).

Mapping changes in glacier area based on repeat glacier extent mapping is straightforward in theory: create maps of glacier ice from two (or more) different images, and compare the computed areas. This can be done either from the raster map or from vector outlines that have been derived from a classified image. Another choice that the analyst must make is whether to compare maps of contiguous glacier ice as a whole, or to compare computed areas on a glacier-by-glacier basis. If the aim is to compute glacier area change for each separate glacier, great care must be taken to ensure that the ice divides (the boundaries between glaciers at their upper reaches) match exactly. If the ice divides do not match, then one can easily obtain the erroneous result that the glacier or glaciers on one side of a divide have grown, while those on the other side have shrunk, even if the total ice area did not change. This kind of investigation is therefore most safely performed over a whole system or mountain range (e.g., Racoviteanu *et al.*, 2008a).

Satellite imagery can also be used to derive length changes. High-resolution satellite imagery has been used to reconstruct glacier front variations over a period of four decades for the well-known Gangotri Glacier of northern India (Bhambri *et al.*, 2012). A study from Jostedalsbreen, Norway, showed that the high spatial variability measured from field data could be observed by LANDSAT (within one pixel, 30 m resolution) for glacier tongues not in cast shadow (Paul *et al.*, 2011).

As satellite imagery only goes back a few decades, it is common to assess glacier changes by comparing with other sources such as aerial photographs, topographic maps or inventory data (Casassa *et al.*, 2002). Glacier outlines derived from topographic maps are a valuable source, but they are usually mapped by non-glaciologists and might contain inaccuracies due to the existence of seasonal snow or misinterpretation of debris cover (Andreassen *et al.*, 2008; Bhambri and Bolch, 2009). Satellite imagery has also been used in several regions to map glacier changes since the Little Ice Age maximum by mapping the glacier forefield, which is spectrally different from the terrain beyond the Little Ice Age maximum moraines (Paul and Kääb, 2005; Baumann *et al.*, 2009).

7.4.6 Volumetric glacier changes

DEMs can be used to measure volume changes of glaciers by subtracting pairs of DEMs. DEMs can be generated by different methods, such as space- and airborne optical stereo data, interferometric SAR (InSAR) data, space and airborne radar and laser altimetry, and topographic maps. There are three nearly global elevation products available today with high enough spatial resolution to study glacier changes:

1 The Shuttle Radar Topography Mission (SRTM) flown in February 2000.
2 ASTER Global DEM, version 2 (GDEM2) based upon stereo scenes acquired from 2000 to the present.
3 The ICESat mission from 2003 to 2009 using spaceborne laser altimetry, or Light Detection and Ranging (LiDAR).

These data sets all have different strengths and limitations. The SRTM DEM has relatively high resolution and covers the globe between 59°S and 60°N latitude, but it includes data voids in mountainous regions of high relief. Radar penetration of snow in the upper reaches of glaciers can lead to elevation biases there (Gardelle *et al.*, 2012). The ICESat elevation estimates are the most accurate (e.g., Nuth and Kääb, 2011), but the spatial coverage is much poorer, with 170 m spacing between points along-track and hundreds or thousands of meters between tracks, depending on latitude. Nevertheless, these data are valuable for obtaining region-wide information (Kääb *et al.*, 2012; Bolch *et al.*, 2013; Gardner *et al.*, 2013).

While the SRTM and standard ASTER DEMs have been proved to be useful as a baseline data set, more accurate DEMs are generally needed for precise volume change determination (e.g., Berthier *et al.*, 2010). Such DEMs can be created from individual image pairs from suitable optical stereo satellite instruments, such as ASTER, SPOT5, ALOS, and Cartosat-1 (e.g., Kääb *et al.*, 2012; Berthier *et al.*, 2007). LiDAR DEMs acquired using airborne systems are generally the most precise, but are limited to smaller areas. One of the largest ice bodies mapped nearly completely with LiDAR to date is Vatnajökull, the largest ice cap of Iceland (Johannesson *et al.*, 2012).

Elevation change of glaciers can be calculated by DEM differencing or repeat altimetry measurements (Figure 7.5). DEM differencing is a popular technique that has been used for decades to assess glacier changes (e.g., Finsterwalder, 1954; Abermann *et al.*, 2009). Changes in the volume of the glacier are converted to mass changes using estimates of density. This method is called the geodetic method

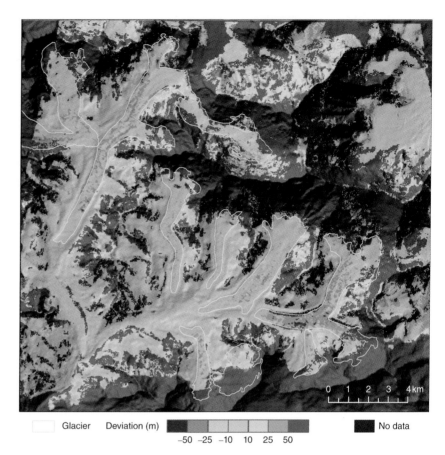

Figure 7.5 Glacier elevation changes measured by subtracting two DEMs in the Mt. Everest area of Nepal. DEMs are based on 1970 Corona and 2007 Cartosat data. Glacier outlines are plotted as light yellow lines (Bolch *et al.*, 2011).

of measuring mass changes and can be used to scale up to the whole glacier the traditional direct mass balance measurements based on field investigations, or to check their accuracy, or to assess changes in unmeasured areas or periods (e.g., Arendt *et al.*, 2002; Cogley , 2009; Fischer, 2010; Andreassen *et al.*, 2012).

7.4.7 Glacier velocity

There are two main techniques for deriving glacier surface flow velocities from remote sensing data:

1 tracking of features on the glacier's surface in sequential imagery, either optical or radar (feature tracking);
2 satellite radar interferometry (InSAR or D-InSAR) (Luckman *et al.*, 2007).

As in all methods for detecting change, feature tracking (Scambos *et al.*, 1992) requires that the input imagery should be accurately co-registered.

Feature tracking software, such as *IMCORR* or *Cosi-corr* (see Heid and Kääb (2012) for a comparison of algorithms), automatically finds matching features from both images and outputs a displacement vector for each location on a grid. For glaciers, these individual features could be ice pinnacles, crevasses, boulders, or a distinctive pattern of debris and ice, which can be identified on both images. Compared to field methods, where a day's work might produce a handful of velocity measurements for a glacier, feature tracking in satellite imagery is much more effective, yielding tens of thousands of vectors.

When feature tracking is applied to radar data, matching is generally done using the intensity images, which are images of the strength of the returned beam, either by tracking large-scale features that move with the ice and are visible in the imagery, or by tracking the small-scale pattern of light and dark in the intensity image that is due to phase coherence from one image to the next over stable surfaces. This latter technique is known as "speckle tracking," and it allows smaller patch sizes to be used compared to feature-tracking (Strozzi *et al.*, 2002; Joughin, 2002). Studies that use feature tracking for velocity measurement include Kääb (2005), Bolch *et al.* (2008), Scherler *et al.* (2008), and Quincey *et al.* (2009).

Interferometric synthetic aperture radar (InSAR) is used for the observation of geometric displacements of the ground or glacier surface. Displacements on the order of centimeters or even millimeters can be measured, depending on the time difference between acquisitions. InSAR is based on the fact that the phase of the radar beam returned to the instrument from a spot on the ground does not change from overflight to overflight, provided the instrument is in the same position, unless the surface is changing. With two overflights, the phase part of the radar images can be subtracted to create an interferogram, which is a depiction of how the phase has changed from one overflight to the next. A single interferogram contains information about changes on the ground (deformation of the surface, or changed dielectric properties), as well as effects from differing satellite positions between overflights (orbital contributions) and topography. If the instrument is in slightly different positions for the two overflights, the effect on the interferogram can be modeled and removed.

To measure surface velocity with this technique, the effect of topography must be removed. This can be done using a DEM from an external source, such as SRTM or GDEM2, in which case only two satellite overflights are required (two-pass method). Alternatively, in the three-pass method, the first two images are used to generate a DEM. The time separation should be short between these two in order to minimize effects of surface deformation and, thus, decorrelation of the signal. The image of the third overflight is then used, together with the first or second image, to retrieve another interferogram. The second interferogram contains not only information about topography, but also the displacement, and by combining the two interferograms, the component of the ice velocity in the direction toward or away from the satellite sensor can be isolated.

The three-pass method is generally known as differential interferometric synthetic aperture radar, or D-InSAR. D-InSAR makes it possible to detect sub-meter changes in images that have a nominal ground resolution of several meters.

Problems with this technique for glacier velocity mapping include difficulty achieving the stability of the phase (phase coherence) necessary to make the method work, data availability, and layover (radar beam shadowing) in regions of high relief.

An example study using this technique is described in Yasuda and Furuya (2013), who studied short-term glacier velocity fluctuations in the Kunlun Shan range. More information about this technique can be found in Eldhuset *et al.* (2003) and Luckman *et al.* (2007).

7.5 Glaciers of the Greenland ice sheet

7.5.1 Surface elevation

The earliest satellite observations of surface elevation change over the Greenland ice sheet were based on radar altimeter data from the ERS-1 (1992–95) and ERS-2 (1995–2003) satellites (Johannessen *et al.*, 2005; Zwally *et al.*, 2005). The results indicated that the ice sheet was growing above the ELA and thinning at lower elevations, with an overall gain in volume. However, both studies indicated more rapid elevation increases at high elevations and less thinning over the margins than did results from repeat aircraft laser altimeter surveys within the same period (Krabill *et al.*, 2000). The differences at lower altitudes, where slopes are steeper, are probably due to the much larger footprint of the radar altimeter ($\approx$20 km) compared with the 1–3 m airborne LiDAR footprint. At higher elevations, the difference may be due to changes in the radar penetration depth being superimposed on surface elevation changes (Thomas *et al.*, 2008).

Since 2002, the ICESat instrument GLAS, a laser altimeter, has been able to sample the faster-flowing outlet glaciers more precisely than the satellite radar altimeter systems, although the crossover points required to detect change are geographically sparse. Data from this instrument have shown that dynamic thinning is far more significant on the fast-flowing marine-terminating outlet glaciers than on the slower flowing parts of the ice sheet, and now reaches all latitudes in Greenland (Figure 7.6; Pritchard *et al.*, 2009). Chapter 8 contains more details on Accumulation and Surface Elevation techniques over Greenland.

7.5.2 Glacier extent

Observations of surface elevation change have focused attention on the frontal behavior, dynamics, and flow rates of the tidewater outlet glaciers in Greenland, where the greatest changes appeared to be concentrated.

The long archive of images from the LANDSAT series of satellites has proved to be particularly useful for monitoring the ice-front positions of these outlet glaciers all around the ice sheet. Manual identification of ice fronts in LANDSAT images, at multi-year intervals from 1972 to 2010, showed that marine-terminating glaciers

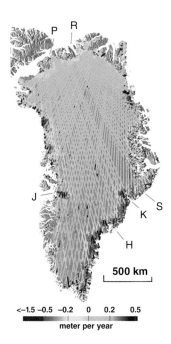

Figure 7.6 Rate of change of surface elevation for Greenland based on IceSAT GLAS data collected after 2002 (Pritchard *et al.*, 2009). Major outlet glaciers are labeled: H – Helheim Glacier; K – Kangerdlussuaq Glacier; J – Jacobshavn Glacier; S – Sortebrae; P – Petermann Glacier, R – Ryder Glacier.

in the south-east and west were generally stable between 1972 and 1985, but began a gradual retreat in 1992, which accelerated between 2000 and 2010 (Howat and Eddy, 2011). A higher temporal resolution time series based on LANDSAT images identified a synchronous re-advance in 2006 in the southeast (Murray *et al.*, 2010). The transition from stability to general retreat, which began in the early 1990s, was coincident with warming trends in land and sea surface temperatures (Howat and Eddy, 2011; Bevan *et al.*, 2012).

Automating the identification of ice-fronts using edge-detection techniques on over 100 000 Moderate Resolution Imaging Spectroradiometer (MODIS) images allowed a ten-year daily time series of ice front positions to be produced for 32 east Greenland glaciers (Seale *et al.*, 2011). In spite of the trade-off between revisit frequency (daily compared to 16 days in the case of LANDSAT) and spatial resolution (250 m, compared to 15 or 30 m for LANDSAT), a seasonal advance and retreat cycle of magnitude proportional to glacier width was identified. Ice fronts can be identified all year using SAR imagery, including during the polar night and when clouds obscure the ice fronts in optical imagery. This capability allows mosaics of the entire ice sheet to be generated for the same time of year at annual intervals, with the earliest coverage available in 1992. Moon and Joughin (2008) used SAR mosaics compiled for 1992, 2000, 2006 and 2007 to compare ice-front retreat rates between epochs and between land and tidewater terminating glaciers.

Occasionally, large individual calving events from Greenland outlet glaciers with floating tongues are observed in near real time in satellite imagery. For example, in August 2010, the Petermann Glacier in northern Greenland calved back by an estimated 28 km (Johannessen *et al.*, 2011). The now multi-decadal archive of optical and SAR imagery allows these events to be placed in a longer term perspective.

7.5.3   *Glacier dynamics*

The retreat of outlet glacier ice fronts around the ice sheet often accompanies, or results in, an acceleration in ice flow via mechanisms which have yet to be fully understood (Joughin *et al.*, 2010). The application of feature-tracking and InSAR techniques has resulted in surface velocity measurements over large numbers of glaciers at multi-year intervals, in addition to high-resolution time series of measurements on individual glaciers.

Undoubtedly, the most dramatic stories for Greenland to be revealed, using remote sensing to measure ice velocities, concerned the rapid doubling or more of flow speeds by three major outlet glaciers. First, in the west, Jakobshavn Isbrae began speeding up in 1998 (Joughin *et al.*, 2004; Luckman and Murray, 2005) and then, in the southeast, Helheim began to accelerate around 2002, and Kangerlussuaq around 2004 (Howat *et al.*, 2005; Luckman *et al.*, 2006). The latter two have since decelerated, but both are still flowing faster than before (Howat *et al.*, 2007; Murray *et al.*, 2010).

In parallel with results for these individual glaciers, the mapping of surface velocities around the whole of Greenland showed a widespread acceleration south of 66°N between 1996 and 2000, which rapidly extended to 70°N by 2000 (Rignot and Kanagaratnam, 2006; Joughin *et al.*, 2010). Mirroring the pattern of ice-front retreat, feature tracking of early LANDSAT 5 images from 1985 has demonstrated that flow speeds were also stable around the ice sheet until temperatures began increasing in the mid-1990s (Bevan *et al.*, 2012). Moon *et al.* (2012) used data from a number of satellites to investigate changes in glacier speed for many of the ≈200 of Greenland's major outlet glaciers, and document complex patterns of velocity change that depend on glacier type (terminating in the ocean versus ice shelves versus land). These results indicate that the changes currently being experienced by Greenland's glaciers are more complex than previously thought. Underlying this complexity, however, appears to be an oceanic driver that is causing major changes in Greenland's outlet glaciers (Walsh *et al.*, 2012).

Comparing annual ice discharge flux for a glacier, based on remotely sensed flow speeds, with surface mass balance, allows an estimate of net mass balance for the catchment to be made. For example, in 1995/96, glaciers in the southeastern part of the ice sheet were shown to be flowing faster than was required to maintain balance (Rignot *et al.*, 2004). This flux-balance method also showed that, between 2000 and 2010, Jakobshavn Isbrae and Kangerdlugssuaq must have been losing mass, while Helheim was apparently gaining mass (Howat *et al.*, 2011).

Feature tracking and InSAR have also been used to investigate short-term variability in ice flow speeds. Examples include the surge of Sortebrae (Pritchard *et al.*, 2005), the mini-surge of Ryder Glacier and seasonal cycles of acceleration and deceleration on outlet glaciers (Luckman and Murray, 2005) and land-terminating sectors of the ice-sheet (Joughin *et al.*, 2008; Palmer *et al.*, 2011). Such short-term accelerations are thought to be driven by seasonal surface melt, percolating to the bed and increasing the basal water pressure, thereby enhancing basal lubrication

(Zwally *et al.*, 2002). Once an efficient subglacial drainage system is re-established, the flow speeds decrease again (e.g., Sundal *et al.*, 2011).

InSAR can be used not only to measure surface velocities, but also to identify parts of a glacier that are moving differently from other parts in a qualitative sense. InSAR analysis has been used, for example, to locate the boundaries of rock glaciers, which typically are spectrally indistinguishable from their surroundings, based on finding areas that are exhibiting creep. As another example, InSAR can be used to identify the location of grounding lines on glaciers with floating tongues. The grounding lines are indicated by the limit of tidal flexure revealed by differencing two velocity-only interferograms. This technique was used in northern Greenland to reveal the inland migration of grounding lines between 1992 and 1996, indicating glacier thinning with a dynamic origin (Rignot *et al.*, 2001).

7.6 Summary

Advances have been made recently in remote sensing of glaciers on a number of fronts, including more complete and accurate glacier inventories, improved glacier mapping techniques, and new insights from gravimetric satellites. Through international cooperative efforts such as the Global Land Ice Measurements from Space (GLIMS) initiative (Raup *et al.*, 2007) and the Global Terrestrial Network for Glaciers (GTN-G; Haeberli *et al.*, 2007), satellite remote sensing of glaciers has led to the ability to produce glacier outlines quickly over large regions, leading to the production of nearly complete global glacier inventories.

Another recent project, spurred by sea level modeling needs and deadlines of the Fifth Assessment Report of the Intergovernmental Panel on Climate Change (IPCC), has led to the production of a nearly globally complete set of glacier outlines, known as the Randolph Glacier Inventory (RGI). This data set lacks the thorough attributes and documentation of data sources of GLIMS data, but it has proved crucial to the IPCC work. For example, the glaciers and ice caps on and near Greenland have been completely mapped for the first time for the RGI (Figure 7.7; Rastner *et al.*, 2012). Both RGI and GLIMS outlines can be downloaded from the GLIMS website (http://glims.org). Glaciers from both GLIMS and the RGI in central Asia and the greater Himalayan region are being used for various ongoing projects (Figure 7.8). As discussed above, simple band ratio techniques can map glacier ice quickly, but debris cover remains a difficulty. More sophisticated techniques, such as texture analysis or the "fuzzy C-means" method (Schowengerdt, 2007), have yielded promising results for a wide range of glacier types (Racoviteanu *et al.*, 2009; Furfaro *et al.*, submitted).

Recent results from a completely different type of measurement, satellite gravimetry, have yielded direct estimates of mass changes in glacier ice. The Gravity Recovery and Climate Experiment (GRACE) system is a pair of satellites, orbiting in tandem, that use laser interferometry to measure the distance between

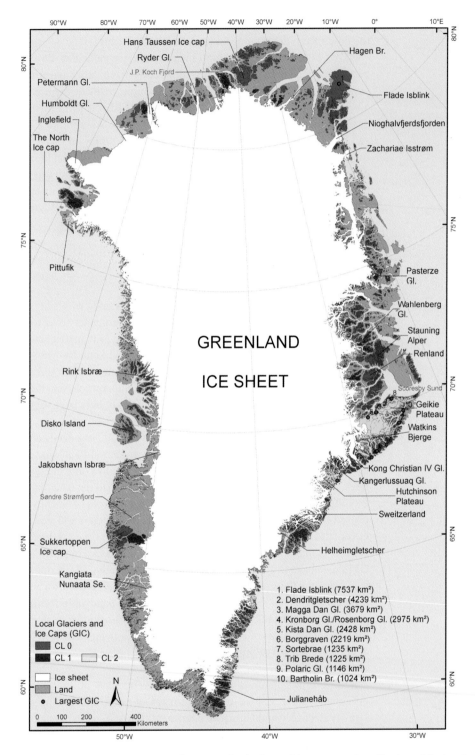

Figure 7.7 Map of glaciers and ice caps surrounding Greenland, obtained by automated application of the band ratio method to optical imagery, mostly LANDSAT (Rastner *et al.*, 2012).

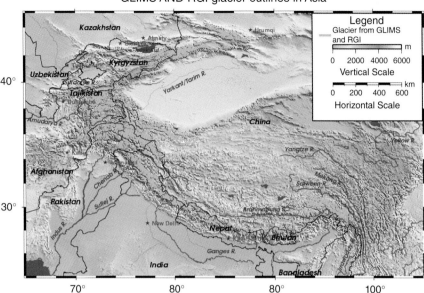

Figure 7.8 Glacier map for Central Asia, produced from GLIMS and RGI data sources using the GMT (Generic Mapping Tools) software.

the two satellites, producing information on changes in the gravitational field (Tapley *et al.*, 2004). One study applied GRACE data to measure glacier ice changes in all major glacierized regions on Earth (Jacob *et al.*, 2012). The spatial resolution of this technique is coarse (some 100 km) and the uncertainties are large compared to the signal in some regions (e.g., Himalaya), as the gravity signal depends on many other factors, including groundwater depletion and fluvial erosion. However, this method produces estimates of mass change that are independent of any other technique. It is probable that future systems will have better spatial resolution, and so will have better signal-to-noise performance (Silvestrin *et al.*, 2012).

In summary, remote sensing methods are capable of measuring many parameters of glaciers and glacier change, leading to greater insight into processes affecting changes in glaciers and hence climate. Field-based measurements are indispensable, as they yield high-precision data and give key insight into processes. However, due to expense and difficult logistics, such measurements are limited to a small number of sites. Remote sensing can cover large numbers of glaciers per image, and some long-term data collections (LANDSAT) are available for free. Algorithms and computational resources are now capable of producing maps of glacier boundaries at useful accuracy over large regions in a matter of days or hours. New sensors will be coming online soon that will continue and extend this capability. Gravitation-sensing satellites can directly measure changes in mass along their orbit tracks, though still at low resolution compared to individual glaciers.

References

Abermann, J., Lambrecht, A., Fischer, A. & Kuhn, M. (2009). Quantifying changes and trends in glacier area and volume in the Austrian Ötztal Alps (1969–1997–2006). *The Cryosphere* **3**, 205–215.

Albert, T.H. (2002). Evaluation of Remote Sensing Techniques for Ice-Area Classification Applied to the Tropical Quelccaya Ice Cap, Peru. *Polar Geography* **26**, 210–226.

Andreassen, L.M., Paul, F., Kääb, A. & Hausberg, J.E. (2008). LANDSAT-derived glacier inventory for Jotunheimen, Norway, and deduced glacier since the 1930s. *The Cryosphere* **2**, 131–145.

Andreassen, L.M., Kjøllmoen, B., Rasmussen, A., Melvold, K. & Nordli, Ø. (2012). Langfjordjøkelen, a rapidly shrinking glacier in northern Norway. *Journal of Glaciology*. Accepted.

Arendt, A.A., Echelmeyer, K.A., Harrison, W.D., Lingle, C.S. & Valentine, V.B. (2002). Rapid wastage of Alaska glaciers and their contribution to rising sea level. *Science* **297**(5580), 382–386.

Atwood, D.K., Meyer, F. & Arendt, A.A. (2010). Using L-band SAR coherence to delineate glacier extent. *Canadian Journal of Remote Sensing* **36**(S1), S186.

Bahr, D.B., Meier, M.F. & Peckham, S.D. (1997). The Physical Basis of Glacier Volume-Area Scaling. *Journal of Geophysical Research* **102**, 20355–20362.

Bajracharya, S.R. & Shrestha, B.R. (Eds.) (2011). *The status of glaciers in the Hindu Kush-Himalayan Region*. ICIMOD, Kathmandu, Nepal.

Baumann, S., Winkler, S. & Andreassen, L.M. (2009). Mapping glaciers in Jotunheimen, South-Norway, during the "Little Ice Age" maximum. *The Cryosphere* **3**, 231–243.

Berthier, E., Arnaud, Y., Vincent, C. & Rémy, F. (2006). Biases of SRTM in high-mountain areas: Implications for the monitoring of glacier volume changes. *Geophysical Research Letters* **33**, L08502, doi:10.1029/2006GL025862.

Berthier, E., Arnaud, Y., Kumar, R., Ahmad, S., Wagnon, P. & Chevallier P. (2007). Remote Sensing Estimates of Glacier Mass Balances in the Himachal Pradesh (Western Himalaya, India). *Remote Sensing of Environment* **108**, 327–338.

Berthier, E., Schiefer, E., Clarke, G.K.C., Menounos, B. & Rèmy F. (2010). Contribution of Alaskan glaciers to sea-level rise derived from satellite imagery. *Nature Geoscience* **3**, 92–95, doi:10.1038/NGEO737.

Bevan, S.L., Luckman, A.J. & Murray, T. (2012). Glacier dynamics over the last quarter of a century at Helheim, Kangerdlugssuaq and 14 other major Greenland outlet Glaciers. *The Cryosphere Discussions* **6**, 1637–1672.

Bhambri, R. & Bolch, T. (2009). Glacier Mapping: A Review with special reference to the Indian Himalayas. *Progress in Physical Geography* **33**(5), 672–704.

Bhambri, R., Bolch, T., Chaujar, R.K. & Kulshreshtha S.C. (2011). Glacier changes in the Garhwal Himalaya, India, from 1968 to 2006 based on remote sensing. *Journal of Glaciology* **57**(203), 543–556.

Bhambri, R., Bolch, T. & Chaujar, R.K. (2012). Frontal recession of Gangotri Glacier, Garhwal Himalayas, from 1965–2006, measured through high resolution remote sensing data. *Current Science* **102**(3), 489–494.

Bishop, M., Bonk, R., Kamp, U. & Shroder, J. (2001). Terrain Analysis and Data Modeling for Alpine Glacier Mapping. *Polar Geography* **25**, 182–201.

Bolch, T., Buchroithner, M.F., Peters, J., Baessler, M. & Bajracharya, S. (2008). Identification of glacier motion and potentially dangerous glacial lakes in the Mt. Everest region/Nepal using spaceborne imagery. *Natural Hazards and Earth System Science* **8**, 1329–1340.

Bolch, T., Menounos, B. & Wheate, R. (2010a). LANDSAT-based Inventory of Glaciers in Western Canada, 1985–2005. *Remote Sensing of Environment* **114**, 127–137.

Bolch, T., Yao, T., Kang, S., Buchroithner, M.F., Scherer, D., Maussion, F., Huintjes, E. & Schneider, C. (2010b). A glacier inventory for the western Nyainqentanglha Range and Nam Co Basin, Tibet, and glacier changes 1976–2009. *The Cryosphere* **4**, 419–433.

Bolch, T., Pieczonka, T. & Benn, D.I. (2011). Multi-decadal mass loss of glaciers in the Everest area (Nepal Himalaya) derived from stereo imagery. *The Cryosphere* **5**, 349–358.

Bolch, T., Sandberg Sørensen, L., Simonssen, S.B., Mölg, N., Machguth, H., Rastner, P. & Paul, F. (2013). Mass loss of Greenland's glaciers and ice caps 2003–2008 revealed from ICESat laser altimetry data. *Geophysical Research Letters* **40**, doi: 10.1029/2012GL054710.

Braithwaite, R.J. (1984). Can the Mass Balance of a Glacier Be Estimated from Its Equilibrium-Line Altitude? *Journal of Glaciology* **30**(106), 364–368.

Brenning, A. (2009). Benchmarking classifiers to optimally integrate terrain analysis and multispectral remote sensing in automatic rock glacier detection. *Remote Sensing of Environment* **113**(1), 239–247.

Casassa, G., Smith, K., Rivera, A., Araos, J., Schnirch, M, & Schneider, C. (2002). Inventory of glaciers in Isla Riesco, Patagonia, Chile, based on aerial photography and satellite imagery. *Annals of Glaciology* **34**, 373–378.

Casey, K.A., Kääb, A. & Benn, D.I. (2012). Geochemical characterization of supraglacial debris via in situ and optical remote sensing methods: a case study in Khumbu Himalaya, Nepal. *The Cryosphere* **6**, 85–100.

Cogley, J.G. (2009). Geodetic and direct mass-balance measurements: comparison and joint analysis. *Annals of Glaciology* **50**(50), 96–100.

Dashora, A., Lohani, B. & Malik, J.N. (2007). A repository of earth resource information --- Corona satellite programme. *Current Science* **92**(7), 926–932.

Eldhuset, K., Andersen, P.H., Hauge, S., Isaksson, E. & Weydahl, D.J. (2003). ERS Tandem InSAR Processing for DEM Generation, Glacier Motion Estimation and Coherence Analysis on Svalbard. *International Journal of Remote Sensing* **24,** 1415–1437.

Evans, D.L., Alpers, W., Cazenave, A., Elachi, C., Farr, T., Glackin, D., Holt, B., Jones, L., Liu, W.T., McCandless, W., Menard, Y., Moore, R. & Njoku, E. (2005).

Seasat – A 25-year legacy of success. *Remote Sensing of Environment* **94**(3), 384–404. ISSN 0034–4257, 10.1016/j.rse.2004.09.011.

Farinotti, D., Huss, M., Bauder, A. & Funk, M. (2009). An estimate of the glacier ice volume in the Swiss Alps. *Global and Planetary Change* **68**(3), 225–231.

Finsterwalder, R. (1954). Photogrammetry and glacier research with special reference to glacier retreat in the eastern Alps. *Journal of Glaciology* **2**(15), 306–315.

Fischer, A. (2010). Glaciers and climate change: Interpretation of 50 years of direct mass balance of Hintereisferner. *Global and Planetary Change* **71**(1–2), 13–26.

Floricioiu, D. & Rott, H. (2001). Seasonal and short-term variability of multifrequency, polarimetric radar backscatter of Alpine terrain from SIR-C/X-SAR and AIRSAR data. *IEEE Transactions on Geoscience and Remote Sensing* **39**(12), 2634–2648.

Furfaro, R., Kargel, J. & Leonard, G. (submitted). Fuzzy C Mean Applications to and Validation of ASTER-Based Landcover Mapping of Root Glacier (Alaska). *The Cryosphere.*

Gardner, A.S., Moholdt, G., Cogley, J.G., Wouters, B., Arendt, A.A., Wahr, J., Berthier, E., Pfeffer, T.W., Kaser, G., Hock, R., Ligtenberg, S.R.M., Bolch, T., Sharp, M.J., Hagen, J.O., van den Broeke, M.R. & Paul, F. (2013). Narrowing the gap: A consensus estimate of glacier mass wastage. *Science.*

Gjermundsen, E., Mathieu, R., Kääb, A., Chinn, T., Fitzharris, B., & Hagen, J.O. (2011). Assessment of multispectral glacier mapping methods and derivation of glacier area changes, 1978–2002, in the central Southern Alps, New Zealand, from ASTER satellite data, field survey and existing inventory data. *Journal of Glaciology* **57**(204), 667–683.

Haeberli, W. (1998). Historical evolution and operational aspects of worldwide glacier monitoring. In: Haeberli, W., Hoelzle, M. & Suter, S. (eds). *Into the second century of worldwide glacier monitoring: prospects and strategies*, 35–51. UNESCO Studies and Reports in Hydrology.

Haeberli, W. & Epifani, F. (1986). Mapping the Distribution of Buried Glacier Ice – an Example from Lago Delle Locce, Monte Rosa, Italian Alps. *Annals of Glaciology* **8**, 78–81.

Haeberli, W., Hoelzle, M., Paul, F. & Zemp, M, (2007). Integrated monitoring of mountain Glaciers as key indicators of global climate change: the European Alps. *Annals of Glaciology* **46**, 150–160.

Hall, D.K., Riggs, G.A. & Salomonson, V.V. (1995). Development of Methods for Mapping Global Snow Cover Using Moderate Resolution Imaging Spectroradiometer Data. *Remote Sensing of Environment* **54**(2), 127–140.

Heid, T. & Kääb, A. (2012). Evaluation of existing image matching methods for deriving glacier surface displacements globally from optical satellite imagery. *Remote Sensing of Environment* **118**, 339–355.

Howat, I.M. & Eddy, A. (2011). Multi-decadal retreat of Greenland's marine-terminating glaciers. *Journal of Glaciology* **57**(203), 389–396.

Howat, I.M., Joughin, I., Tulaczyk, S. & Gogineni, S. (2005). Rapid retreat and acceleration of Helheim glacier, east Greenland. *Geophysical Research Letters* **32**, L22502.

Howat, I.M., Joughin, I. & Scambos, T.A. (2007). Rapid changes in ice discharge from Greenland outlet glaciers. *Science* **315**(5818), 1559–1561.

Howat, I.M., Ahn, Y., Joughin, I., van den Broeke, M.R., Lenaerts, J.T.M. & Smith, B. (2011). Mass balance of Greenland's three largest outlet glaciers, 2000–2010. *Geophysical Research Letters* **38**, L12501.

Jacob, T., Wahr, J., Pfeffer, T.W. & Swenson, S. (2012). Recent contributions of glaciers and ice caps to sea level rise. *Nature* **482**, 514–518.

Johannessen, O.M., Khvorostovsky, K., Miles, M.W. & Bobylev, L.P. (2005). Recent ice-sheet growth in the interior of Greenland. *Science* **310**, 1013–1016.

Johannessen, O.M., Babiker, M. & Miles, M.W. (2011). Petermann Glacier, North Greenland: massive calving in 2010 and the past half century. *The Cryosphere Discussions* **5**(1), 169–181.

Johannesson, T., Björnsson, H., Magnusson, E., Gudmundsson, S., Palsson, F., Sigurdsson, O., Thorsteinsson, T. & Berthier, E. (2012). Ice-volume changes, bias-estimation of mass-balance measurements and changes in subglacial lakes derived by LiDAR-mapping of the surface of Icelandic glaciers. *Annals of Glaciology* **54**(63).

Joughin, I. (2002). Ice-sheet Velocity Mapping: a Combined Interferometric and Speckle-tracking Approach. *Annals of Glaciology* **34**, 195–201.

Joughin, I., Abdalati, W. & Fahnestock, M. (2004). Large fluctuations in speed on Greenland's Jakobshavn Isbrae glacier. *Nature* **432**, 608–610.

Joughin, I., Howat, I., Alley, R.B., Ekstrom, G., Fahnestock, M., Moon, T., Nettles, M., Truffer, M., and Tsai, V.C. (2008). Ice-front variation and tidewater behavior on Helheim and Kangerdlugssuaq glaciers, Greenland. *Journal of Geophysical Research* **113**, F01004.

Joughin, I., Smith, B.E., Howat, I.M., Scambos, T. & Moon, T. (2010). Greenland flow variability from ice-sheet-wide velocity mapping. *Journal of Glaciology* **56**(197), 415–430.

Krabill, W., Abdalati, W., Frederick, E., Manizade, S., Martin, C., Sonntag, J., Swift, R., Thomas, R., Wright, W. & Yungel, J. (2000). Greenland ice sheet: High-elevation balance and peripheral thinning. *Science* **289**, 428–430.

Kääb, A. (2005). Combination of SRTM3 and Repeat ASTER Data for Deriving Alpine Glacier Flow Velocities in the Bhutan Himalaya. *Remote Sensing of Environment* **94**, 463–474.

Kääb, A., Chiarle, M., Raup, B. & Schneider, C. (2007). Climate Change Impacts on Mountain Glaciers and Permafrost. *Global and Planetary Change* **56**, vii–ix.

Kääb, A. (2008). Glacier Volume Changes Using ASTER Satellite Stereo and ICESat GLAS Laser Altimetry. A Test Study on Edgeøya, Eastern Svalbard. *IEEE Transactions on Geoscience and Remote Sensing* **46**, 2823–2830.

Kääb, A. (2010). The role of remote sensing in worldwide glacier monitoring.In: Pellika, P. (ed). *Remote Sensing of Glaciers*, 285–296. CRC Press.

Kääb, A., Berthier E., Nuth, C., Gardelle, J. & Arnaud, Y. (2012). Contrasting patterns of early twenty-first-century glacier mass change in the Himalayas. *Nature* **488**(7412), 495–498.

Karimi, N., Farokhnia, A., Karimi, L., Eftekhari, M. & Ghalkhani, H. (2012). Combining optical and thermal remote sensing data for mapping debris-covered glaciers (Alamkouh Glaciers, Iran). *Cold Regions Science and Technology* **71**, 73–83.

Kieffer, H., Kargel, J.S., Barry, R. *et al.* (2000). New Eyes in the Sky Measure Glaciers and Ice Sheets. *Eos Transactions, American Geophysical Union* **81**(24), 265–271.

Lambrecht & Kuhn (2007). Glacier Changes in the Austrian Alps During the Last Three Decades, Derived from the New Austrian Glacier Inventory. *Annals of Glaciology* **46**, 177–184.

Lemke, P., Ren, J., Alley, R.B., Allison, I., Carrasco, J., Flato, G., Fujii, Y., Kaser, G., Mote, P., Thomas, R.H. & Zhang, T. (2007). Observations: Changes in Snow, Ice and Frozen Ground. In: Solomon, S., Qin, D., Manning, M., Chen, Z., Marquis, M., Averyt, K.B., Tignor, M. & Miller, H.L. (eds). *Climate Change 2007: The Physical Science Basis.* Contribution of Working Group I to the Fourth Assessment Report of the Intergovernmental Panel on Climate Change. Cambridge, UK and New York, NY, USA: Cambridge University Press.

Luckman, A., & Murray, T. (2005). Seasonal variations in velocity before retreat of Jakobshavn Isbrae, Greenland. *Geophysical Research Letters* **32**, L08501.

Luckman, A., Murray, T., de Lange, R., & Hanna, E. (2006). Rapid and synchronous ice-dynamic changes in east Greenland. *Geophysical Research Letters* **33**, L03503.

Luckman, A., Quincey, D. & Bevan, S. (2007). The potential of satellite radar interferometry and feature tracking for monitoring flow rates of Himalayan glaciers. *Remote Sensing of Environment* **111**(2–3), 171–181.

Moholdt, G., Nuth, C., Hagen, J.O. & Kohler, J. (2010). Recent elevation changes of Svalbard glaciers derived from ICESat laser altimetry. *Remote Sensing of Environment* **114**, 2756–2767.

Mool, P., Bajracharya S.R., Joshi S.P., Sakya K. & Baidya A. (2002). *Inventory of glaciers, glacial lakes and glacial lake outburst floods monitoring and early warning systems in the Hindu-Kush Himalayan region, Nepal.* International Center for Integrated Mountain Development (ICIMOD) tech. pub., 227 pp. Kathmandu, Nepal. ISBN 92 9115 345 1.

Moon, T. & Joughin, I. (2008). Changes in ice front position on Greenland's outlet glaciers from 1992 to 2007. *Journal of Geophysical Research* **113**, F02022.

Moon, T., Joughin I., Smith, B. & Howat, I. (2012). 21st-Century evolution of Greenland outlet glacier velocities. *Science* **336**(6081), 576–578.

Murray, T., Scharrer, K., James, T.D., Dye, S.R., Hanna, E., Booth, A.D., Selmes, N., Luckman, A., Hughes, A.L.C., Cook, S. & Huybrechts, P. (2010). Ocean regulation hypothesis for glacier dynamics in southeast Greenland and implications for ice sheet mass changes. *Journal of Geophysical Research* **115**, F03026.

Nagler, T. & Rott, H. (2000). Retrieval of wet snow by means of multitemporal SAR data. *IEEE Transactions on Geoscience and Remote Sensing* **38**(2), 754–765, doi:10.1109/36.842004

Narama, C., Kääb A., Duishonakunov, M. & Abdrakhmatov K. (2010). Spatial variability of recent glacier area changes in the Tien Shan Mountains, Central Asia, using Corona (≈1970), LANDSAT (≈2000), and ALOS (≈2007) satellite data. *Global and Planetary Change* **71**(1–2), 42–54.

Nuth, C. & Kääb A. (2011). Co-registration and bias corrections of satellite elevation data sets for quantifying glacier thickness change. *The Cryosphere* **5**, 271–290.

Oerlemans, J. (2005). Extracting a Climate Signal from 169 Glacier Records. *Science* **308**, 675–677.

Palmer, S., Shepherd, A., Nienow, P. & Joughin, I. (2011). Seasonal speedup of the Greenland ice sheet linked to routing of surface water. *Earth and Planetary Science Letters* doi.org/10.1016/j.epsl.2010.12.037.

Paul, F. & Kääb, A. (2005). Perspectives on the production of a glacier inventory from multispectral satellite data in Arctic Canada: Cumberland Peninsula, Baffin Island. *Annals of Glaciology* **42**, 59–66.

Paul, F. & Linsbauer, A. (2012). Modeling of glacier bed topography from glacier outlines, central branch lines, and a DEM. *International Journal of Geographical Information Science* **1**, 1–18.

Paul, F., Kääb, A., Maisch, M., Kellenberger, T. & Haeberli, W. (2002). The new remote-sensing derived Swiss glacier inventory: I. *Methods. Annals of Glaciology* **34**, 355–361.

Paul, F., Huggel, C. & Kääb, A. (2004). Combining Satellite Multispectral Image Data and a Digital Elevation Model for Mapping Debris-covered Glaciers. *Remote Sensing of Environment* **89**, 510–518.

Paul, F., Barry, R.G., Cogley, J.G., Frey, H., Haeberli, W., Ohmura, A., Ommanney, C.S.L., Raup, B., Rivera, A., & Zemp, M. (2009). Recommendations for the Compilation of Glacier Inventory Data from Digital Sources. *Annals of Glaciology* **50**, 119–126.

Paul, F., Andreassen, L.M. & Winsvold, S.H. (2011). A new glacier inventory for the Jostedalsbreen region, Norway, from LANDSAT TM scenes of 2006 and changes since 1966. *Annals of Glaciology* **52**(59), 153–162.

Pritchard, H., Murray, T., Luckman, A., Strozzi, T. & Barr, S. (2005). Glacier surge dynamics of Sortebræ, east Greenland, from synthetic aperture radar feature tracking. *Journal of Geophysical Research* **110**, F03005.

Pritchard, H.D., Arthern, R.J., Vaughan, D.G. & Edwards, L.A. (2009). Extensive dynamic thinning on the margins of the Greenland and Antarctic ice sheets. *Nature* **461**(7266), 971–975.

Quincey, D.J., Luckman, A. & Benn, D. (2009). Quantification of Everest region glacier velocities between 1992 and 2002, using satellite radar interferometry and feature tracking. *Journal of Glaciology* **55**(192), 596–606.

Rabatel, A., Dedieu, J.P. & Vincent, C. (2005). Using remote-sensing data to determine equilibrium-line altitude and mass-balance time series: validation on three French glaciers, 1994–2002. *Journal of Glaciology* **51**(175), 539–546.

Racoviteanu, A.E., Arnaud, Y., Williams, M.W. & Ordonez, J. (2008a). Decadal Changes in Glacier Parameters in the Cordillera Blanca, Peru, Derived from Remote Sensing. *Journal of Glaciology* **54**, 499–510.

Racoviteanu, A., Williams, M.W. & Barry, R. (2008b). Optical remote sensing of glacier mass balance: a review with focus on the Himalaya. *Sensors*, Special issue: Remote sensing of the environment, 3355–3383.

Racoviteanu, A.E., Paul, F., Raup, B., Singh Khalsa, S.J. & Armstrong, R. (2009). Challenges and recommendations in mapping of glacier parameters from space: results of the 2008 Global Land and Ice Measurements from Space (GLIMS) workshop, Boulder, Colorado, USA. *Annals of Glaciology* **53**, 53–69.

Radić, V. & Hock, R. (2010). Regional and global volumes of glaciers derived from statistical upscaling of glacier inventory data. *Journal of Geophysical Research*, Earth Surface **115**, doi:10.1029/2009JF001373.

Ranzi, R., Grossi G., Iacovelli, L. & Taschner, S. (2004). *Use of multispectral ASTER images for mapping debris-covered glaciers within the GLIMS Project.* Proceedings of the International Geoscience and Remote Sensing Symposium 2004, 1144–1147.

Raper, S.C.B. & Braithwaite, R.J. (2005). The Potential for Sea Level Rise: New Estimates from Glacier and Ice Cap Area and Volume Distributions. *Geophysical Research Letters* **32**, doi: 10.1029/2004GL021981.

Raup, B., Kääb, A., Kargel, J.S., Bishop, M.P., Hamilton, G., Lee, E., Paul, F., Rau, F., Soltesz, D., Singh Khalsa, S.J., Beedle, M. & Helm, C. (2007). Remote Sensing and GIS Technology in the Global Land Ice Measurements from Space (GLIMS) Project. *Computers and Geosciences* **33**, 104–125.

Rastner, P., Bolch, T., Mölg, N., Machguth, H., Le Bris, R. & Paul, F. (2012). The first complete inventory of the local glaciers and ice caps on Greenland. *The Cryosphere* **6**, 1483–1495, www.the-cryosphere.net/6/1483/2012/, doi:10.5194/tc-6-1483-2012.

Richards, J.A. (1986). *Remote Sensing Digital Image Analysis*. Springer-Verlag Inc.

Rignot, E. & Kanagaratnam, P. (2006). Changes in the velocity structure of the Greenland ice sheet. *Science* **311**(5763), 986–990.

Rignot, E., Gogineni, S., Joughin, I. & Krabill, W. (2001). Contribution to the glaciology of northern Greenland from satellite radar interferometry. *Journal of Geophysical Research* **106**(D24), 34007–30019.

Rignot, E., Braaten, D., Gogineni, S.P., Krabill, W.B. & McConnell, J.R. (2004). Rapid ice discharge from southeast Greenland glaciers. *Geophysical Research Letters* **31**, L10401.

Rott, H. (1994). Thematic studies in Alpine areas by means of polarimetric SAR and optical imagery. *Advances in Space Research* **14**(3), 217–226.

Sabins, F.F., Jr. (2007). *Remote Sensing: Principles and Applications*, 3rd Edition. Long Grove, Illinois, USA. Waveland Press, Inc.

Salomonson, V.V. & Appel, I. (2004). Estimating fractional snow cover from MODIS using the normalized difference snow index. *Remote Sensing of Environment* **89**, 351–360.

Scambos, T.A., Dutkiewicz, M.J., Wilson, J.C. & Bindschadler, R.A. (1992). Application of Image Cross-Correlation to the Measurement of Glacier Velocity Using Satellite Image Data. *Remote Sensing of Environment* **42**, 177–186.

Scherler, D., Leprince, S. & Strecker, M.R. (2008). Glacier-surface velocities in alpine terrain from optical satellite imagery – accuracy improvement and quality assessment. *Remote Sensing of Environment* **112**(10), 3806–3819.

Schowengerdt, R.A. (2007). *Remote Sensing: Models and Methods for Image Processing*, 3rd Edition. San Diego, CA, Academic Press.

Seale, A., Christoffersen, P., Mugford, R.I. & O'Leary, M. (2011). Ocean forcing of the Greenland ice sheet: Calving fronts and patterns of retreat identified by automatic satellite monitoring of eastern outlet glaciers. *Journal of Geophysical Research* **116**, F03013.

Sidjak, R.W. & Wheate, R.D. (1999). Glacier Mapping of the Illecillewaet Icefield, British Columbia, Canada, Using LANDSAT TM and Digital Elevation Data. *International Journal of Remote Sensing* **20**, 273–284.

Silvestrin, P., Aguirre, M., Massotti, L., Leone, B., Cesare, S., Kern, M. & Haagmans, R. (2012). The Future of the Satellite Gravimetry After the GOCE Mission. In: Kenyon S. *et al.* (eds) *Geodesy for Planet Earth*, International Association of Geodesy Symposia **136**(2), 223–230, doi: 10.1007/978-3-642-20338-1_27.

Strozzi, T., Luckman, A., Murray, T., Wegmuller, U. & Werner, C.L. (2002). Glacier Motion Estimation Using SAR Offset-Tracking Procedures. *IEEE Transactions on Geoscience and Remote Sensing* **40**, 2384–2391.

Sundal, A.V., Shepherd, A., Nienow, P., Hanna, E., Palmer, S. & Huybrechts, P. (2011). Melt-induced speed-up of Greenland ice sheet offset by efficient subglacial drainage. *Nature* **469**(7331), 521–524.

Tapley, B.D., Bettadpur, S., Ries, J.C., Thompson, P.F. & Watkins, M.M. (2004). GRACE measurements of mass variability in the Earth system. *Science* **305**(5683), 503–505.

Thomas, R., Davis, C., Frederick, E., Krabill, W., Li, Y., Manizade, S. & Martin, C. (2008). A comparison of Greenland ice-sheet volume changes derived from altimetry measurements. *Journal of Glaciology* **54**(185), 203–212.

Walsh, K.M., Howat, I.M., Ahn, Y. & Enderlin, E.M. (2012). Changes in the marine-terminating glaciers of central east Greenland, 2000–2010. *The Cryosphere* **6**, 211–220.

Yasuda, T. & Furuya, M. (2013). Short-term glacier velocity changes at West Kunlun Shan, Northwest Tibet, detected by Synthetic Aperture Radar data. *Remote Sensing of Environment*, **128**, 87–106.

Zemp, M., Hoelzle, M. & Haeberli, W. (2009). Six decades of glacier mass-balance observations: a review of the worldwide monitoring network. *Annals of Glaciology* **50**, 101–111.

Zwally, H.J., Abdalati, W., Herring, T., Larson, K., Saba, J. & Steffen, K. (2002). Surface melt-induced acceleration of Greenland ice-sheet flow. *Science* **51**, 218–222.

Zwally, H.J., Giovinetto, M.B., Li, J., Cornejo, H.G., Beckley, M.A., Brenner, A.C., Sabba, J.L. & Li, D. (2005). Mass changes of the Greenland and Antarctic ice sheets and shelves and contributions to sea-level rise: 1992–2002. *Journal of Glaciology* **51**, 509–527.

Acronyms

ASTER	Advanced Spaceborne Thermal Emission and Reflection Radiometer
DN	Digital Numbers
ELA	Equilibrium Line Altitude
ERTS	1 Earth Resources Technology Satellite
ETM+	Enhanced Thematic Mapper Plus
GCP	ground control points
GIFOV	Ground instantaneous field of view
GLAS	Geoscience Laser Altimetry System
GLIMS	Global Land Ice Measurements from Space
GRACE	The Gravity Recovery and Climate Experiment
GTN-G	Global Terrestrial Network for Glaciers
ICESat	Ice, Cloud, and land Elevation Satellite
IPCC	Intergovernmental Panel on Climate Change
IRS	Indian Remote Sensing
LDCM LANDSAT	Data Continuity Mission
LIDAR	Light Detection and Ranging
MSS	Multispectral scanner
NDSI	normalized difference snow index
NIR	Near Infrared
RGI	Randolph Glacier Inventory
SPOT	Satellite Pour l'Observation de la Terre
SRTM	Shuttle Radar Topography Mission
SWIR	short-wave infrared
TM	Thematic Mapper
UTM	Universal Transverse Mercator

Websites cited

http://glims.org
http://landsat.gsfc.nasa.gov/

8 Remote sensing of accumulation over the Greenland and Antarctic ice sheets

Lora Koenig[1], Richard Forster[2], Ludovic Brucker[1,3] and Julie Miller[2]

[1] NASA / Goddard Space Flight Center, Greenbelt, USA
[2] University of Utah, Salt Lake City, USA
[3] Universities Space Research Association, Columbia, USA

Summary

Earth's sea level fluctuations are predominantly determined by the amount of ice contained in the Antarctic and Greenland ice sheets. Variation in that amount of ice is still a critical unknown, and a complete understanding of past, present, and future variations of mass balance over both ice sheets is essential to understanding and modeling sea level fluctuations.

As the only input term to the mass balance of the Greenland and Antarctic ice sheets, snow accumulation is a critical process to measure accurately. The vast size of the ice sheets and the spatial variability in accumulation rates make it impossible to record enough accumulation measurements from *in situ* methods. Modeling efforts are advancing and can estimate accumulation at tens-of-kilometer grid spacing, but sensitivity to boundary forcing conditions and a paucity of validation data leave these data sets less than ideal. Therefore, remote sensing of ice sheet accumulation is an important technique to develop.

Several promising techniques have been explored, using various portions of the electromagnetic spectrum, but none of them are at a point where accumulation can be measured in an operational mode. Spaceborne active and passive microwave remote sensing have received the most attention, due to their responsiveness from snow and firn properties related to accumulation rates, and their ability to operate independently of sunlight and through cloud cover. Spatial and temporal variability in liquid water and refrozen subsurface ice structures, however, prevents the use of a single broadly applied algorithm for snow accumulation retrievals, even at microwave wavelengths.

Other techniques show promise under specific circumstances, such as altimetry, gravity, and airborne sensors, warranting continued study. This chapter explains that determining the spatio-temporal variability of accumulation over the ice sheets remains one of the most

Remote Sensing of the Cryosphere, First Edition. Edited by M. Tedesco.
© 2015 John Wiley & Sons, Ltd. Published 2015 by John Wiley & Sons, Ltd.
Companion Website: www.wiley.com/go/tedesco/cryosphere

daunting tasks facing scientists. This task will only be accomplished by utilizing a variety of remote sensing techniques and sensors in conjunction with models and *in situ* measurements. The chapter describes remote sensing methods, from both spacecraft and aircraft, for determining accumulation, as well as, *in situ* measurements and model results.

8.1 Introduction to accumulation

Earth's sea level fluctuations are predominantly driven by the amount of ice contained in the Greenland and Antarctic ice sheets (GrIS and AIS, respectively). Variation in that amount of ice is a critical unknown which hinders accurate sea level prediction. Monitoring the variability in ice content is done by estimating the mass balance of the GrIS and AIS, with snow accumulation being the only input to mass balance. Therefore, measuring the present and past net accumulation of snow over the Earth's ice sheets is essential for understanding the current and previous contributions of the ice sheets to sea level changes.

The large size and remote-harsh environment of the ice sheets, along with the spatio-temporal variability in accumulation magnitude, make it impossible to measure ice sheet-wide accumulation from *in situ* measurements alone. Consequently, remote sensing of accumulation is required to assess mass balance changes of the ice sheets.

The mass balance of an ice sheet is defined as the result of all mass gain (termed accumulation) minus all mass loss (termed ablation). Figure 8.1 is a schematic of an ice sheet, showing the two main zones used to describe mass balance, the accumulation and ablation zones. The accumulation zone is a region of mass gain, where snowfall and snow loading by wind (termed positive wind redistribution) outweigh ablation processes. The ablation zone is a region of mass loss and is dominated by the processes of ice calving and flow (termed ice dynamics), melt, snow striping by wind (termed negative wind redistribution), evaporation, and sublimation. The accumulation zone is higher in altitude than the ablation zone (Figure 8.1).

Within the accumulation zone, there are two important sub-zones – the percolation zone and the dry-snow zone. The percolation zone experiences some surface melt, usually annually, producing water that percolates vertically into the snow and refreezes – still representing an overall mass gain. The dry-snow zone rarely, if ever, experiences melt. The warmer GrIS has a much larger percolation zone than the AIS. The transition from the accumulation zone to the ablation zone is defined as the equilibrium line, where mass is neither gained nor lost. The amount of ablation and accumulation must both be quantified in order to calculate changes in the total mass of the ice sheet over time, therefore quantifying the ice sheets contribution to sea level.

The terminology for accumulation varies across disciplines and can sometimes be confusing. In remote sensing, it is very important to understand the specific physical quantity that a sensor is measuring, how the measurement relates to a geophysical property, and the proper terminology for comparing satellite retrieval data to *in situ* and modeled data. Historically, satellite data were used in the dry-snow

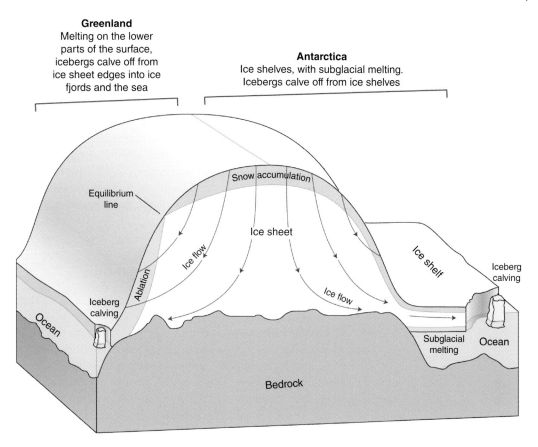

Figure 8.1 Ice sheet schematic, showing the zones of accumulation and ablation. The equilibrium line is the altitude where mass is neither gained nor lost, and it marks the transition from the accumulation zone (mass gain) to the ablation zone (mass loss). Different remote sensing techniques are used to calculate the mass over the entire ice sheet (Hugo Alhenius UNEP/GRID-Arendal).

zone of the ice sheets (Benson, 1962), and calibrated with snow pit and/or ice core data. Because of these early studies in remote sensing, the term "accumulation" describes the net amount of mass added to an ice sheet.

The majority of satellite and ice core data cannot distinguish the process by which the snow was deposited (e.g., precipitation or wind redistribution). With the development of atmospheric models, however, the term accumulation was more specifically separated into precipitation and evaporation – accumulation being defined as precipitation minus evaporation. Coupling atmospheric models to snow/firn models provided even more specificity, and total ice sheet surface mass balance (SMB) was quantified. SMB is defined as precipitation minus evaporation, sublimation, and melt and plus or minus wind (eolian) redistribution of snow. All of these physical quantities vary spatially and temporally, making them difficult quantities to measure directly. In this chapter, we will discuss how current satellites sensors, such as microwave radiometers, are capable of deriving accumulation, while other sensors are used to monitor only one component of SMB, such as laser altimeters detecting wind distribution of snow.

In this chapter, we will show that determining the spatio-temporal variability of accumulation and/or quantifying any of the individual processes of SMB over the ice sheets remain daunting tasks. Completing these tasks will likely require utilizing a variety of remote sensing techniques and sensors (radiometers, scatterometers, altimeters, and gravimeters) in conjunction with models and *in situ* measurements. Here, we describe and review remote sensing methods, from both spacecraft and aircraft, used for determining accumulation, as well as *in situ* measurements and model results. Chapter 6 discusses the remote sensing methods to calculate ablation, needed to calculate a full mass balance assessment of an ice sheet.

8.2 Spaceborne methods for determining accumulation over ice sheets

Precipitation and snowfall are weather phenomena that change daily across the globe and over the ice sheets. Monitoring precipitation requires comprehensive spatial coverage and high temporal resolution, best achieved by polar-orbiting satellites. Over the polar regions, where orbits converge, polar-orbiting satellites offer numerous daily observations, providing an ideal monitoring frequency and spatial coverage for accumulation retrievals. In theory, satellite retrievals of accumulation rate can be derived from passive and active microwave sensors, laser altimeters, and gravimeters. Difficulties primarily arise in using satellites to determine accumulation from two complications:

1 Satellite sensors used to derive accumulation are influenced by other geophysical and glaciological properties (e.g., temperature, snow grain size, density, roughness) inter-locked to accumulation.
2 Definitive measurements of spatial and temporal variations of accumulation from *in situ* sources are sparse and are influenced by microscale (smaller than the sensors field of view) processes, such as wind redistribution and melt water retention among others, making them difficult to directly compare with satellite observations (McConnell *et al.*, 2000; Eisen *et al.*, 2008).

Though much effort has been put into deriving accumulation from satellite measurements since the 1970s, no single method gives a fully accurate spatial and temporal picture of accumulation. Accumulation retrievals were pioneered by microwave radiometry. Passive microwave remote sensing has been used most extensively to derive long-term accumulation rates using frequencies between ≈ 6 and ≈ 37 GHz (Zwally, 1977; Vaughan *et al.*, 1999; Winebrenner *et al.*, 2001; Arthern *et al.*, 2006).

More recent sensors, operating at lower frequencies (1.4 GHz), such as those on board the European Space Agency's (ESA) Soil Moisture and Ocean Salinity (SMOS) mission and the joint National Aeronautic and Space Administration (NASA)/Argentina's Comisión Nacional de Actividades Espaciales (CONAE) Aquarius mission, offer new potential for passive microwave retrievals, since they record emission emanating from deeper layers within the ice sheet. Additional satellite methods show promise in deriving accumulation including gravimetry and altimetry (e.g., Horwath *et al.*, 2012), but these methods require additional information about ice dynamics and, for altimetry, firn densification rates (Arthern and Wingham, 1998; Li and Zwally, 2002).

8.2.1   *Microwave remote sensing*

Microwave remote sensing methods are the most commonly used for determining accumulation over the ice sheets. Microwave measurements, either passive (emission) or active (backscatter), are sensitive to snow properties (e.g., temperature, grain size, and density) related to accumulation variations. For instance, ice sheet regions with lower accumulation rates and temperatures have larger grain size which, in general, lowers the snow's emissivity and brightness temperature compared to higher accumulation, smaller grain size locations (Gow, 1969, 1971; Brucker *et al.*, 2010).

Microwaves penetrate snow/firn from a few centimeters to hundreds of meters, depending on the frequency (Surdyk, 2002). This penetration, along with microwave wavelengths of similar scale (cm scale) to the snow/firn layering structure and grain size, creates scattering and reflections of the waves related to accumulation. Although microwave remote sensing methods have been most successful in deriving accumulation patterns, the fact remains that they are not a direct measurement of accumulation, but rather a proxy related to the snow/firn properties. Passive and active remote sensing methods, therefore, will always require *in situ* measurements and/or radiative transfer modeling to establish accumulation algorithms.

Although microwave radiometry has been used since the 1970s, accurately parameterizing the physical interactions between the radiation and the ever-changing snow/firn structure remains difficult. A leading challenge is to couple snow metamorphism models with radiative transfer models to accurately quantify scattering from snow grains of different sizes and shapes (Brucker *et al.*, 2011a). Nevertheless, significant progress has been achieved since the 1990s, with the development of physical radiative transfer models to calculate the interaction and propagation of microwave radiation within the firn (e.g., Wiesmann and Mätzler, 1999; Tsang and Kong, 2001; Picard *et al.*, 2009; Brucker *et al.*, 2011b; Picard *et al.*, 2013). These developments contributed to the retrievals of glaciological properties that will ultimately provide better parameters for extracting the accumulation variable from the microwave signal.

8.2.1.1 Passive microwave remote sensing

The passive microwave record is the longest, temporally consistent satellite dataset over the ice sheets capable of investigating changes in accumulation. Table 8.1 gives a summary of the passive microwave sensors used for accumulation retrievals over the ice sheet. The length and continuity of these data provides considerable motivation for developing accumulation algorithms that utilize the high temporal resolution of the passive microwave data.

Passive microwave sensors, such as the series of Special Sensor Microwave Imager Sounder (SSM/I and SSMIS) and Advanced Microwave Scanning Radiometer (AMSR-E, AMSR2), have been used in numerous accumulation studies, because the microwave radiation, naturally emitted by the snow/firn does not require solar illumination, and the long-wavelength emission is weakly influenced by the dry polar atmosphere. These characteristics allow microwave measurements of the ice sheet through clouds and during the dark polar winters.

Temperature-dependent radiation is emitted by the snow/firn from depths of centimeters at high frequencies (e.g., 89 GHz) to tens of meters at lower frequencies (e.g., 6 GHz). As the emitted radiation travels towards the satellite sensor, it is both scattered and absorbed within the snow and ice layers and is reflected at layer interfaces. The magnitude of the scattering, absorption and reflection contains information about the snow and ice properties the radiation is traveling through and, thus, can be used to monitor both temperature and snow/firn properties that relate directly to accumulation rates (see also Chapter 5 on snow depth and snow water equivalent retrievals). To a first approximation, assuming an isothermal medium, recorded microwave brightness temperature can be approximated by the emissivity of snow/firn multiplied by its physical temperature (Ulaby *et al.*, 1981).

Zwally (1977) first related accumulation to the passive microwave signal by showing that emissivity changes over firn were caused by changes in scattering

Table 8.1 Major passive microwave sensors used for cryospheric remote sensing.

Passive microwave sensor	Frequencies (GHz)	Operational period
ESMR	19	Dec. 1972 – May 1977[1]
SMMR	6.6, 10.7, 18.0, 21.0, 37.0	Oct. 1978 – Aug. 1987[2]
SSM/I	19.3, 22.2, 37.0, 85.5	July 1987 – present[3]
AMSR-E	6.9, 10.7, 18.7, 23.8, 36.5, 89.0	May 2002 – Oct. 2011[4]
SSMIS	19.3, 22.2, 37.0, 91.7	Oct. 2003 – present[3]
AMSR2	6.9, 10.7, 18.7, 23.8, 36.5, 89.0	May 2012 – present

[1] Parkinson *et al.*, 1999;
[2] Gloerson *et al.*, 1990;
[3] Maslanik and Stroeve, 1990;
[4] Ashcroft and Wentz, 2006.

properties related to grain size which, in turn, was governed by accumulation rate. Since this initial work, there have been numerous studies using passive microwave sensors to derive accumulation rates in both Greenland and Antarctica (Zwally and Giovinetto, 1995; Vaughan *et al.*, 1999; Zwally *et al.*, 2000; Winebrenner, *et al.*, 2001; Bindschadler *et al.*, 2005; Arthern *et al.*, 2006; Koenig *et al.*, 2007).

Three primary methods have been proposed to relate accumulation to microwave radiometry. The first method uses the higher frequency channels (e.g., ≈19 GHz) and takes advantage of the fact that the emissivity in the Rayleigh-Jeans approximation changes, as scattering increases or decreases in the Rayleigh/Mie scattering regimes (Zwally, 1977). This method has been used to estimate accumulation directly from algorithms calibrated with *in situ* measurements.

The second method takes advantage of the fact that vertically polarized radiation, near the Brewster angle (of ≈53° for the snow/air interface), is preferentially transmitted through layer boundaries, while the horizontally polarized radiation is more reflected. Winebrenner *et al.* (2001) showed that the polarization ratio could be used to determine accumulation using lower microwave frequencies (≈6.9 GHz), and this work was furthered by Arthern *et al.* (2006). These two methods have been used to determine long-term accumulation averages over the ice sheets and to produce maps of the spatial variations of accumulation. Figure 8.2 shows the Arthern *et al.* (2006) map of Antarctic accumulation, derived from passive microwave AMSR-E data and ice surface temperatures from the Advanced Very High Resolution Radiometer (AVHRR).

A third method for monitoring accumulation using passive microwave data is based on the extinction-diffusion time model developed by Winebrenner *et al.* (2004). The model is a convolution equation, relating surface temperatures to brightness temperatures, and it depends on one characteristic time scale that is physically related to the microwave penetration depth and the thermal diffusivity of the firn. Koenig *et al.* (2007) showed that this characteristic time scale, a model parameter, varied spatially and temporally, and that those variations over the West Antarctic ice divide region were related to long-term mean accumulation rates. This method showed promise for deriving temporally varying accumulation rates, although it requires annual *in situ* data, generally derived from ice cores, which are currently not available to develop an algorithm capable of determining time-varying accumulation rates from passive sensors.

Though long temporal subsets of the passive microwave signal have been linked to long-term accumulation rates using all three methods, a fundamental question remains: can passive microwave remote sensing monitor temporal changes in accumulation? The answer is most likely "yes", but temporal variations discovered in the microwave signal are left invalidated with respect to accumulation, due to the lack of spatially distributed annual or near-annual accumulation measurements (Koenig *et al.*, 2007).

Point-source ice core data provide annual accumulation measurements, but do not have the spatial correlation to characterize the large passive microwave pixels (of approximately 13 km × 15 km at 85 GHz, up to approximately 43 km × 69 km at 19.35 GHz) necessary for algorithm development. The increase in new shallow

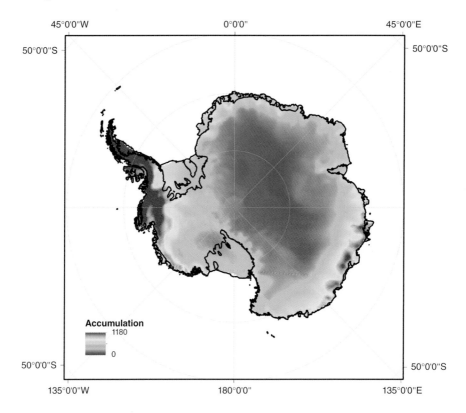

Figure 8.2 Arthern *et al.* (2006) map of Antarctic accumulation in kg/m² /a (or mm of water equivalent per year), derived from passive microwave AMSR-E data at 6.9 GHz and ice surface temperatures from the Advanced Very High Resolution Radiometer (AVHRR). (Data from British Antarctic Survey).

ice cores collected in Antarctica by the International Trans Antarctic Scientific Expedition (ITASE) teams has greatly increased the ground truth data, and merits revisiting studies on how well passive microwave can monitor accumulation. Additionally, new modeled data of accumulation, described in Section 8.4, merits comparison with passive microwave data.

A drawback of passive microwave remote sensing is that accumulation estimates can only be derived from the dry-snow zone, where melt does not occur. In regions that experience melt, passive microwave retrieval of temperature and accumulation are difficult for two reasons.

- First, when meltwater is present, there is a large microwave dielectric constant difference between ice and liquid water. The water creates an absorption-dominated regime, raising the brightness temperature compared to the scattering dominated regime of dry snow.
- Second, after the liquid water refreezes, a melt-freeze layer is created, along with larger snow grains that are not related to accumulation. The melt-freeze layer and larger snow grains confuse algorithms developed for dry-snow zone scattering parameters, resulting in incorrect accumulation retrievals.

Magand *et al.* (2008) did a detailed study of microwave accumulation retrievals in a large sector of Antarctica, and found that while there was reasonable agreement with *in situ* measurements in the dry-snow zone, there were large discrepancies in the coastal regions where melt occurred, biasing microwave-retrieved accumulation. The large interior region of the AIS, excluding the Antarctic Peninsula, some ice shelves and coastal regions, has minimal melt, thus providing good potential for passive microwaves to measure accumulation.

On the GrIS, however, the melt restriction on passive microwave retrievals is increasingly a concern. The dry-snow zone of the GrIS is decreasing as melt extent increases, due to warming temperature trends in the Arctic (Abdalati and Steffen, 2001; ACIA, 2005; Tedesco *et al.*, 2011). It is known that temporal variation of accumulation in Southeast Greenland, where accumulation rates are very high, dominate the accumulation signal for the entire GrIS (van de Broeke *et al.*, 2009; Burgess *et al.*, 2010), yet passive microwave sensors are unable to determine even a long-term accumulation field in this area, due to intense seasonal melt. Additionally, in July of 2012, almost the entire GrIS experienced melt, well into the traditional dry-snow zone (Nghiem *et al.*, 2012). As melt extent increases, passive microwave remote sensing will have more limited capabilities for measuring accumulation rates over the GrIS.

8.2.1.2 *Active microwave remote sensing*

Active microwave remote sensing, like passive, has also been considered for its potential use in estimating accumulation rates. Active microwave sensors (Table 8.2) or radar, used for accumulation estimates over the ice sheets, fall into two categories – synthetic aperture radar (SAR) and scatterometers. Both measure backscatter, derived using the radar equation using the power returned back to the radar after scattering from the surface and within the upper layers of snow/firn. The relationship between radar backscatter and snow accumulation is complex, and research has been limited to a few studies focusing on either the dry-snow zone in Greenland and Antarctica, or the percolation zone of Greenland.

Algorithms to retrieve snow accumulation rates in the dry-snow zone and in the percolation zone are based on both theoretical models and empirical relationships derived from comparisons between active microwave signatures and *in situ* measurements of snow accumulation rates. The relationship between snow accumulation and radar backscatter is inverse in both the dry snow zone and the percolation zone; however, the scattering response results from different physical mechanisms:

- Within the dry-snow zone, larger snow grains develop when snow accumulation rates are low, resulting in relatively high backscatter measurements, and smaller snow grains develop when snow accumulation rates are high, resulting in relatively low backscatter measurements.
- Within the percolation zone, a layer of accumulating snow overlying a strong scattering layer decreases backscatter measurements over time. Low snow accumulation rates in the percolation zone result in relatively small backscatter decreases and high accumulation rates result in relatively large backscatter decreases (Figure 8.3).

Table 8.2 Major active microwave sensors used for cryospheric remote sensing.

Active microwave sensor	Frequencies (GHz)	Operational period
SASS	14.6	Jul. 1978 – Oct. 1978
ERS 1/2	5.3	Jan. 1992 – May 1996
NSCAT	14.0	Sep. 1996 – Jun. 1997
QuikSCAT	13.4	Jul. 1999 – Nov. 2009
SeaWinds	13.4	Apr. 2003 – Oct. 2003
ASCAT	5.3	May 2007 – present
OSCAT	13.5	Sep. 2009 – present

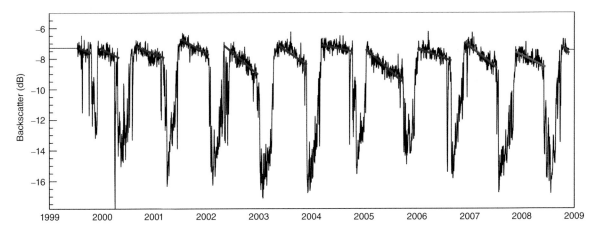

Figure 8.3 Active microwave snow accumulation signatures over the percolation zone of the Greenland ice sheet. A layer of accumulating snow of increasing depth, overlying a scattering layer created by melt metamorphosis of snow and firn layers, decreases the backscatter measurement over time, resulting in the observed decreasing winter season backscatter signatures (red lines). Vertically polarized backscatter data was acquired from NASA's Ku-band scatterometer SeaWinds on QuikSCAT, and obtained from the Scatterometer Climate Record Pathfinder (SCP) (www.scp.byu.edu).

Long and Drinkwater (1994) first observed the inverse relationship between accumulation rate and Ku-band backscatter measurements, acquired from NASA's first scatterometer (SASS) aboard the Seasat-A satellite, across a southwest transect of the dry-snow zone of the GrIS. Jezek and Gogineni (1992) observed a similar relationship in comparisons between L-band backscatter measurements acquired from the SAR aboard Seasat-A, and C-band backscatter measurements acquired from ESA's SAR aboard the European Remote Sensing Satellite (ERS-1).

The inverse relationship between radar backscatter and snow accumulation in the GrIS dry-snow zone has been investigated in further detail, using both SAR and scatterometers. Differences between large-scale mappings of backscatter measurements over the GrIS are related to the satellite's observation geometry and the associated radar's antenna scanning pattern and spatial response, which results in differing spatial and temporal resolutions. The relatively narrow swath width of a typical SAR scene (25 km) requires mosaicking of multiple orbits over several weeks to compile a complete, relatively high spatial resolution (25 m) mapping of backscatter measurements over the GrIS.

In contrast, scatterometers typically have wide swaths ($\approx$500–1800 kilometers), which provide complete ice sheet coverage as frequently as twice daily. The resulting mapping of backscatter measurements from scatterometers has a relatively low spatial resolution ($\approx$25–50 kilometers). Thus, spatial resolution is sacrificed for temporal resolution. In general, SAR provides single, higher spatial resolution mosaics observed on weekly time-scales. Scatterometry provides lower spatial resolution images on a near-daily time scale, allowing for the construction of higher temporal resolution time series. The relationships derived between radar backscatter and snow accumulation from both SAR and scatterometer studies, however, are similar.

Forster *et al.* (1999) used a coupled snow metamorphosis/backscatter model and backscatter measurements acquired from the ERS-1 SAR mosaic by Fahnestock *et al.* (1993) to map accumulation in Greenland. Results of this study indicate that Rayleigh scattering from snow grains dominates the microwave response at high incidence angles ($>30°$), and further suggests that backscatter measurements have increased sensitivity to changes in snow accumulation rates when the average snow accumulation rate for the region is low. Forster *et al.* (1999) concluded that an inverse relationship exists between radar backscatter and snow accumulation rates, and suggested a secondary relationship with mean annual temperature. A subsequent study by Munk *et al.* (2003) extended the work of Forster *et al.* (1999) by generating a look-up table which uniquely maps backscatter measurements and mean annual temperatures to snow accumulation rates in the GrIS dry-snow zone.

A study by Drinkwater *et al.* (2001) derived an empirical relationship between the change in backscatter measurements with incidence angle and *in situ* point measurements of snow accumulation rates from firn cores in the dry-snow zone. They used C-band backscatter measurements from ESA's scatterometer onboard the European Remote Sensing Satellite (ERS-2), and Ku-band backscatter measurements from NASA's NSCAT scatterometer aboard the Advanced Earth Observing Satellite (ADEOS-1). The derived exponential relationship was compared to a coupled snow metamorphosis/backscatter model similar to Forster *et al.* (1999).

Results suggest that C-band backscatter measurements are more sensitive to snow layering and buried scattering layers within the boundary zone between the dry-snow zone and the percolation zone, given the longer wavelength and increased penetration depth, which allows for interaction with more annual layers. Ku-band backscatter measurements are more sensitive to volume scattering from recently accumulated snow, given the shorter wavelength, which allows for increased interaction with individual snow grains. Flach *et al.* (2005) extended the work by Drinkwater *et al.* (2001) and developed an inversion technique to retrieve time-varying geophysical surface properties (i.e., snow grain size, density, layer thicknesses, and snow accumulation rate) to be used in predicting retracking errors in altimetry elevation measurements from ESA's CryoSat mission.

Another type of study in the GrIS dry-snow zone, by Oveisgharan and Zebker (2007), introduced the first backscatter model relating interferometric synthetic aperture radar (InSAR) measurements of coherence and backscatter to snow grain size and snow accumulation rates. This study suggested that, because larger snow grain size and deeper annual layers both increase backscatter measurements, to accurately model snow accumulation rates, a depth hoar layer must be introduced. Model results were validated in a small area of the dry-snow zone of the GrIS, using InSAR data from ERS-2.

Two studies focusing on using scatterometer data to estimate snow accumulation rates in the percolation zone are those by Wismann *et al.* (1997) and Nghiem *et al.* (2005). In the percolation zone, subsurface ice layers formed by melt-freeze metamorphoses create a strong scattering layer within the snowpack. Wismann *et al.* (1997) conducted the first study that observed decreasing winter season backscatter signatures in the percolation zone, and then described the relationship between backscatter measurements and snow accumulation rates as inverse and approximately linear in the decibel (dB) domain. This study used a two-layer backscatter model (i.e., accumulating snow layer overlying a scattering layer) and Monte-Carlo simulations to estimate the attenuation rate of the backscatter measurement as a function of the depth of the accumulated snow layer. This model was validated using C-band backscatter measurements from ERS-1 and ERS-2. The annual decrease in the backscatter measurement was used as a snow accumulation metric, and a large-scale mapping of annual snow accumulation rates was constructed. It was then visually compared to results in Ohmura and Reeh (1991) and found to be in general agreement. Limitations identified by the study include uncertainty in the conversion of the attenuation rate of the backscatter measurement to the rate of increasing snow depth and the uncertainty in snow density, given the lack of *in situ* measurements.

The technique used by Wismann *et al.* (1997) was subsequently adapted by Nghiem *et al.* (2005). This study derived a linear (dB) backscatter approximation (Nghiem *et al.*, 1990) that defined decreasing winter season backscatter signatures as a function of the attenuation rate of the backscatter measurement and the depth of the accumulating snow layer. An attenuation coefficient was derived, using *in situ* snow height recordings from two sonic height instruments. Using Ku-band backscatter measurements from NASA's SeaWinds scatterometer on the QuikSCAT satellite, the derived backscatter approximation, and the derived attenuation coefficient, a snow depth was calculated and was used to construct a large-scale mapping of winter season snow accumulation rates in the percolation zone of the GrIS.

A limitation identified by Nghiem *et al.* (2005) is the use of fixed winter season dates, which does not provide a complete estimate of winter season snow accumulation rates. The coarse temporal sampling of ERS- scatterometer data used by Wismann *et al.* (1997) resulted in an implicit approximation of snow accumulation rates over the melt season, which most likely resulted in errors in snow accumulation estimates in their study. The frequent temporal sampling of scatterometer data acquired from QuikSCAT requires an alternative approach.

Nghiem *et al.* (2005) used fixed dates as a simple method to separate stable winter season backscatter signatures from more complex melt season backscatter signatures. To identify the winter season accurately in a backscatter time series, freeze-up (i.e., the final melt event of the melt season) and melt onset (i.e., the initial melt event of the melt season) dates must be detected on a pixel-by-pixel basis, which requires a more sophisticated retrieval algorithm than the one used by Nghiem *et al.* (2005).

A second limitation is the use of a single attenuation coefficient over the entire study region. Spatially, the geophysical surface properties of the snowpack are modified by atmospheric conditions, topography, and latitude, which influence the attenuation coefficient. Thus, it is unlikely that the attenuation coefficient remains constant over the GrIS.

Recent work by Miller *et al.* (2011) builds on work by Wismann *et al.* (1997) and Nghiem *et al.* (2005) by comparing QuikSCAT decreasing winter season backscatter signatures with modeled snow accumulation within GrIS the percolation, wet snow, and ablation zones. A simple empirical relationship between winter season backscatter decreases and spatially calibrated Polar MM5 snow accumulation data (Burgess *et al.*, 2010) is derived, with a strong negative correlation coefficient. Results indicate that decreasing winter season backscatter signatures are a function of the depth of the accumulating snow layer and the stratigraphy of the underlying ice facie, which influences the scattering mechanism. This technique shows potential in the higher elevations of the percolation zone.

One of the few studies that test the ability of active remote sensing to estimate accumulation for a portion of the AIS was Dierking *et al.* (2012). They used a coupled snow metamorphosis/backscatter model, developed from empirical parameterizations of firn grain size and density, as functions of depth, surface temperature, and accumulation rate. Their model made significant progress over previous models by incorporating dense media radiative transfer (DMRT), as well as two internal firn layers per accumulation year. This was tested with measured accumulation rates, Envisat ASAR C-band wide-swath mode images, and QuikSCAT Ku-band scatterometer data from Dronning Maud Land, Antarctica. They found C-band to be more sensitive to accumulation rate changes than Ku-band, and also found that the spatial variability in firn properties needs to be considered when comparing accumulation rates from different positions on the ice sheet.

8.2.2 *Other remote sensing techniques and combined methods*

Both radar and laser altimetry are used over the ice sheets to investigate mass balance (See, for example, Chapter 9 for more information). Altimetry measures the surface height of the ice sheet. The change in surface height over time (dh/dt) is controlled by the surface mass balance, mass balance at the base of the ice sheet, and the vertical (firn densification) and horizontal flow velocities (Hagen and Reeh, 2004). Theoretically, altimeters can detect accumulation by monitoring the surface elevation change. This, however, has yet to be fully realized, because firn compaction rates are similar to accumulation over much of the ice sheets, and quantifying densification is difficult (Arthern and Wingham, 1998; Helsen *et al.*, 2008; Pritchard *et al.*, 2009).

Bindschadler *et al.* (2005) used a combination of passive microwave and Ice Cloud and land Elevation Satellite (ICESat) laser altimetry data to estimate accumulation events. They used one of the highest frequency channels (85 GHz) to detect new snowfall, using a Hibert-Huang transform. This technique allowed for new snowfall events and areas to be detected by a change in the surface emission due to the presence of new snow grains at the surface. This method then used altimetry data, showing the surface height change over the area for the time period detected by the passive microwave sensor, to determine the volume of snowfall. Although promising, results from this technique can be influenced by surface temperatures and atmospheric conditions. It also cannot distinguish between surface emission changes due to precipitation or wind redistribution. Limitations aside, this method is capable of delineating possible areas of new snowfall with coincident altimetry and passive microwave data.

Palm *et al.* (2011) also took advantage of a combined technique using LiDAR data from ICESat and from the Cloud-Aerosol LiDAR with Orthogonal Polarization (CALIOP), along with visible data from the Moderate Resolution Imaging Spectroradiometer (MODIS), to detect and quantify blowing snow events in Antarctica. While blowing snow is only a small component of total Antarctic SMB, it can be regionally important. Satellite retrievals like those made by Palm *et al.*, (2011) will be very useful for comparison with regional atmospheric models that incorporate blowing snow (Lenaerts *et al.*, 2012).

Gravity measurements over the ice sheets provide a measurement of total mass gain or loss. The Gravity Recovery and Climate Experiment (GRACE) satellite, launched in 2002, provides monthly mass measurements over the ice sheets at a very low spatial resolution (grids over the ice sheet are ≈50 km; see Chapter 10). GRACE measurements clearly show the seasonal mass fluctuations over the ice sheets (Velicogna and Wahr, 2006; Luthcke *et al.*, 2006; Llubes *et al.*, 2007). While accumulation is only one component of the mass change measured by GRACE, in portions of the accumulation zone where ice surface velocities do not account for a significant dynamic component of the mass balance, the differences in mass gained and lost over a year are a measure of the net accumulation rate. While remote sensing gravity measurements show promise for determining snow accumulation over the ice sheets, additional research directed in this area is required.

Horwath *et al.* (2012) combined Envisat radar altimetry data with GRACE data to compare the inter-annual surface mass balance variations, relating volume changes and mass changes respectively, in Antarctica. They showed that both methods had similar patterns, and that a specific accumulation event could be identified and quantified in West Antarctica.

Measurements of precipitation are also possible with remote sensing methods using high frequency radars used to profile clouds, such as CloudSat. To date, snowfall retrievals have met with limited success over the ice sheet. However, Liu (2008) was able to determine global precipitation rates using CloudSat, including those over the ice sheets. The CloudSat-derived precipitation fields over the ice sheets, however, have not been validated. The launch of the Global Precipitation Mission (GPM) in 2014, carrying more advanced Ka- and Ku-band precipitation radars and a multi-channel radiometer, promises future improvements of precipitation measurements on a global scale, including over the cryosphere.

While visible data cannot be used to measure accumulation directly, it has been used to delineate the accumulation zone over the Greenland ice sheet (Benson and Box, 2008). The major problem with using visible imagery at the poles is that it cannot be used during the months of polar darkness or through clouds.

8.3 Airborne and ground-based measurements of accumulation

8.3.1 Ground-based

Accumulation measurements on the ice sheets are made using a variety of techniques. Accumulation can be measured at point locations using snow stakes, ultrasonic sensors, snow pits and firn/ice cores. Accumulation can also be measured along transects using ground penetrating radar (GPR), providing better spatial coverage for comparison with satellite footprints (see Eisen *et al.* (2008) for a detailed explanation of each method).

For the AIS, Vaughan *et al.* (1999) created one of the first comprehensive compilations of surface accumulation measurements. The compilation is spatially extensive, with over 1800 data points throughout Antarctica. This data set reports the time frame over which the ground measurements were taken, but uses the long-term temporal average, not distinguishing annual accumulation rates. Vaughan *et al.* (1999) used a passive microwave background to interpolate between the compilation of *in situ* measurement, and determined the average accumulation rate over the AIS to be ≈ 17 cm/yr of water equivalent.

Eisen *et al.* (2008) furthered the compilations of *in situ* accumulation measurements in Antarctica by compiling a comprehensive review of spatial and temporal accumulation measurements in East Antarctica, as well as an assessment of the quality of measurement techniques. This review compiled accumulation data from numerous measurement techniques, including stakes, ultrasonic sounders, snow pits, radar profiling and firn and ice cores, including those gathered on ITASE traverses. This review was a large step forward in compiling the necessary data for remote sensing calibration and validation. It also made the important distinction that not all *in situ* measurements are equally accurate – an important caveat when creating accumulation algorithms for satellite sensors.

Over the GrIS, NASA's Program for Regional Climate Assessment (PARCA) started the first major assessment of accumulation rates. McConnell *et al.* (2000) undertook a major coring effort to characterize the accumulation area above 2000 m (mostly included in the dry-snow zone). These cores, known as the PARCA cores, are a great example of the spatial sampling of accumulation needed for satellite algorithm validation and calibration. The major drawbacks of these cores is that they only measure accumulation rates up to 1998, before many of the Earth-observing satellites were in orbit, and at altitudes below 2000 m there are very few measurements (Mosely-Thompson *et al.*, 2001). Furthermore, in the ablation zone of the GrIS, there are no extensive *in situ* accumulation measurements in either time or space.

Bales *et al.* (2001, 2009) continued efforts by compiling ice sheet accumulation measurements over Greenland from ice cores and snow pits, and updating approximately a decade later. Data from these studies shows the annual accumulation rate over the GrIS to be ≈ 30 cm yr^{-1} water equivalent. The ice core and pit data were krigged, without using a passive microwave background, to create a long-term map of accumulation over Greenland. The Bales *et al.* (2001, 2009) method is a good independent method for comparing with satellites but, again, it lacks temporal variations, because the ice cores and pits used to determine the spatial extent did not overlap in time.

High/very high-frequency (100 to 900 MHz) ground penetrating radar (GPR) data provide temporal measurements of accumulation along a line, increasing the spatial resolution over other accumulation measurements (Kanagaratnam *et al.*, 2004). Radars detect isochronal layers in the firn. When the layers are dated, in conjunction with ice cores, they can be used to determine long-term accumulation rates over linear distances (Spikes *et al.*, 2004). Spikes *et al.* (2004) and Arcone *et al.* (2005) showed that shallow 400 MHz radar data, gathered during the US ITASE transverse over the West Antarctic Ice Sheet (WAIS), contained isochronal layers with datable accuracy of less than one year.

Spikes *et al.* (2004) used the radar data to obtain accumulation rates by picking a radar echoed layer dated at 1966, and four additional layers dating from previous years, over one of the eight transects collected in 2000. The radar had a vertical resolution of ≈ 35 cm, capable of detecting near-annual or possibly annual layering if the accumulation rate is high. Additional measurements from high frequency radars were collected in East Antarctica by the US-Norwegian ITASE traverse by Müller *et al.* (2010) and by Anshütz *et al* (2008), providing thousands of kilometers of needed spatial and temporal accumulation rates at decadal time scales. In Greenland, a 400 MHz GPR was used to measure accumulation rates in the high accumulation area of the southeast GrIS by the Arctic Circle Traverse in 2010 (ACT-10) (Miège *et al.*, 2013).

In addition to high/very high frequency radars, ultra/super-high-frequency (UHF) (2–20 GHz) frequency-modulated continuous wave (FMCW) radars have been shown to map stratigraphic layers over ice sheets and ice caps at nearly annual rates (Richardson *et al.*, 1997; Marshall and Koh, 2008; Langley *et al.*, 2008). These radars provide high vertical resolutions with large bandwidth (≈5 cm, given a 2.5 GHz bandwidth in snow), and are capable of determining annual accumulation rates.

Gogineni *et al.* (2007) showed the utility of a UHF wide band radar, operating up to S-band, to image internal layering in the near surface firn near Summit, Greenland. Similarly, Rink *et al.* (2006) showed that a 12–18 GHz (X-band) radar tested at WAIS divide and Summit, Greenland, could map stratigraphic layers in snow pits, determine accumulation rates and follow layers for many kilometers. These studies have all proved the usefulness of UHF to image internal layers and determine annual accumulation rate, but a comprehensive study of annual accumulation rates over a large spatial area has yet to be conducted.

Two UHF radars, developed by the University of Kansas Center for Remote Sensing of Ice Sheets (CReSIS), called the Snow radar and the Ku-band radar, collected data as part of the Satellite Era Accumulation Traverse (SEAT) in 2010 and 2011. Figure 8.4 shows a radar echogram from the Ku-band radar during the SEAT traverse. The high vertical resolution provides images of the annual layers as well as sub-annual firn structure.

Ground-based or *in situ* measurement will always be intricately tied to remote sensing efforts over the ice sheets. Ground measurements alone are incapable of monitoring the spatial and temporal variations of accumulation of the large ice sheets yet, at the same time, remote sensing algorithms providing the spatio-temporal coverage require *in situ* measurements for development and calibration. A symbiotic relationship must exist.

8.3.2 *Airborne*

As previously mentioned, radar echograms (Figure 8.4), revealing firn internal layers, provide quality, spatially-extensive, *in situ*, accumulation data. Radar measurements can be extended over ground based applications by flying on specially equipped aircraft. Hawley *et al.* (2006) first showed that airborne radar at 13.2 GHz (Ku-band) could determine annual layers in the dry-snow zone of the GrIS down to 12 m, when compared to dating based on annual density peaks. This study confirmed the ability of UHF radars to monitor density defined annual layers, and it pointed to the future possibility of extensive monitoring of accumulation rates from air and space.

In 2009, NASA launched the largest airborne mission monitoring the polar regions, called Operation Ice Bridge (OIB) (Koenig *et al.*, 2010). OIB is expected to continue until the launch of ICESat-2 in 2017 (see Chapter 15). OIB includes three radars, developed by CReSIS, capable of determining accumulation rates over the ice sheets. The three radars are the Ku-band (13–17 GHz) radar, the Snow radar (2–8 GHz) and the Accumulation radar (600–900 MHz). The Ku-band and Snow radar have, for the first time, extensively mapped the firn layering structure

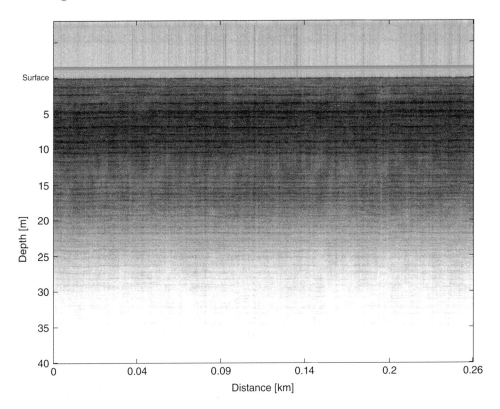

Figure 8.4 CReSIS Ku-band radar echogram collected during the SEAT traverse in West Antarctica. The high vertical resolution provides images of the annual layers, as well as sub annual firn structure. Large bandwidth FMCW radars, like the Ku-band radar, provide an air- and ground-based method for determining accumulation rates over the ice sheets.

over both the ablation and accumulation zones of the Greenland and Antarctic ice sheets. These radar data are currently being analyzed, and they show great promise in providing spatial and temporal accumulation measurements. Radars of this type are not yet ready for space-based platforms, but they still provide a major improvement for remote sensing of the spatio-temporal patterns of accumulation.

8.4 Modeling of accumulation

Because no remote sensing method currently gives the full temporal variation and spatial extent of accumulation on an operational basis, models are often used to produce maps of accumulation rate and to determine ice sheet mass balance (e.g., Rignot *et al.*, 2011). Three primary categories of models exist. Global climate models (GCMs) have a coarse resolution, but provide global coverage. Regional atmospheric models have better spatial resolution, ranging from 50 km to 5 km, more sophisticated atmospheric physics, and are forced on the margins of the analysis boundary by atmospheric reanalysis data sets. Reanalyses are assimilation of satellite meteorological observations into models of increasing complexity. Over the ice sheets, regional atmospheric models and reanalyses are used to determine SMB (e.g., van de Berg *et al.*, 2006; Ettema *et al.*, 2009; Lenaerts *et al.*, 2012b; Bromwich *et al.*, 2011; Cullather and Bosilovich, 2011; Hanna *et al.*, 2011; Fettweis, 2007). Figure 8.5 shows the accumulation of Greenland from a regional atmospheric model, defined as snowfall minus evaporation and sublimation.

While regional models and reanalysis provide spatial and temporal information, the collection of models operated do not agree on the magnitude of accumulation, net precipitation and net evaporation. Bromwich *et al.* (2011) shows that the reanalysis range in trends is equivalent to 1 mm of sea level rise for the AIS. Additionally, recent regional models of accumulation over Antarctica are significantly higher than the passive microwave retrievals of accumulation, most likely due to the model's ability to better resolve high accumulation areas on the coast of

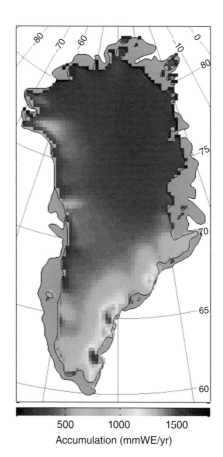

Figure 8.5 Greenland average accumulation rate from (1958–2012), calculated from the regional atmospheric model MAR, in mm of water equivalent per year. Greenland accumulation is dominated by high snowfall in the southeastern coastal region (data courtesy of M. Tedesco).

Antarctica (Lenaerts *et al.*, 2012b). In Greenland, two recent studies of accumulation also found large differences in the model outputs. Chen *et al.* (2011) found that the range in average accumulation between five reanalysis models was ≈15 to ≈30 cm/yr, while Cullather and Bosilovich (2011) compared reanalysis data to regional climate models with a range in accumulation from ≈34 to ≈42 cm/yr average accumulation.

An additional hybrid technique for determining accumulation was developed by Burgess *et al.* (2010). This technique combines the direct measurement of point data from firn cores with the comprehensive spatial extent of a regional climate model. They develop spatially calibrated solid precipitation output (i.e. snowfall) acquired from the Fifth Generation Mesoscale Model modified for polar climates (Polar MM5) (Box *et al.*, 2004, 2006; Bromwich *et al.*, 2001; Cassano *et al.*, 2001). The spatial calibration uses *in situ* point measurements of snow accumulation rates, acquired from firn cores and coastal meteorological stations, with spatial interpolation of regionally derived linear correction functions. The resulting large-scale mapping of snow accumulation rates on the GrIS is linked to *in situ* measurements uniform in space and time. However, the calibration factors are still limited by the original uncertainty estimates in the point data (Figure 8.6). Their results show that an average accumulation rate is ≈34 cm/year, which is 21% higher than a previous measurement (Calanca *et al.*, 2000).

Some discrepancies in modeled accumulation are expected. The regional models often use different forcings at the lateral boundaries, model runs do not always overlap in time, and the resolution of the model grids are different. These discrepancies aside, it is clear that the range in the modeled accumulation must be narrowed over the GrIS and AIS, and further study is needed. Models of SMB, however, have provided the first maps of seasonal and annual variations in

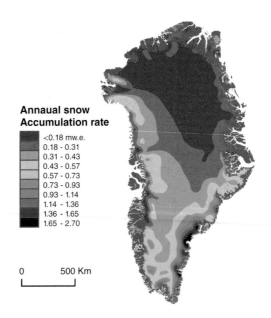

Figure 8.6 Greenland average accumulation rate from 1958–2008 from the ice core calibrated Polar MM5 model.

accumulation over the ice sheet, a feat not yet accomplished by remote sensing. It is clear that in the future both models and remote sensing methods will be used together to improve estimates of SMB.

8.5   The future for remote sensing of accumulation

In this chapter, methods to monitor ice sheet SMB from remote sensing, *in situ* measurements, and models have been discussed. These methods are used to understand the ice sheet past and current mass balance and, thus, to determine present sea level contributions. Most of the time, these methods are used independently. Future studies, producing decade-long SMB time series over both the AIS and GrIS, would benefit from combining many of the methods presented here. Such an approach would favor the use of the most accurate method for a given region – for instance, using radiometer retrievals in the dry-snow zone and radar retrievals in the percolation zone. Considering surface processes alone is not enough for predicting future sea level rise. The movement of ice, called ice dynamics, has incredible potential for rapidly increasing the rate at which ice is transferred to the ocean, causing sea level rise. The combination of satellite sensors, and models of both the atmosphere and ice dynamics, will provide finer details to investigate the relationship between ice dynamics and SMB near the ice sheet margins.

It has also been mentioned that the contribution of blowing snow on SMB, producing scour and redistribution of snow (especially in Antarctica), is very likely a major SMB component. However, quantitative studies are extremely rare and are spatially and temporally limited. Determining the magnitude of mass transport of blowing snow (and sublimation) to better understand the fraction of total SMB appears to be one of the challenges of the coming decades. To achieve this challenge, remote sensing, *in situ* measurements, and model simulations will be analyzed to obtain the finest spatial and temporal resolution possible, as well as the best sensitivity to snow and atmospheric properties. Of note, these studies will also need to determine the distance along which snow is transported by wind, and the snow volume moved from the ice sheet to the surrounding sea ice and, eventually, to the ocean.

SMB depends on both snow accumulation and terms of mass loss, both of which are part of complex feedback mechanisms. For instance, the well-known albedo feedback is accelerating, especially in the Arctic, affecting global temperatures through the surface radiative balance. Warming temperatures over the Earth's polar regions change the atmospheric humidity, cloud cover, and precipitations. Feedbacks associated with changes in cryosphere, ocean, and atmosphere will affect the ice sheet SMB, but a consensus of their respective magnitude has not yet been reached. Future remote sensing studies will face the challenge of quantifying the impacts on SMB of aerosols, rising temperatures, changing precipitation, and cloud type and cover. Satellite sensors must be capable of monitoring all of these processes.

8.6 Conclusions

As the only input term to the mass balance of the Greenland and Antarctic ice sheets, snow accumulation is a critical process to measure. The vast size of the ice sheets, and spatial variability in accumulation rates, render *in situ* measurement alone useless for determining accurate ice-sheet-scale accumulation. Modeling efforts are progressing and can estimate accumulation at tens-of-kilometer grid spacing, but models are sensitive to boundary forcing conditions, and a paucity of validation data leaves these data sets less than ideal. Therefore, remote sensing of ice sheet accumulation is an important technique to develop.

There are several promising techniques, using various portions of the electromagnetic spectrum, but none of them are at a point where accumulation can be measured in an operational mode. Spaceborne active and passive microwave remote sensing have received the most attention, due to their responsiveness to snow and firn properties related to accumulation rates, and due to their ability to operate independently of sunlight and through cloud cover. Spatial and temporal variability in liquid water and refrozen subsurface ice structures prevents the use of a single broadly applied algorithm for snow accumulation retrievals, even at microwave wavelengths. Other techniques show promise under specific circumstances, such as altimetry, gravity, and airborne sensors, warranting continued study. Combining several different techniques across the electromagnetic spectrum in specific regions of the ice sheets seems to be the most likely way forward for remote sensing of snow accumulation.

References

Abdalati, W. & Steffen, K. (2001). Greenland ice sheet melt extent: 1979–1999. *Journal of Geophysical Research* **106**(D24), 33983–33988.

Anschütz, H., Sinisalo, A., Isaksson, E., *et al.* (2011). Variation of accumulation rates over the last eight centuries on the East Antarctic Plateau derived from volcanic signals in ice cores. *Journal of Geophysical Research – Atmospheres* **116**, doi:10.1029/2011JD015753.

Arcone, S., Spikes V. & Hamilton, G. (2005). Stratigraphic variation within polar firn caused by differential accumulation and ice flow, interpretation of a 400 MHz short-pulse radar profile from West Antarctica. *Journal of Glaciology* **51**(174), 407–422, doi: 10.3189/172756505781829151.

Arctic Climate Impact Assessment (ACIA) (2005). An Introduction to the Arctic Climate Impact Assessment, In: Symon, C. *et al.* (eds). *Arctic Climate Impact Assessment*, pp. 1–29. Cambridge, Cambridge University Press.

Arthern, R. & Wingham, D. (1998). The natural fluctuations of firn densification and their effect on the geodetic determination of ice sheet mass balance. *Climatic Change* **40**(3–4), 605–624, doi: 10.1023/A:1005320713306.

Arthern, R.J., Winebrenner, D.P. & Vaughan, D.G. (2006). Antarctic snow accumulation mapped using polarization of 4.3 cm wavelength microwave emission. *Journal of Geophysical Research – Atmospheres* **111**(D6), doi: 10.1029/2004JD005667.

Ashcroft, P. & Wentz, F. (2006). *AMSR-E/Aqua L2A Global Swath Spatially-Resampled Brightness Temperatures V002*. Boulder, Colorado USA, National Snow and Ice Data Center, Digital media.

Bales, R., McConnell, J., Mosley-Thompson, E. & Lamorey, G. (2001). Accumulation map for the Greenland Ice Sheet, 1971–1990. *Geophysical Research Letters* **28**(15), 2967–2970, doi: 10.1029/2000GL012052.

Bales, R., Guo, Q., Shen, D., *et al.* (2009). Annual accumulation for Greenland updated using ice core data developed during 2000–2006 and analysis of daily coastal meteorological data. *Journal of Geophysical Research – Atmospheres* **114**, doi: 10.1029/2008JD011208.

Benson, C.S. (1962). *Stratigraphic Studies in the snow and firn of the Greenland ice sheet, Research Report 70*. U.S. Army Cold Regions Research and Engineering Laboratory, Hanover, NH.

Benson, R. & Box, J. (2008). *MODIS-derived Greenland ice sheet equilibrium line altitude 2000–2008, comparison with surface melt and accumulation variability*. AGU Fall Meeting, San Francisco, CA.

Bindschadler, R., Choi, H., Shuman, C. & Markus, T. (2005). Detecting and measuring new snow accumulation on ice sheets by satellite remote sensing. *Remote Sensing of Environment* **98**(4), 388–402, doi: 10.1016/j.rse.2005.07.014.

Box, J., Bromwich, D. & Bai, L. (2004). Greenland ice sheet surface mass balance 1991–2000, Application of Polar MM5 mesoscale model and *in situ* data. *Journal of Geophysical Research – Atmospheres* **109**(D16), doi: 10.1029/2003JD004451.

Box, J., Bromwich, D., Veenhuis, B., *et al.* (2006). Greenland ice sheet surface mass balance variability (1988–2004) from calibrated polar MM5 output. *Journal of Climate* **19**(12), 2783–2800, doi: 10.1175/JCLI3738.1.

Bromwich, D., Chen, Q., Bai, L., Cassano, E. & Li, Y. (2001). Modeled precipitation variability over the Greenland ice sheet. *Journal of Geophysical Research – Atmospheres* **106**(D24), 33891–33908, doi: 10.1029/2001JD900251.

Bromwich, D., Nicolas, J. & Monaghan, A. (2011). An Assessment of Precipitation Changes over Antarctica and the Southern Ocean since 1989 in Contemporary Global Reanalyses. *Journal of Climate* **24**(16), 4189–4209, doi: 10.1175/2011JCLI4074.1.

Brucker, L., Picard, G. & Fily, M. (2010). Snow grain-size profiles deduced from microwave snow emissivities in Antarctica. *Journal of Glaciology* **56**(197), 514–526.

Brucker, L., Royer, A., Picard, G., Langlois, A. & Fily, M. (2011a). Hourly simulations of the microwave brightness temperature of seasonal snow in Quebec, Canada, using a coupled snow evolution-emission model. *Remote Sensing of Environment* **115**(8), 1966–1977.

Brucker, L., Picard, G., Arnaud, L., Barnola, J.M., Schneebeli, M., Brunjail, H., Lefebvre, E. & Fily, M. (2011b). Modeling time series of microwave brightness temperature at Dome C, Antarctica, using vertically resolved snow temperature and microstructure measurements. *Journal of Glaciology* **57**(201), 171–182.

Burgess, E., Forster, R., Box, J., *et al.* (2010). A spatially calibrated model of annual accumulation rate on the Greenland Ice Sheet (1958–2007). *Journal of Geophysical Research – Earth Surface* **115**, doi:10.1029/2009JF001293.

Calanca, P., Gilgen, H., Ekholm, S. & Ohmura, A. (2000). Gridded temperature and accumulation distributions for Greenland for use in cryospheric models. *Annals of Glaciology* **31**, 118–120, doi: 10.3189/172756400781820345.

Cassano, J., Box, J., Bromwich, D., Li, L. & Steffen, K. (2001). Evaluation of polar MM5 simulations of Greenland's atmospheric circulation. *Journal of Geophysical Research – Atmospheres* **106**(D24), 33867–33889, doi: 10.1029/2001JD900044.

Chen, L., Johannessen, O., Wang, H. & Ohmura, A. (2011). Accumulation over the Greenland Ice Sheet as represented in reanalysis data. *Advances in Atmospheric Sciences* **28**(5), 1030–1038, doi: 10.1007/s00376–010-0150-9.

Cullather, R. & Bosilovich, M. (2011). The Moisture Budget of the Polar Atmosphere in MERRA. *Journal of Climate* **24**(11), 2861–2879, doi: 10.1175/2010JCLI4090.1.

Dierking, W., Linow, S. & Rack, W. (2012). Toward a robust retrieval of snow accumulation over the Antarctic ice sheet using satellite radar. *Journal of Geophysical Research – Atmospheres* **117**, doi:10.1029/2011JD017227.

Drinkwater, M., Long, D. & Bingham, A. (2001). Greenland snow accumulation estimates from satellite radar scatterometer data. *Journal of Geophysical Research – Atmospheres* **106**(D24), 33935–33950, doi:10.1029/2001JD900107.

Eisen, O., Frezzotti, M., Genthon, C., Isaksson, E., Magand, O. van den Broeke, M.R., Dixon, D.A., Ekaykin, A., Holmlund, P., Kameda, T., Karlöf, L., Kaspari, S., Lipenkov, V.Y., Oerter, H., Takahashi, S. & Vaughan, D.G. (2008). Ground-based measurements of spatial and temporal variability of snow accumulation in east Antarctica. *Reviews of Geophysics* **46**(2), doi:Rg2001 10.1029/2006rg000218.

Ettema, J., van den Broeke, M., van Meijgaard, E., *et al.* (2009). Higher surface mass balance of the Greenland ice sheet revealed by high-resolution climate modeling. *Geophysical Research Letters* **36**, doi: 10.1029/2009GL038110.

Fahnestock, M., Bindshadler, R., Kwok, R. & Jezek, K. (1993). Greenland ice-sheet surface-properties and ice dynamics from ERS-1 SAR imagery. *Science* **262**(5139), 1530–1534, doi: 10.1126/science.262.5139.1530.

Fettweis, X. (2007). Reconstruction of the 1979–2006 Greenland ice sheet surface mass balance using the regional climate model MAR, *The Cryosphere* **1**, 21–40, doi:10.5194/tc-1-21-2007.

Flach, J., Partington, K., Ruiz, C., Jeansou, E. & Drinkwater, M. (2005). Inversion of the surface properties of ice sheets from satellite microwave data. *IEEE Transactions on Geoscience and Remote Sensing* **43**(4), 743–752, doi: 10.1109/TGRS.2005.844287.

Forster, R., Rignot, E., Isacks, B. & Jezek, K. (1999). Interferometric radar observations of Glaciares Europa and Penguin, Hielo Patagonico Sur, Chile. *Journal of Glaciology* **45**(150), 325–337, doi: 10.3189/002214399793377220.

Gloerson, P., Cavalieri, D., Campbell, W.J. & Zwally, J. (1990). *Nimbus-7 SMMR Polar Radiances and Arctic and Antarctic Sea Ice Concentrations*. Boulder, Colorado USA, National Snow and Ice Data Center. Digital Media.

Gogineni, S., Braaten, D., Allen, C.J., *et al.* (2007). Polar radar for ice sheet measurements (PRISM). *Remote Sensing of Environment* **111**(2–3). 204–211 doi:10.1016/j.rse.2007.01.022.

Gow, A. (1969). On the rates of growth of grains and crystals in southern polar firn. *Journal of Glaciology* **8**(3), 241–252.

Gow, A. (1971). *Depth-time-temperature relationships of ice crystal growth in polar glaciers, Research Report 300*. US Army Cold Regions Research and Engineering Laboratory, Hanover, NH.

Hagan, J. & Reeh, N. (2004). Chapter 2, *In situ* measurement techniques, land ice. In: Bamber, J. & Payne, A. (eds). *Mass Balance of the Cryosphere*, pp. 11–41. University Press, Cambridge.

Hanna, E., Huybrechts, P., Cappelen, J., Steffen, K., Bales, R.C., Burgess, E., McConnell, J.R., Steffensen, J.P., Van den Broeke, M., Wake, L., Bigg, G., Griffiths, M. & Savas, D. (2011). Greenland Ice Sheet surface mass balance 1870 to 2010 based on Twentieth Century Reanalysis, and links with global climate forcing. *Journal of Geophysical Research – Atmospheres* **116**(D24), doi:10.1029/2011JD016387.

Hawley, R.L., Morris, E.M., Cullen, R., Nixdorf, U., Shepherd, A.P. & Wingham D.J. (2006). ASIRAS airborne radar resolves internal annual layers in the dry-snow zone of Greenland. *Geophysical Research Letters* **33**(4), doi:10.1029/2005gl025147.

Helsen, M., van den Broeke, M., van de Wal, R., *et al.* (2008). Elevation changes in Antarctica mainly determined by accumulation variability. *Science* **320**(5883), 1626–1629, doi: 10.1126/science.1153894.

Horwath, M., Legresy, B., Remy, F., Blarel, F. & Lemoine, J.M. (2012). Consistent patterns of Antarctic ice sheet interannual variations from ENVISAT radar altimetry and GRACE satellite gravimetry. *Geophysical Journal International* **189**(2), 863–876, doi: 10.1111/j.1365-246X.2012.05401.x.

Jezek, K.C. & Gogenini, P. (1992). Microwave remote sensing of the Greenland ice sheet, Remote Sensing Society Newletter. *IEEE Transactions on Geoscience and Remote Sensing* (**85**), 6–10.

Kanagaratnam, P., Gogineni, S., Ramasami, V. & Braaten, D. (2004). A wideband radar for high-resolution mapping of near-surface internal layers in glacial ice. *IEEE Transactions on Geoscience and Remote Sensing* **42**(3), 483–490, doi: 10.1109/TGRS.2004.823451.

Koenig, L.S., Steig, E.J., Winebrenner, D.P. & Shuman, C.A. (2007). A link between microwave extinction length, firn thermal diffusivity, and accumulation rate in West Antarctica. *Journal of Geophysical Research – Earth Surface* **112**(F3), doi:F03018 10.1029/2006jf000716.

Koenig, L., Martin, S., Studinger, M. & Sonntag, J.G. (2010). Polar Airborne Observations Fill Gap in Satellite Data. *EOS* **91**(38), 333–334, doi: 10.1029/2010EO380002.

Langley, K., Hamran, S.E., Hogda, K.A., *et al.* (2008). From glacier facies to SAR backscatter zones via GPR. *IEEE Transactions on Geoscience and Remote Sensing* **46**(9), 2506–2516, doi: 10.1109/tgrs.2008.918648.

Lenaerts, J. & van den Broeke, M. (2012). Modeling drifting snow in Antarctica with a regional climate model, 2. Results. *Journal of Geophysical Research – Atmospheres* **117**, doi:10.1029/2010JD015419.

Lenaerts, J., van den Broeke, M., van de Berg, W., van Meijgaard, E. & Munneke, P. (2012). A new, high-resolution surface mass balance map of Antarctica (1979–2010) based on regional atmospheric climate modeling. *Geophysical Research Letters* **39**, doi:10.1029/2011GL050713.

Li, J., Wang, W., Zwally, J. & Wolff, E. (2002). Interannual variations of shallow firn temperature at Greenland summit. *Annals of Glaciology* **35**(35), 368–370.

Liu, G. (2008). Deriving snow cloud characteristics from CloudSat observations. *Journal of Geophysical Research – Atmospheres* **113**(D18), doi:10.1029/2007JD009766.

Llubes, M., Lemoine, J. & Remy, F. (2007). Antarctica seasonal mass variations detected by GRACE. *Earth and Planetary Science Letters* **260**(1–2), 127–136, doi: 10.1016/j.epsl.2007.05.022.

Long, D. & Drinkwater, M. (1994). Greenland ice-sheet surface-properties observed by the Seasat–A scatterometer as enhanced resolution. *Journal of Glaciology* **40**(135), 213–230.

Luthcke, S., Zwally, H., Abdalati, W., *et al.* (2006). Recent Greenland ice mass loss by drainage system from satellite gravity observations. *Science* **314**(5803), 1286–1289, doi: 10.1126/science.1130776.

Magand, O., Picard, G., Brucker, L., Fily, M. & Genthon, C. (2008). Snow melting bias in microwave mapping of Antarctic snow accumulation. *The Cryosphere* **2**(2), 109–115.

Marshall, H.P. & Koh, G. (2008). FMCW radars for snow research. *Cold Regions Science and Technology* **52**(2), 118–131, doi: 10.1016/j.coldregions.2007.04.008.

Maslanik, J. & Stroeve, J. (1990). *DMSP SSM/I-SSMIS Daily Polar Gridded Brightness Temperatures.* Boulder, Colorado USA: National Snow and Ice Data Center. Digital media.

McConnell, J., Mosley-Thompson, E., Bromwich, D., Bales, R. & Kyne, J. (2000). Interannual variations of snow accumulation on the Greenland Ice Sheet (1985–1996), new observations versus model predictions. *Journal of Geophysical Research – Atmospheres* **105**(D3), 4039–4046, doi: 10.1029/1999JD901049.

Miège, C., Forster, R.R., Box, J.E., Burgess, E.W., McConnell, J.R. & Spikes, V.B. (2013). Validation and spatial variability of Southeast Greenland high accumulation rates from firn cores and surface-based radar. *Annals of Glaciology* **54**(63).

Miller, J., Forster, R., Long, D., Schröder, R., McDonald, K. & Box, J. (2011). *Snow accumulation rate retrieval across the Greenland ice facies using SeaWinds on QuikSCAT.* American Geophysical Union, Fall Meeting 2011, abstract #C41E-0467, San Francisco, CA.

Mosley-Thompson, E., McConnell, J., Bales, R., *et al.* (2001). Local to regional-scale variability of annual net accumulation on the Greenland ice sheet from

PARCA cores. *Journal of Geophysical Research – Atmospheres* **106**(D24), 33839–33851, doi: 10.1029/2001JD900067.

Müller, K., Sinisalo, A., Anschutz, H., *et al.* (2010). An 860 km surface mass-balance profile on the East Antarctic plateau derived by GPR. *Annals of Glaciology* **51**(55), 1–8.

Munk, J., Jezek, K., Forster, R. & Gogineni, S. (2003). An accumulation map for the Greenland dry-snow facies derived from spaceborne radar. *Journal of Geophysical Research – Atmospheres* **108**(D9), doi:10.1029/2002JD002481.

NASA Press Release (2012). *Satellites see unprecedented Greenland ice sheet surface melt.* http://www.nasa.gov/topics/earth/features/greenland-melt.html.

Nghiem, S.V., Kong, J.A. & Shin, R.T. (1990). *Study of polarimetric response of sea ice with layered random medium model.* IEEE, College Park, MD, USA.

Nghiem, S., Steffen, K., Neumann, G. & Huff, R. (2005). Mapping of ice layer extent and snow accumulation in the percolation zone of the Greenland ice sheet. *Journal of Geophysical Research – Earth Surface* **110**(F2), doi:10.1029/2004JF000234.

Nghiem, S.V., D.K. Hall, T.L. Mote, M. Tedesco, M.R. Albert, K. Keegan, C.A. Shuman, N.E. DiGirolamo & G. Neumann (2012). The extreme melt across the Greenland ice sheet in 2012. *Geophysical Research Letters* **39**, L20502.

Ohmura, A. & Reeh, N. (1991). New Precipitation and accumulation maps for Greenland. *Journal of Glaciology* **37**(125), 140–148.

Oveisgharan, S. & Zebker, H. (2007). Estimating snow accumulation from InSAR correlation observations. *IEEE Transactions on Geoscience and Remote Sensing* **45**(1), 10–20, doi: 10.1109/TGRS.2006.886196.

Palm, S., Yang, Y., Spinhirne, J. & Marshak, A. (2011). Satellite remote sensing of blowing snow properties over Antarctica. *Journal of Geophysical Research – Atmospheres* **116**, doi:10.1029/2011JD015828.

Parkinson, C., Comiso, J. & Zwally, H.J. (1999). *Nimbus-5 ESMR Daily Polar Gridded Brightness Temperatures.* Boulder, Colorado, USA, National Snow and Ice Data Center, Digital media.

Picard, G., Brucker, L., Fily, M., Gallée, H. & Krinner, G. (2009). Modeling time-series of microwave brightness temperature in Antarctica. *Journal of Glaciology* **55**(191), 537–551.

Picard, G., Brucker, L., Roy, A., Dupont, F., Fily, M., Royer, A. & Harlow, C. (2013). Simulation of the microwave emission of multi-layered snowpacks using the Dense Media Radiative transfer theory: the DMRT-ML model. *Geoscientific Model Development* **6**, 1061–1078, doi:10.5194/gmd-6-1061-2013.

Pritchard, H., Arthern, R., Vaughan, D. & Edwards, L. (2009). Extensive dynamic thinning on the margins of the Greenland and Antarctic ice sheets. *Nature* **461**(7266), 971–975, doi: 10.1038/nature08471.

Richardson, C., Aarholt, E., Hamran, S.-E., Holmlund, P. & Isaksson, E. (1997). Spatial distribution of snow in western Dronning Maud Land, East Antarctica, mapped by a ground-based snow radar. *Journal of Geophysical Research – Solid Earth* **102**(B9), 20343–20353.

Rignot, E., Mouginot, J. & Scheuchl, B. (2011). Ice Flow of the Antarctic Ice Sheet. *Science* **333**(6048), 1427–1430, doi: 10.1126/science.1208336.

Rink, T., Kanagaratnam, P., Braaten, D., Akins, T., Gogineni, S. & IEEE (2006). *A Wideband Radar for Mapping Near-Surface Layers in Snow.* Paper presented at IEEE International Geoscience and Remote Sensing Symposium (IGARSS), Denver, CO, Jul 31–Aug 04.

Spikes, V., Hamilton, G., Arcone, S., Kaspari, S., Mayewski, P. & Jacka, J. (2004). Variability in accumulation rates from GPR profiling on the West Antarctic plateau. *Annals of Glaciology* **39**, 238–244, doi: 10.3189/172756404781814393.

Surdyk, S. (2002). Using microwave brightness temperature to detect short-term surface air temperature changes in Antarctica, An analytical approach. *Remote Sensing of Environment* **80**(2), 256–271.

Tedesco, M., Fettweis, X., van den Broeke, M., *et al.* (2011). The role of albedo and accumulation in the 2010 melting record in Greenland. *Environmental Research Letters* **6**(1), doi:10.1088/1748-9326/6/1/014005.

Tsang, L. & Kong, J.A. (2001). *Scattering of Electromagnetic Waves, Vol. 3, Advanced Topics.* Wiley Interscience, 413 pages.

Ulaby, F., Moore, R. & Fung, A. (1981). *Microwave Remote Sensing Active and Passive, Vol. 2, Microwave Remote Sensing Fundamentals and Radiometry.* Addison-Wesley Publishing Company, 456 pages.

van de Berg, W., van den Broeke, M., Reijmer, C. & van Meijgaard, E. (2006). Reassessment of the Antarctic surface mass balance using calibrated output of a regional atmospheric climate model. *Journal of Geophysical Research – Atmospheres* **111**(D11), doi:10.1029/2005JD006495.

van den Broeke, M., Bamber, J., Ettema, J., *et al.* (2009). Partitioning Recent Greenland Mass Loss. *Science* **326**(5955), 984–986, doi: 10.1126/science.1178176.

Vaughan, D.G., Bamber, J.L., Giovinetto, M., Russell, J. & Cooper, A.P.R. (1999). Reassessment of net surface mass balance in Antarctica. *Journal of Climate* **12**(4), 933–946.

Velicogna, I. & Wahr, J. (2006). Measurements of time-variable gravity show mass loss in Antarctica. *Science* **311**(5768), 1754–1756, doi: 10.1126/science.1123785.

Wiesmann, A. & Matzler, C. (1999). Microwave emission model of layered snowpacks. *Remote Sensing of Environment* **70**(3), 307–316, doi: 10.1016/S0034-4257(99)00046-2.

Winebrenner, D., Steig, E., Schneider, D. & Jacka, J. (2004). Temporal co-variation of surface and microwave brightness temperatures in Antarctica, with implications for the observation of surface temperature variability using satellite data. *Annals of Glaciology* **39**, 346–350, doi: 10.3189/172756404781813952.

Winebrenner, D.P., Arthern, R.J. & Shuman, C.A. (2001). Mapping Greenland accumulation rates using observations of thermal emission at 4.5-cm wavelength. *Journal of Geophysical Research – Atmospheres* **106**(D24), 33919–33934.

Wismann, V., Winebrenner, D., Boehnke, K., Arthern, R. & IEEE (1997). *Snow accumulation on Greenland estimated from ERS scatterometer data.*

Igarss '97 – 1997 International Geoscience and Remote Sensing Symposium, Proceedings Vols I – Iv, 1823 – 1825.

Zwally, H., Giovinetto, M. & Rothrock, D. (1995). Accumulation in Antarctica and Greenland derived from passive-microwave data, A comparison with contoured compilations. *Annals of Glaciology* **21**, 123 – 130.

Zwally, H., Giovinetto, M. & Steffen, K. (2000). Spatial distribution of net surface mass balance on Greenland. *Annals of Glaciology* **31**, 126 – 132, doi: 10.3189/172756400781820318.

Zwally, J. (1977). Microwave emissivity and accumulation rate of polar firn, *Journal of Glaciology* **18**(79), 195 – 214.

Acronyms

ACT-10	Arctic Circle Traverse 2010
ADEOS-1	Advanced Earth Observing Satellite
AIS	Antarctica ice sheet
AMSR-E and AMSR2	Advanced Microwave Scanning Radiometer
ASAR	Advanced Synthetic Aperture Radar
AVHRR	Advanced Very High Resolution Radiometer
CALIOP	Cloud-Aerosol LIdar with Orthogonal Polarization
CONAE	Argentina's Comisión Nacional de Actividades Espaciales
CReSIS	University of Kansas' Center for Remote Sensing of Ice Sheets
dB	Decibel
DMRT	Dense Media Radiative Transfer
ERS-1	European Remote Sensing Satellite
ERS-2	European Remote Sensing Satellite
ESA	European Space Agency
FMCW	Frequency-modulated continuous wave
GCM	Global climate model
GPM	Global Precipitation Mission
GPR	Ground penetrating radar
GRACE	Gravity Recovery and Climate Experiment satellite
GrIS	Greenland ice sheet
ICESat	Ice Cloud and land Elevation Satellite
InSAR	Interferometric Synthetic Aperture Radar
ITASE	International Trans Antarctic Scientific Expedition
MODIS	Moderate Resolution Imaging Spectroradiometer
NASA	National Aeronautic and Space Administration
NSCAT	National Aeronautic and Space Administration scatterometer
OIB	Operation Ice Bridge

PARCA	National Aeronautic and Space Administration's Program for Regional Climate Assessment
QuikSCAT	Quick scatterometer
SAR	Synthetic Aperture Radar
SASS	Satellite-borne scatterometer
SEAT	Satellite Era Accumulation Traverse
SMB	Surface mass balance
SMOS	Soil Moisture and Ocean Salinity
SSM/I and SSMIS	Special Sensor Microwave Imager Sounder
UHF	Ultra/super-high-frequency
WAIS	West Antarctic Ice Sheet

Website cited

http://www.antarctica.ac.uk//bas_research/data/online_resources/snow_accumulation/ (British Antarctic Survey)

9 Remote sensing of ice thickness and surface velocity

Prasad Gogineni and Jie-Bang Yan

Center for Remote Sensing of Ice Sheets, University of Kansas, Lawrence, USA

Summary

This chapter provides an introduction to radar sounding and imaging of the polar ice sheets and the application of Interferometric Synthetic Aperture Radar (InSAR) for ice dynamics and topography measurements. It begins with a brief review of the electrical properties of ice that are important to the design of radars for sounding ice and interpreting results. We then develop expressions from the radar equation for received radar signals from various types of targets. We show that radar sensitivity is directly proportional to the power-aperture product, which is determined by the combined performance of power amplifiers, the antenna array, and signal processing. We also discuss the trade-offs between radar resolution and the signal-to-noise ratio, and the approaches used to improve the signal-to-noise ratio via hardware and software techniques. We explain the use of Synthetic Aperture Radar (SAR) processing and pulse compression in a radar for improving along-track resolution and obtaining the high sensitivity required to sound ice around ice-sheet margins. We also provide a brief overview of antenna arrays and their use on both short- and long-range aircraft.

In addition to the theoretical aspects, we show several examples of actual data collected over the interior ice sheet that contain surface and bed echoes, as well as reflections from internal layers. We also provide results over outlet glaciers for illustrating the challenges associated with sounding fast-flowing glaciers such as Jakobshavn.

Finally, we provide a brief overview of the principle of InSAR operation, the steps involved in generating InSAR images, and measurement errors. We have also referenced a few examples to show that InSAR systems have previously been used to collect accurate information on ice sheet topography and ice flow dynamics.

9.1 Introduction

The application of radars to the sounding of ice sheets began with the pioneering work of Mr. Amory Waite, which led to the first successful demonstration

Remote Sensing of the Cryosphere, First Edition. Edited by M. Tedesco.
© 2015 John Wiley & Sons, Ltd. Published 2015 by John Wiley & Sons, Ltd.
Companion Website: www.wiley.com/go/tedesco/cryosphere

in the late 1950s. Waite was also credited with conducting the first airborne ice measurements in the early 1960s. Since then, the application of radars to glaciology has expanded substantially and radar has become an invaluable tool in the study of ice sheets and glaciers. Nowadays, radars with the capability to sound ice and image the ice bed are being operated on a variety of aircraft (Leuschen *et al.*, 2010; Gogineni *et al.*, 2012).

In addition, satellite-borne Synthetic Aperture Radars (SARs) have made a significant impact on the field of glaciology (Jezek *et al.*, 2003; Rignot and Kanagaratnam, 2006; Joughin *et al.*, 2010). SAR has been used to map the melt zones of the Greenland ice sheet (Fahnestock *et al.*, 1993), surface topography, and surface velocity (Kwok and Fahnestock, 1996). Furthermore, Canadian RADARSAT, in collaboration with NASA, was used to map the Antarctic continent over a period of three months. This map will serve as a baseline to document changes in the ice sheet's retreat and growth. Data collected during the RADARSAT mapping mission were also used to produce surface velocities of key glaciers. Recently, a complete surface velocity map of the Antarctic ice sheet was produced by Rignot *et al.* (2011). Radars have played, and will continue to play, a major role in the study of the polar ice sheets.

This chapter gives a brief introduction to radar sounding and imaging of the polar ice sheets and Interferometric Synthetic Aperture Radar (InSAR). It provides a concise summary of the principle of operation of a typical radar sounder, discusses the concepts of pulse compression and advanced radars used to sound and image ice, and presents sample results from different ice-sheet areas. It also includes a brief overview of radars used to map near-surface internal layers in polar firn. Finally, it discusses the principle of InSAR operation and its application to ice sheets.

9.1.1 Electrical properties of glacial ice

Ice thickness is determined by measuring the time delay associated with the propagation of an electromagnetic pulse through the ice. Electromagnetic waves are absorbed, reflected and scattered while propagating through the ice to its bed. Normally, both scattering at the frequencies used to sound and image ice, and energy lost through reflections caused by internal ice layers, are negligible. Thus, most of the radar signal attenuation is caused by absorption. The electromagnetic wave propagating through the ice at depth z can be expressed as:

$$E(z) = E_0 e^{-\gamma z} \tag{9.1}$$

where:

E_0 = electric field intensity at the surface in V/m
γ = propagation constant = $\alpha + j\beta$, where $j = \sqrt{-1}$
α = attenuation constant in nepers/m

β = phase constant in radians

z = depth below the surface

$E(z)$ = electric field at z.

Glacial ice is a mixture of pure ice and air and a small amount of impurities, including aerosols and ash. It is formed from the annual accumulation of snow that is compressed by the weight of overlying snow. Dust, ash and other trace gases present in the atmosphere at the time of snowfall are incorporated into the ice. Glacial ice can be considered a low-loss dielectric at certain frequencies (1 to 1000 MHz) at which radars for sounding and imaging ice sheets and mapping internal layers are operated. The presence of these impurities causes small changes in the dielectric properties of the ice. In general, the real part of the dielectric constant of cold ice remains nearly constant at about 3.15, and the imaginary part varies with the acidity and dust content of the ice. The imaginary part is also a strong function of temperature. The electric properties of ice are documented by Evans (1963) and Fujita *et al.* (2000).

The real and imaginary parts of the propagation constant can be expressed in terms of the material electrical and magnetic properties as

$$\gamma = \alpha + j\beta = j\omega\sqrt{\mu\varepsilon} = j\omega\sqrt{(\mu\varepsilon'(1 - j\varepsilon''/\varepsilon'))} \tag{9.2}$$

For low-loss dielectric, we can approximate and simplify Equation 9.2 as:

$$\approx j\omega\sqrt{(\mu\varepsilon')}(1 - j\varepsilon''/[(2\varepsilon)]') \tag{9.3}$$

Equating real and imaginary parts on both sides, we can determine attenuation and propagation constants:

$$\alpha = \omega * \varepsilon''/2 * \sqrt{(\mu/\varepsilon')} \tag{9.4}$$

$$\beta = \omega\sqrt{\mu\varepsilon'} \tag{9.5}$$

where:

μ is the permeability of the medium

ε is the permittivity of the medium

$\omega = 2\pi f$ is the frequency in radians.

The imaginary part of the dielectric constant of pure ice as a function of temperature was measured and documented by Fujita *et al.* in 2000. The imaginary part of the dielectric constant decreases is a function of frequency, with broad minima around 1 GHz, that depends on temperature. In the absence of scattering, attenuation as a function of frequency remains fairly constant. However, additional losses resulting from acidic and other impurities will increase the attenuation as a function of frequency. Thus, radars operating at frequencies higher than about 250 MHz must be designed to overcome higher losses.

9.2 Radar principles

9.2.1 Radar sounder

The radar sounder transmits a high-power electromagnetic pulse toward the ice, as illustrated in Figure 9.1(a). The radar measures the time delay between the transmission of the pulse and the received echoes from the ice surface and ice bed. The time difference between the two received echoes is converted into ice thickness as:

$$\partial t = \tau_2 - \tau_1 = \frac{2R_2}{c} - \frac{2R_1}{c} \tag{9.6}$$

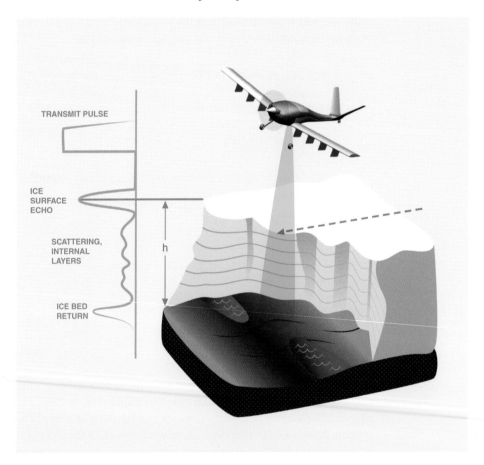

TRANSMIT PULSE

ICE SURFACE ECHO

SCATTERING, INTERNAL LAYERS

ICE BED RETURN

h

Figure 9.1 (a) Measurement of time delays to the ice surface and ice bed. (b) Radar echogram generated by combining lines of radar data collected along a flight line.

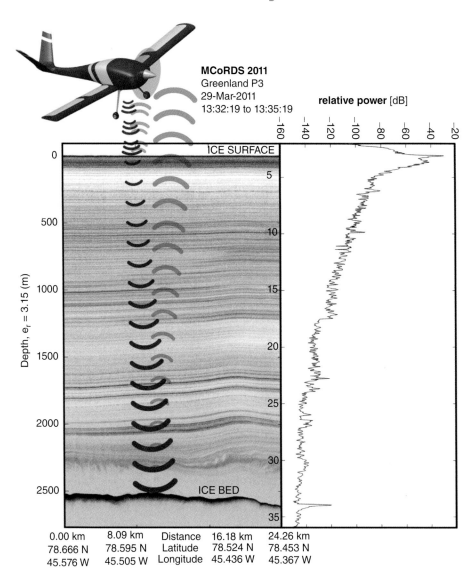

MCoRDS 2011
Greenland P3
29-Mar-2011
13:32:19 to 13:35:19

Figure 9.1 (*continued*)

where:

R_1 = range to the ice surface from the radar antenna phase center

R_2 = electrical range to the ice-bed interface and ice thickness, $h = v_p \partial t$, where

$\quad v_p$ = velocity of wave propagation in ice

Normally, pulses are transmitted as the radar-carrying platform moves along a predetermined path, and the received signals are recorded along with geo-location information. The received signals are processed to generate an image, which is referred to as an echogram, as shown in Figure 9.1(b). The echogram in the figure

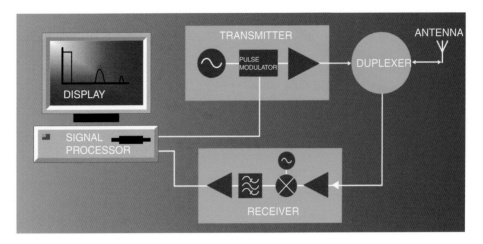

Figure 9.2 The block diagram of a typical radar sounder.

is generated from data collected over the Greenland ice sheet interior with a radar operating at 195 MHz on a NASA P-3 aircraft. The echogram shows clear echoes from the ice surface, ice bed, and internal ice layers.

Figure 9.2 shows a block diagram of a simple radar for sounding ice sheets. It consists of a transmitter, a duplexer, an antenna, a receiver, and a processor. The transmitter generates a high-power pulse. This pulse is supplied to the antenna through the duplexer, which isolates the receiver from the high-power pulse during transmission. The antenna radiates the pulse toward the target (ice) and captures reflected signals from the ice surface and ice bed. The receiver amplifies and filters the received signals. The processor digitizes and processes the received signals, first to estimate the time difference between the surface and bed echoes, and then to generate estimates of ice thickness. The early radars were primarily incoherent systems that used high peak power to overcome the loss in ice. Now, a majority of the radars being used are coherent systems with low peak power and high sensitivity, which is required to sound and image ice sheets. The high sensitivity is obtained with the application of advanced digital signal processing techniques.

9.2.2 Radar equation

The radar equation used to design and analyze a system for sounding and imaging ice depends on the type of radar, the target response, and the processing techniques employed to improve radar performance. Radar returns from the ice-bed interface of glacial ice can be treated as reflections from a planar interface and as backscattered signals from rough interfaces. The radar equations needed for these two cases are different, and the use of the wrong equation will result in a large error, as will be illustrated later. Most textbooks provide a detailed development of the radar equation. Here, we will start with the equation for a planar reflector located at a range R, and modify it for sounding ice.

Power received from a planar interface is given by

$$P_r = \frac{P_T G_T G_r \lambda^2 |\Gamma|^2}{(4\pi)^2 (2R)^2} \qquad (9.7)$$

where P_r = received power; P_T = transmitted power; G_T = transmitter antenna gain; G_R = receiver antenna gain; and Γ = reflection coefficient of the planar interface.

We can modify Equation 9.7 for ice sounding by accounting for the two-way transmission loss through the air-firn interface and assuming that the aircraft is flying at a height of h_a meters above ice with a thickness of h_i meters:

$$P_r = \frac{P_T G_T A_e (1 - |\Gamma_{af}|^2) |\Gamma_b|^2}{4\pi (2h)^2 L_{is}} \qquad (9.8)$$

where:

Γ_{af} = reflection coefficient at the air-snow interface
Γ_b = reflection coefficient at the ice bed interface
L_{is} = loss through the ice
$h = h_a + h_i$.

The receive-antenna gain is related to its effective aperture as:

$$G_R = \frac{4\pi A_e}{\lambda^2} \qquad (9.9)$$

The received signal must be higher than the receiver noise for reliable detection. It is useful to express the power received in terms of a signal-to-noise ratio. When the receiver performance is limited by thermal noise, the noise power can be expressed as:

$$P_n = KTBF \qquad (9.10)$$

where:

K = Boltzmann constant
T = receiver temperature
B = receiver bandwidth
F = receiver noise figure.

$$\left(\frac{S}{N}\right) = \frac{P_r}{P_n} = \frac{P_T G_T A_e (1 - |\Gamma|^2)^2 |\Gamma_b|^2}{4\pi (2h)^2 L_{is} KTBF} \qquad (9.11)$$

The above equation shows that radar performance is directly proportional to the power-aperture product and inversely proportional to the receiver bandwidth. To detect weak echoes from ice, we must have a high-power aperture product or a narrow bandwidth. There is a trade-off between resolution and signal-to-noise (S/N) ratio improvement. The range resolution is also inversely proportional to the bandwidth and is given by:

$$\partial r = \frac{c}{2B} \tag{9.12}$$

Thus, decreasing bandwidth to improve the S/N ratio results in poor resolution. The signal-to-noise ratio can also be increased by a factor of N for coherent integration of N pulses, whereas it only increases by the square root of N for incoherent integration (Skolnik, 1981). The difference in processing gain between coherent and incoherent integration is not large for a small number of pulses, but becomes significant for large values of N. Equation 9.11 can be rewritten for a coherent radar with an integration gain of CN as:

$$\left(\frac{S}{N}\right)_N = \frac{P_r}{P_n} = \frac{P_T G_T A_e (1 - |\Gamma_{af}|^2)^2 |\Gamma_b|^2 C_N}{4\pi (2h)^2 L_{is} KTBF} \tag{9.13}$$

Bandwidth is approximately equal to the inverse of the pulse width. We can substitute for B in equation 9.13 to express S/N ratio in terms of the energy in the pulse.

$$\left(\frac{S}{N}\right)_N = \frac{P_r}{P_n} = \frac{P_T T_p G_T A_e (1 - |\Gamma_{af}|^2)^2 |\Gamma_b|^2 C_N}{4\pi (2h)^2 L_{is} KTF} \tag{9.14}$$

The above equation demonstrates that pulse energy must be large to obtain a large signal-to-noise ratio.

As a first example, let us compute the S/N ratio for a coherent radar with the following parameters: transmit power, $P_T = 200$ W; transmit and receive antenna gain = 30; receiver bandwidth = 20 MHz; receiver noise figure = 2; Receiver temperature = 290 K; reflection coefficient at the interface = 0.1; wavelength = 2 m, ice thickness = 300 m; height of the aircraft above the ice surface = 500 m; two-way ice loss = 20 dB/km; and the number of pulses integrated, $N = 100$. Effective receiver antenna aperture can be determined using equation 9.9:

$$A_e = \frac{G_r \lambda^2}{4\pi} = \frac{30 \times 2^2}{4\pi} = 9.55 \text{ m}^2$$

We can compute the transmission, assuming the dielectric constant of firn as 2:

$$T_p = (1 - |\Gamma_{af}|^2)^2 = \left(1 - \left|\frac{\sqrt{\varepsilon_{af}} - 1}{\sqrt{\varepsilon_{af}} + 1}\right|^2\right)^2 = \left(1 - \left|\frac{\sqrt{2} - 1}{\sqrt{2} + 1}\right|^2\right)^2 = 0.94$$

We can compute noise power using equation 9.10:

$$P_n = 1.38 \times 10^{-23} \times 290 \times 20.16^6 \times 2 = 1.6 \times 10^{-13} \quad \text{W} = -97.96 \text{ dBm}$$

Substituting the above values into equation 9.11 gives a S/N of 27.4 dB. The S/N ratio will be about 17.4 dB for the same number of pulses.

For sounding thicker, more lossy ice, we can further increase radar sensitivity by increasing peak power. However, the use of very high peak power pulses is not desirable, due to cable and power supply limitations, particularly for airborne applications. One effective way to increase peak power is through the use of pulse compression, in which a long coded pulse is transmitted and a received pulse is decoded to generate a short pulse. This method of detecting weak targets can be accomplished without sacrificing resolution. We will discuss the concept of pulse compression and issue range sidelobes in the next section.

The equation for a pulse-compression radar must be modified as:

$$\left(\frac{S}{N}\right)_N = \frac{P_T G_T A_e (1 - |\Gamma_{af}|^2)^2 |\Gamma_b|^2 C_N C_{pc}}{4\pi(2h)^2 L_{is} KTBF} \tag{9.15}$$

where C_{pc} = pulse compression gain.

$$C_{pc} = k\frac{\tau_u}{\tau_c} \tag{9.16}$$

where k is a constant less than 1 to account for weighting to reduce range sidelobes, and τ_u and τ_c are uncompressed and compressed pulse widths, respectively.

With $k = 0.5$, the compression gain when a 10 μs pulse is compressed to a 50 ns pulse is 100. Thus the S/N ratio is larger than that obtained with coherent integration in the earlier example. Let us substitute for C_{pc} in equation (9.15):

$$\left(\frac{S}{N}\right)_N = \frac{P_T G_T A_e (1 - |\Gamma_{af}|^2)^2 |\Gamma_b|^2 C_N k\frac{\tau_u}{\tau_c}}{4\pi(2h)^2 L_{is} KTBF} \tag{9.17}$$

$B\tau_c \approx 1$, thus $\left(\frac{S}{N}\right)_N = \dfrac{P_T G_T A_e (1 - |\Gamma_{af}|^2)^2 |\Gamma_b|^2 C_N k\tau_u}{4\pi(2h)^2 L_{is} KTBF}$ (9.18)

Equation 9.18 shows that there is no improvement in the signal-to-noise ratio by decreasing the bandwidth of a pulse compression radar for a planar interface.

Now let us consider the case of a distributed target with a radar. The use of a distributed radar equation is important for designing systems operating as unfocused or focused synthetic aperture radar (SAR) to improve along-track resolution. We will start the discussion with the radar equation for a point target (Skolnik, 1981).

Power received from a point target located at range R is given by:

$$P_r = \frac{P_T G_T G_R \lambda^2 \sigma}{(4\pi)^3 R^4} \tag{9.19}$$

where σ is the radar cross-section of the target in m^2.

A distributed target consists of a number of randomly distributed scatterers in the illuminated spot and is characterized by a radar cross-section per unit area as:

$$\sigma = \sigma^0 A \tag{9.20}$$

A radar depth sounder is a radar altimeter for the ice-bed interface and the illuminated area:

$$A = \pi c \tau_{ch} \tag{9.21}$$

Substituting equations 9.20 and 9.21 into 9.19 and accounting for air-firn transmission loss and ice attenuation, the power received from the ice bed is given by:

$$P_r = \frac{P_T G_T G_R \lambda^2 \pi c \tau_c h (1 - |\Gamma_{af}|^2)^2 \sigma^0}{(4\pi)^3 h^4 L_{is}} \tag{9.22}$$

$$= \frac{P_T G_T G_R \lambda^2 \pi c \tau_c (1 - |\Gamma_{af}|^2)^2 \sigma^0}{(4\pi)^3 h^3 L_{is}} \tag{9.23}$$

The signal-to-noise ratio, including pulse compression gain and coherent integration gain, is:

$$\left(\frac{S}{N}\right)_N = \frac{P_T G_T G_R \lambda^2 \pi c \tau_c (1 - |\Gamma_{af}|^2)^2 \sigma^0 C_N k \tau_u}{(4\pi)^3 h^3 L_{is} KTBF} \tag{9.24}$$

As a second example, let us compute the signal-to-noise ratio for a distributed target with the same parameters as those used in the earlier example. Let us assume a root mean square (rms) height of 20 cm for the ice bed with a slope of 1°. The radar back-scattering coefficient for the ice-bed interface under a geometric optics approximation is given by:

$$\sigma^0(\theta) = \frac{|\Gamma(0)|^2 e - \dfrac{\tan^2(\theta)}{2s^2}}{2s^2 \cos^4(\theta)} \tag{9.25}$$

Using the above equation (9.25), we can determine the backscatter coefficient at normal incidence angle as

$$\sigma^0(0) = 12.1 \text{ dB/m}^2$$

$$\text{S/N} = 29.8 \text{ dB}.$$

The signal-to-noise ratio for a distributed target is about 19 dB lower than that for a planar reflector. This example illustrates the need to consider bed characteristics in designing a radar to sound different areas of an ice sheet. This is also important when analyzing and interpreting radar data, particularly for determining bed conditions.

9.3 Pulse compression

As mentioned in the previous section, radar sensitivity for detecting weak targets can be enhanced by increasing pulse energy. This can be accomplished by increasing the peak power or using a longer pulse. Increasing peak power can cause dielectric breakdown of cables carrying the pulse to the transmit antenna, and this will also require power supplies capable of supplying large peak currents. Because resolution is directly proportional to pulse width, increasing pulse width degrades resolution. Pulse compression is a technique used to transmit a long pulse to obtain the high energy needed to detect weak targets, and to compress the pulse on reception so that a fine resolution can be achieved.

The theory and design of chirp radars was discussed in a paper in open literature as early as 1960 (Klauder et al., 1960), and the pulse-compression technique is now widely used in long-range radars. Most of the advanced airborne and spaceborne radar sounders use pulse compression to obtain the sensitivity required to detect weak echoes from sub-surface targets (Li et al., 2012; Watters et al., 2006). Here, we provide only a brief introduction to pulse compression using a linearly-chirped signal. A more complete discussion of pulse compression using chirped and coded waveforms, including design and analysis, is available in several classical textbooks on radars (Skolnik, 1981; Stimson, 1998; Rihaczek and Hershkowitz, 2000).

In a pulse-compression radar, a long chirped or coded pulse is transmitted, and a received signal is decoded or "dechirped" to generate a short pulse. The main disadvantage of pulse compression is the range sidelobes generated during the dechirping process. The range sidelobes exist for a time duration equal to the uncompressed pulse width on both sides of the main target peak. The range sidelobes of a strong return from a large target can mask the weak echoes of nearby weak targets. Most of the recent research on pulse compression techniques focuses on designing and developing hardware and algorithms to obtain low range sidelobes.

To understand the concept of pulse compression and its application to radar sounders/imagers, we will consider a chirped signal. The transmitter frequency is increased linearly as a function of time over the bandwidth needed to obtain fine resolution, as shown in the upper two sub-plots in Figure 9.3.

The received signal is passed through a dispersive filter. The delay decreases as a function of frequency, as shown on the bottom left corner. The lower frequencies transmitted earlier are delayed longer than the higher frequencies. This results in a compressed pulse with a $\sin x / x$-type time or range response, as shown on the lower right side of the figure. The main peak is located at about zero relative time, and the first range sidelobes are located at relative times of about ± 0.5. The amplitude of these sidelobes is only about 13 dB below the main peak. The spectral response of the transmit waveform or received signal can be shaped using weighting functions to reduce sidelobes. Table 9.1 provides a summary of a few typical weighting functions.

The use of weighting to reduce sidelobes results in broadening of the main lobe and the loss of resolution. Thus, choosing a weighting function is a trade-off between resolution and sidelobes. A Dolph-Chebyshev window provides the best possible resolution for a given sidelobe level. The application of a suitable window to shape the amplitude of the chirp signal results in low range sidelobes near the main lobe. However, the presence of Fresnel ripples causes large sidelobes to appear on both sides of the main lobe at $\pm \frac{\tau_u}{2}$, which limits the fall-off rate of

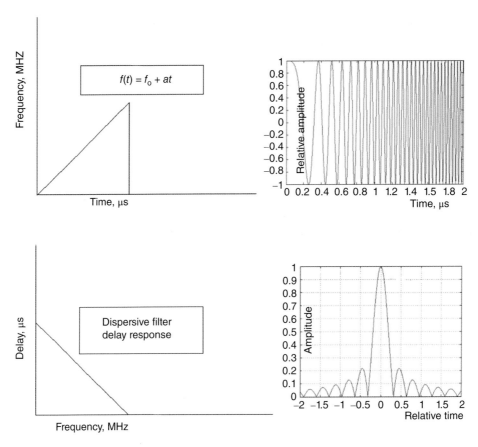

Figure 9.3 Illustration of the concept of pulse compression with a linearly chirped signal. (a) The instantaneous frequency of a chirp increases with time. (b) A time-domain plot of a chirp signal. (c) Response of a dispersive filter. (d) Compressed signal in time-domain.

Table 9.1 Comparison of a few windows used to reduce range sidelobes.

Window	Sidelobe level (dB)	Resolution	S/N loss dB	Fall-off rate dB/octave	Comments
Rectangular	13	0.88	0	6	Reference
Hanning	32	1.44	1.4	18	
Hamming	43	1.30	1.34	6	
Tukey, taper = 0.1					
Flat top	44	2.94		6	
Dolph-Chebyshev	41	1.17	1.2	0	
Blackman	58	1.68		18	

distant sidelobes. The Fresnel ripples can be reduced by tapering the start and end of the time-domain chirp signal (Kowatsch and Stocker, 1982). Misaridis and Jensen (2005) also investigated the application of amplitude tapering of the transmit waveform, coupled with windowing of the received signal for an ultrasound imaging system to obtain extremely low range sidelobes.

The Multichannel Coherent Radar Depth Sounder/Imager, described in Section 9.5, developed by the Center for Remote Sensing of Ice Sheets (CReSIS) for sounding and imaging ice sheets, adapts the above approach and employs amplitude-tapered transmit waveforms with a 10% Tukey window. Figure 9.4 shows a time-domain transmit-chirp for a 10 μs waveform used to sound thick ice and map internal layers at depth. The time-domain signal is passed through a delay line with a delay of 36 μs to simulate a point target at a range of 18 μs into the receiver. The received signal's spectrum is modified using a Blackman window, and it is compressed by multiplying it with a conjugated reference chirp signal. The resulting compressed time-domain response shows nearby sidelobes about 60 dB below the main-lobe peak, and distant sidelobes are reduced by more than 80 dB (Li, 2009).

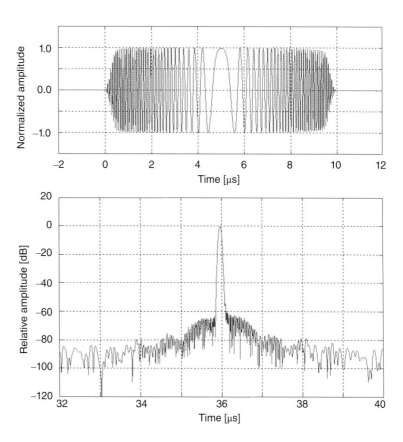

Figure 9.4 MCoRDS/I time-domain waveform (top) and impulse response of the system simulated using a delay line (bottom).

9.4 Antennas

The antennas are an integral part of a radar sounder/imager. The careful design and optimization of antennas is essential to achieving the high sensitivity needed to sound ice over the most challenging areas of the ice sheets – fast-flowing glaciers and ice-sheet margins. An antenna is a transducer between guided waves, in transmission lines or waveguides, and waves propagating in free-space or media, such as ice.

Antennas can be classified into four broad categories: wire, aperture, slot, and planar. The dipole, loops, and modifications of these are examples of wire antennas. The dipole and an array of dipoles are the primary antennas used for sounding and imaging ice with radars operating in the VHF (Very High Frequency) part of the spectrum. Aperture antennas, such as Transverse ElectroMagnetic (TEM) horns and slotted-antennas, are used for radars operating in the microwave part of the spectrum. These radars are used to map near-surface internal layers and sound shallow low-loss ice. The purpose of this section is to provide a brief introduction to a dipole antenna and a linear array. A more extensive discussion on the theory and application of different types of antennas is available in Stutzman and Thiele (1998) and Balanis (2005).

An antenna is characterized by a few basic parameters: its directivity and gain, input impedance and efficiency. The directivity of an antenna is the ratio of maximum radiation intensity in a particular direction to the average radiation intensity of an isotropic radiator. The gain is a product of efficiency and directivity. Input impedance consists of both resistive and reactive components. The real part consists of the loss resistance and the radiation resistance which is a hypothetical resistance that dissipates the same amount of power as is radiated by the antenna. The radiation resistance must be as high as possible compared to the loss resistance, which accounts for Ohmic losses in the conductors used to build the antenna. For efficient power transfer, the radiation resistance must be close to the characteristic impedance of the transmission line feeding the antenna. The antenna pattern shows spatial distribution of radiated energy and is used to determine antenna beamwidths. The efficiency, which characterizes how effective the antenna is in converting a guided wave into a propagating wave, is the ratio of power radiated by the antenna to the total power supplied to it.

Let us consider a half-wavelength dipole oriented along the z-axis to illustrate antenna parameters. The magnetic vector potential at this point can be computed from the current distribution on the wire. The retarded magnetic vector potential and current distribution are given as:

$$A = \int_{-l/2}^{l/2} \frac{\mu[I]}{4\pi r_1} dz \tag{9.26}$$

where:

$r_1 = r - z \cos(\theta)$

$[I] = I_0 e^{-j\beta r_1}$

A is the vector potential at r_1 from a current-carrying wire of length l.

$$I(z) = I_0 \sin\left[\beta\left(\frac{l}{2} - z\right)\right] \quad \text{for } z > 0 \tag{9.27}$$

$$I(z) = I_0 \sin\left[\beta\left(\frac{l}{2} + z\right)\right] \quad \text{for } z < 0 \tag{9.28}$$

Substituting equations 9.27 and 9.28 into 9.26 and carrying out the integration, we can obtain the z-component of A_z as:

$$A_z \approx \frac{\mu I_0 e^{j*\beta*r}}{2\pi r} \frac{\cos\left(\frac{\pi}{2}\cos(\theta)\right)}{\sin(\theta)} \tag{9.29}$$

Using the relationship, we can determine the electric, E and magnetic, H fields as:

$$H_\varphi \approx \frac{I_0 e^{j*\beta*r}}{2\pi r} \frac{\cos\left(\frac{\pi}{2}\cos(\theta)\right)}{\sin(\theta)} \tag{9.30}$$

$$E_r = \eta_0 H_\phi \tag{9.31}$$

The orientation of the dipole is critical for reducing reflections from the sidewalls of the glacier during airborne measurements. The orientation of the dipole axis perpendicular to the flight direction provides nulls at $\pm 90°$. However, the gain reduction provided by a single dipole might not be sufficient to reduce strong reflections from a glacier's sidewalls, as discussed in the section on array and SAR processing.

We can compute power density at point r using the Poynting vector:

$$P = \frac{1}{2}\text{Re}(E \times H) \tag{9.32}$$

Next we can determine average power flowing through cross-sectional areas, given by

$$P_{av} = \oiint P \cdot ds = \frac{\eta_0 I_0^2}{4\pi} \int_0^{\frac{\pi}{2}} \frac{\left[\cos\left(\frac{\pi}{2}\cos(\theta)\right)\right]^2}{\sin(\theta)} d\theta \approx 365 I_0^2 \tag{9.33}$$

$$P_{av} = \frac{1}{2}I_0^2 R_{rad} \tag{9.34}$$

Using equations 9.33 and 9.34, we can determine the radiation resistance $R_{rad} =$ 73 Ohms for a half-wave dipole.

We can also compute the directivity and effective antenna aperture for a half-wave dipole.

$$\text{Directivity, } D = \frac{\text{Maximum radiation intensity}}{\text{Average radiation intensity}} = \frac{\dfrac{\eta_0 I_0^2}{4\pi}}{\dfrac{365 I_0^2}{4\pi r^2}} = 1.64 \tag{9.35}$$

Assuming 100% efficiency, we can compute the effective aperture size of a half-wave dipole as:

$$A_e = \frac{1.64\lambda^2}{4\pi} \text{ m}^2 \tag{9.36}$$

Thus, for a half-wave dipole operating at a frequency of 195 MHz, the effective aperture size is $A_e = 0.31$ m^2. The directivity can be increased by a factor of four by placing the ground plane at a distance of a quarter-wave behind the antenna, which is typical of wing-mounted dipole antennas.

Antenna arrays are used to improve directivity and radar sensitivity to detect long-range weak targets, and to apply advanced signal processing techniques to reduce the surface clutter that can mask weak echoes from the ice bed of fast-flowing glaciers. An antenna array is a geometrical arrangement of elemental radiators to increase antenna gain and perform electronic steering of its beam. Antenna arrays can take many shapes, such as linear, planar, circular, and conformal. For most sounding radars, the antenna array is in the form of a linear array mounted under the fuselage or wings of an aircraft. Six folded dipoles are mounted under each wing of a Twin-Otter aircraft to operate the CReSIS radar sounder/imager, as shown on the top of Figure 9.5(a). Seven modified bowtie dipoles under the fuselage, and four under each wing, are used to operate another radar sounder/imager on the NASA P-3 aircraft, shown on the bottom of Figure 9.5(c). We use a large number of cross-track antenna elements to reduce surface clutter, as will be explained in the next section. We will analyze an array and its performance factors below.

Let us consider a uniform line array of N isotropic elements arranged with a spacing of d meters between elements and with an angle between the axis of the array and the line joining the first element to the observation point P. The radiation pattern of an array is a product of the element pattern and the array factor. The array factor can be computed by summing contributions from each element with the appropriate propagation delay as:

$$A_f(\theta) = \sum_{n=0}^{N-1} e^{-jn\beta d \cos(\theta)} \tag{9.37}$$

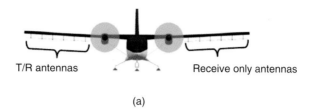

(a)

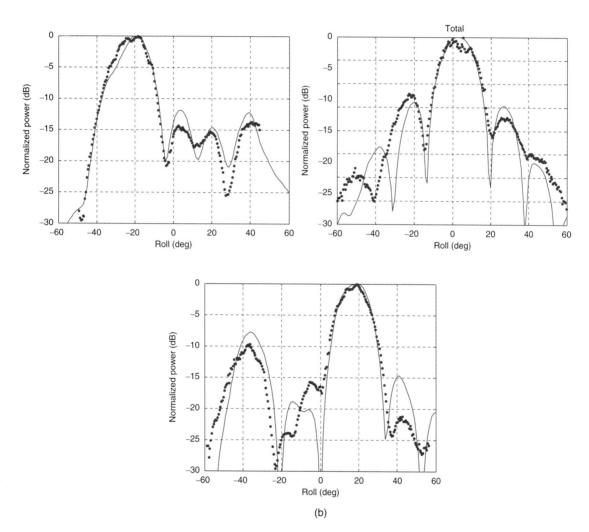

(b)

Figure 9.5 (a) An antenna array consisting of folded dipoles on the wings of the Twin Otter along with the fuselage-mounted antennas for other high resolution radars. (b) A comparison between in-flight measured and simulated radiation patterns of the Twin Otter array (red line: simulated data with Ansys HFSS®; blue dot: measured data). (c) A large antenna array of 15 elements flown on the NASA P-3 aircraft. (Yan et al., 2012. Reproduced with permission of the IEEE.)

Figure 9.5 (*continued*)

(c)

The above equation can be simplified as:

$$A_f(\theta) = \frac{1 - e^{-jN\beta d \cos(\theta)}}{1 - e^{-j\beta d \cos(\theta)}} \tag{9.38}$$

$$A_f(\theta) = \frac{\sin\left(\dfrac{N\beta d \cos(\theta)}{2}\right)}{\sin\left(\dfrac{\beta d \cos(\theta)}{2}\right)} \tag{9.39}$$

Let $x = (\beta d \cos\theta / 2)$

$$A_f(\theta) = \frac{\sin(Nx)}{\sin(x)} \tag{9.40}$$

For the uniform line-array of dipoles, a pattern is obtained by multiplying the array factor with the dipole field pattern, computed earlier as:

$$E_T(\theta) = \frac{\eta_0 I_0 e^{-j\beta r}}{2\pi r} \frac{\cos\left(\dfrac{\pi}{2}\cos(\theta)\right)}{\sin(\theta)} \frac{\sin\left(\dfrac{N\beta d \cos(\theta)}{2}\right)}{\sin\left(\dfrac{\beta d \cos(\theta)}{2}\right)} \tag{9.41}$$

The power radiated, directivity, and beamwidths can be determined using the above equation. An N-element array will have $N-2$ minor lobes, so the six-element will have four minor lobes. The directivity of the N-element dipole array increases by a factor of N, hence by a factor of six for the six-element array. The null-to-null beamwidth is approximately equal to $2\lambda/L$, where L is the length of the array. For a six-element array, it is about 38°. The simplified analysis above does not include a progressive phase shift or delay that can be applied to steer the beam in a desired direction.

Figure 9.5(b) shows a measured and simulated collinear array used by CReSIS for measurements over the Byrd Glacier in East Antarctica (Yan *et al.*, 2012). The measurements were made by flying over a relatively smooth surface while the aircraft rolled from over five or more cycles. The simulated patterns were generated using the High-Frequency Structure Simulator (Ansys HFSS®). The disagreement between the measured and simulated patterns, when the beam is steered to the left, occurs because the aircraft control surfaces that existed during the flight are not included in the simulation. However, the overall agreement between the measured and simulated results is excellent, with a mean deviation of less than 2.7 dB for 85% of the data.

9.5   Example results

Researchers at the University of Kansas have been developing coherent radars for sounding ice sheets for over a decade, and a large volume of data has been collected with these radars (Gogineni *et al.*, 1998, 2012). In this section, we will look at several examples collected with early versions of the University of Kansas coherent radar depth sounder over both the interior and the margins of the ice sheets. A detailed description of this radar is available in Gogineni *et al.* (1998) and Chuah (1997), so only a brief summary is included here. We will use the results from this system to show the need for more advanced radars with Synthetic Aperture and array processing.

The early versions of the KU radar depth sounder were developed to operate at a center frequency of 150 MHz, with a bandwidth of about 20 MHz. These radars used a Surface Acoustic Wave (SAW) expander to generate a chirped pulse of about 2 µs in duration at a pulse repetition frequency of about 9 kHz. The radars also used a SAW compressor to compress received signals to an effective pulse width of about 60 ns. These radars were operated with two four-element dipole arrays on a P-3 aircraft and two 4–5 element dipole arrays mounted under each wing of a Twin Otter aircraft. One of the arrays was used for transmission and the other for reception. The received data were digitized with two-channel 8-bit A/D converters during the early phase and with 12-bit A/D converters later. To keep the data rate low, 32–256 pulses were pre-summed in hardware before storing them on a disc for further processing. Data were further integrated over one Fresnel zone and detected to generate radar echograms.

Figure 9.6(b) shows a radar echogram over the interior ice. The segment of the flight line for which data are shown is highlighted in green in Figure 9.6(a). The ice surface return is the first dark line at about the zero range. The signals above this line are the feed-through signal between the transmit and receive antenna and the range sidelobes of the compressed signal. The early versions of the radar were operated with an analog sensitivity time control (STC) circuit set to reduce strong echoes from the surface. This caused weak signals from internal layers below about 400 m to be masked.

The echogram shows strong and clear echoes from internal layers between 500 m and 1500 m. The reflections from the internal layers are the result of dielectric discontinuities caused by changes in density, conductivity and crystal fabric orientation. The density changes are reported to be the main source of reflections in the upper few hundred meters of ice. Conductivity changes are the primary source of reflections in the middle, and crystal fabric orientations are the primary source near the bed (Fujita *et al.*, 2000).

There is a range window between 1600–1800 m, with no or very weak echoes. This zone of weak echoes is reported to contain ice from the last glacial maximum. Echoes from a few internal layers start appearing at about 1800 m, and these are reported to be related to Dansgaard-Oeschger events, which are rapid climate

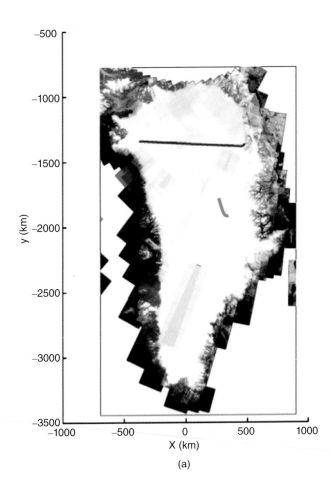

Figure 9.6 (a) Red: a flight line from west to east over the Greenland ice sheet. Green: one of the flight lines flown during the 1999 field season. (b) Radar echogram for the red flight line segment is highlighted in red. (c) Radar echogram for the green flight line.

(a)

Data frame ID: 19990512_01_010

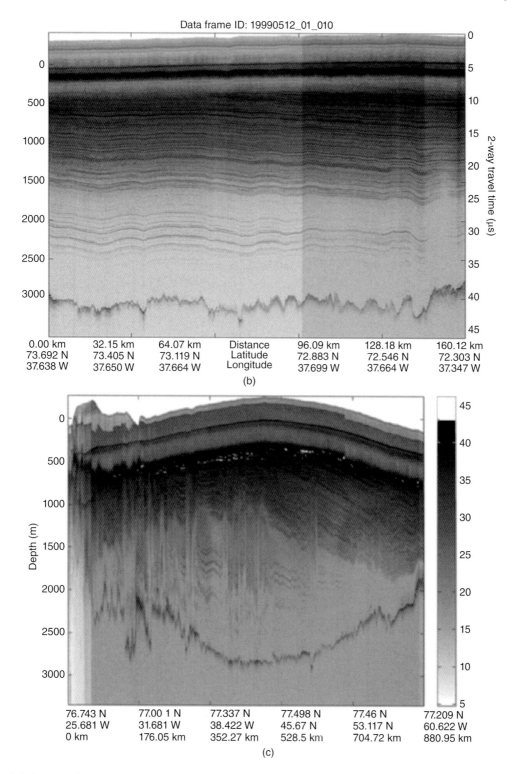

Figure 9.6 (*continued*)

fluctuations, specifically climate warming, over a short period of time, i.e. 30–40 years. The echoes from layers below 2400 m are very weak and are not visible in the echogram. This is partly because the radar did not have adequate sensitivity to map very weak echoes from layers below this depth at the time the data were collected, and also because the ice close to the bed is disturbed.

Figure 9.6(a) also shows a flight path (highlighted in red) going from west to east across the northern and southern parts of the Greenland ice sheet. The flight line did not go over the edge on both sides of the ice sheet. The ice thickness varied from 1500 m at the edge to about 3000 m at the center of the ice sheet. There is a rapid change in ice thickness over the first 170 km of the flight line, with a gradual increase in ice thickness to about 3 km in the middle. This is followed by a gradual decrease to about 1500 m over the next 400 km on the other side of the maximum.

The ice in the northern part of Greenland has low loss, and sounding ice in the interior is not a major challenge. In the southern part of Greenland, here defined as below 70° North, the ice thickness varies from about 1500 m at the edges to about 2400 m in the middle. The signal from 2400 m thick ice in the south is much weaker than that from 3000 m thick ice in the north; the ice in the south is more lossy, due to higher reflection loss caused by thicker annual melt layers and the presence of warmer ice at the bed. Echoes from internal layers below about 1.5 km, which are used to delineate different periods of time, are not observable in this echogram or others from the south.

Now we will look at data collected over outlet glaciers both in the north and the south. Figure 9.7 shows data collected across Jakobshavn glacier in the south and a glacier in the north. The received signal amplitude as a function of range, referred to as an A-scope, for a range-line for the ice in the channel and outside channel, is shown on the right. This type of plot is normally referred to as an A-scope. The returns in the A-scope for Jakobshavn are normalized with respect to bed return from ice outside the channel. The ice-bed echoes for Jakobshavn glacier, which have a S/N ratio of only about 3 dB, are barely discernible. The additional 1.5 km of ice in the channel caused about 67 dB more loss than the 1 km thick ice located outside the channel. The extra loss, as well as surface clutter (backscattered signal, as explained below) from the rough surface typical of fast-flowing glaciers, makes sounding these areas a major challenge.

The ice loss is also high for a few outlet glaciers in the northern and southern parts of Greenland. The A-scopes show that an additional 1.2 km of ice caused an extra 60 dB of loss. As previously stated, ice-bed echoes from fast-flowing glaciers are very weak, mainly due to the extra loss caused by the warm ice near the bed, which is typical of these glaciers. In addition, these weak bed echoes are often masked by signals backscattered from extremely rough glacier surfaces. Because of these factors, radar sounding of fast-flowing glaciers and ice-sheet margins is a major technical challenge.

In the interior, where the ice surface is relatively smooth, the off-vertical signals from antenna sidelobes that illuminate the ice surface are reflected away from

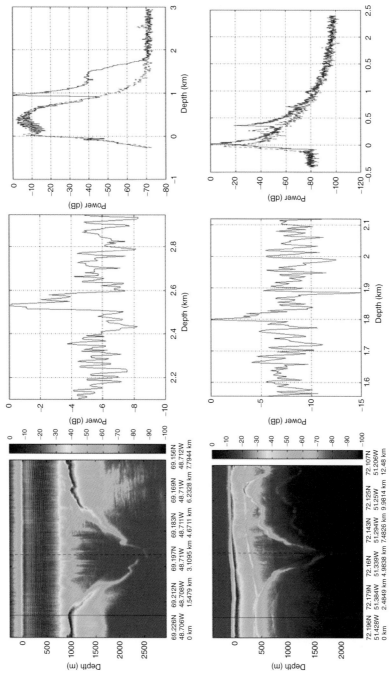

Figure 9.7 Radar echogram of data over Jakobshavn (top) and a glacier in the northern Greenland (bottom). The middle A-scopes show the S/N ratio for the bed return. The A-scopes on the right is a comparison of bed echoes for outside (blue) and inside the channel (red). A-scopes are plotted in range units.

the radar. The ice-bed echoes are strong and clearly visible, allowing us to obtain unambiguous estimates of ice thickness. Over fast-flowing glaciers and margins, however, the ice surface is very rough, due to crevassing and other effects. High-sensitivity radars are required to overcome large attenuation losses, and high spatial resolution is required to reduce surface clutter.

Unfortunately, the size of the antenna array that can be accommodated on aircraft is limited, because of the need to operate the radars in the HF and VHF range of the spectrum. We can obtain high spatial resolution by synthesizing a large aperture in the along-track direction by taking advantage of the Doppler frequency resulting from the forward motion of the aircraft; this is known as Synthetic Aperture Radar (SAR). However, SAR is ineffective at reducing clutter in the cross-track direction, because the line orthogonal to the flight direction has zero Doppler. We need a large cross-track array to reduce clutter. Since it is difficult to operate a very large array on an aircraft, we must use high-resolution array-processing techniques to reduce cross-track clutter. The concepts of SAR and array processing are discussed in the next section.

9.6 SAR and array processing

A complete discussion of SAR and array processing is beyond the scope of this chapter, but we will provide a brief introduction to these topics. An interested reader should consult the many excellent textbooks and literature available on this topic (Stimson, 1998; Van Trees, 2005). As stated earlier, SAR uses the forward motion of the aircraft to synthesize the long antenna needed to improve resolution. The resolution obtained with SAR can be derived using array theory, a Doppler filter and a matched filter concepts.

We will use the Doppler filter concept to derive an expression for the resolution of fully-focused SAR. The radar used to collect data for SAR processing transmits and collects complex received signals from different points along its flight path, as shown in Figure 9.8(a).

The radar receives reflected signals from the target at different locations along its flight path, from the time the target enters the leading edge of the antenna beam to the time the target exits the trailing edge. To simplify the analysis, let us consider a continuous wave (CW) transmit signal as:

$$V_T(t) = e^{-j\omega_0 t} \tag{9.42}$$

The received signal from a target at a range R that corresponds to time delay, $\tau = \frac{2R}{c}$, is given by:

$$V_T(t) = e^{-j\omega_0(t-\tau)} \tag{9.43}$$

R is related to range R_0 and aircraft location x at any instant t by:

$$R = \sqrt{R_0^2 + x^2} \tag{9.44}$$

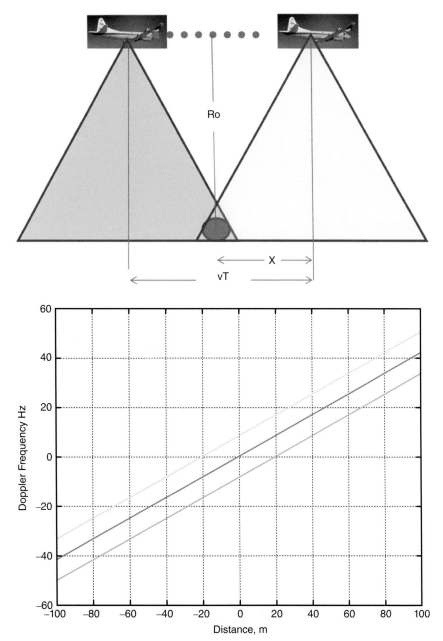

Figure 9.8 (a) Conceptual illustration of a Synthetic Aperture Radar. (b) Doppler frequency as a function of the along-track distance in a synthesized aperture.

We simplify the above equation for $R_o \gg x$ as:

$$R \approx R_0 + \frac{x^2}{2R_0} \tag{9.45}$$

$$V_r(t) = A V_0 e^{-j\omega\left(t - \frac{4\pi R_0}{\lambda} - \frac{x^2}{\lambda R_0}\right)} \tag{9.46}$$

$$\phi(t) = \omega t - \frac{4\pi R_0}{\lambda} - \frac{2\pi x^2}{\lambda R_0} \tag{9.47}$$

We rewrite the above equation in terms of distance from the center of the aperture to its end point and measurement location as

$$\phi(t) = \omega t - \frac{4\pi R_0}{\lambda} - \frac{2\pi(x - x_0)^2}{\lambda R_0} \tag{9.48}$$

$$\omega_r(t) = \frac{d\phi}{dt} = \omega_0 - \frac{4\pi(x - x_0)}{\lambda R_0}\frac{dx}{dt} \tag{9.49}$$

Thus, the Doppler frequency,

$$f_d(t) = \frac{2(x - x_0)}{\lambda R_0}v \tag{9.50}$$

Figure 9.8(b) shows Doppler frequency computed for three targets located within the illuminated aperture. By using three tracking filters, we can separate the three targets.

Now we can derive the along-track SAR resolution using above equation. The along-track resolution is directly related to the filter bandwidth as:

$$r_a = \frac{\lambda R_0 \Delta f_d}{2v} \tag{9.51}$$

where Δf_d is the Doppler-tracking filter bandwidth. The length of the aperture that can be synthesized is related to the velocity of the aircraft and the time of observation of the target as:

$$L = vT \tag{9.52}$$

The observation time and filter bandwidth are related $vT \approx 1$. The length of the aperture is related to the real-antenna beamwidth which, in turn, is determined by the physical aperture size as:

$$L = \frac{R_0 \lambda}{D} \tag{9.53}$$

where D is the antenna effective length in the along-track direction.

Substituting the above equation in the expression for v and Doppler bandwidth, we obtain an expression for resolution as:

$$r_a = \frac{D}{2} \tag{9.54}$$

Thus, the smaller the antenna size, the finer the resolution. The use of a smaller antenna results in a wide beam, which effectively increases observation time. Increased observation time means a narrow Doppler bandwidth and a finer resolution. However, as discussed in the section on radar equations, radar sensitivity is directly proportional to the power-aperture product, so there is a trade-off between resolution and sensitivity.

The advent of fast and cheap computers has made generating a fully-focused SAR image from coherent radar data reasonably easy, though not trivial, using an

FFT-based algorithm. There are a number of algorithms to process SAR data, each with its own advantages and disadvantages. A time-domain back propagation algorithm provides the best results, but it is computationally intensive (Gorham *et al.*, 2006; Kusk and Dall, 2010). Frequency domain processing algorithms, such as the frequency-wavenumber migration algorithm, are fast and easy to implement, and are more widely used to process radar depth sounder data (Legarsksy *et al.*, 1999; Gogineni *et al.*, 2001; Peters *et al.*, 2007; Hélière *et al.*, 2007).

The basic steps involved in SAR processing of radar data are as follows:

1 compensate data for motion and system errors;
2 pulse compress data;
3 range migrate pulse-compressed data using the frequency-wavenumber algorithm to generate a SAR image;
4 perform incoherent integration to reduce fading and improve S/N ratio;
5 generate a radar echogram.

Examples of SAR-processed radar echograms can be found in earlier figures (Figures 9.6(b), 9.6(c), and 9.7). The echogram can also be de-trended in the post-processing stage to enhance layers. With improved radars, we can map echoes from internal layers within 200–300 m of ice-bed.

9.7 SAR Interferometry

9.7.1 Introduction

SAR Interferometry, or Interferometric SAR (InSAR), is a technique that exploits two or more antennas at different spatial sampling locations to determine the surface elevation profile or the topography of a target. The basic idea is to estimate the signal path difference from multiple antenna locations to the target, which is directly related to the target height. The accuracy of the estimation can be on the order of the operating wavelength. At microwave frequencies, the data resolution obtained from space could be as precise as less than a meter. There are many different variations of InSAR systems that can provide high-quality measurements of terrain or surface reflectivity changes over time.

The first application of radar interferometry was filed by Rogers and Ingalls in 1969 to resolve the ambiguity between the north and south poles' reflections in Venus radar mapping (Rogers and Ingalls, 1969). Three years later, Zisk reported the topographic measurement of the moon's surface via radar interferometer (Zisk, 1972). Later, in 1974, Graham demonstrated airborne Earth topographic mapping with his synthetic interferometer radar (Graham, 1974). Using two vertically-separated antennas, Graham was able to simultaneously record the backscattered radar signal in which the phase information was processed to reveal the elevation of the terrain. The accuracy of the data collected could be used to compile 1 : 250,000 scale class B maps given the technology employed at that time.

In 1987, Goldstein and Zebker performed a radar measurement of ocean surface current flow with repeat-track interferometry, and the phase difference between the radar images taken at different times was used to estimate the motion of water current (Goldstein and Zebker, 1987). Since then, InSAR has been applied to and revolutionized a wide range of military and scientific areas, including surveillance, mapping, vegetation surveying, structural analysis, natural disaster monitoring, etc., due to its capability to operate day and night under virtually all weather conditions.

9.7.2 Basic theory

A conventional, single SAR system can only measure the propagation delay from a target that is directly related to the range r of the target, as shown in Figure 9.9(a). Interferometric SAR, on the other hand, is able to reconstruct the target elevation profile by taking advantage of angular diversity.

Assuming an aerial platform carries two antennas separated by a baseline d (and $d \ll r$), as depicted in Figure 9.9(b), the two antennas can be treated as coherent point sources, and the phase difference ϕ due to the path delay from the antennas to the target is given by:

$$\phi = \frac{2n\pi}{\lambda}\delta r = \frac{2n\pi d}{\lambda}\sin(\theta - \beta) \tag{9.55}$$

where λ is the operating wavelength of the radar, and $n = 1$ for single-pass systems or 2 for multi-pass systems or ping-pong mode (which will be explained later). The measured phase difference can then be used to evaluate the look angle θ, and the elevation of the target z can be evaluated using:

$$z = h - r\cos\theta \tag{9.56}$$

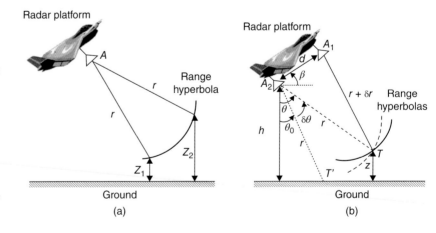

Figure 9.9 (a) A conventional single SAR system; a single measurement of the range r would result in ambiguity in the elevation of the target. (b) An InSAR system with two separated antennas in the along-track direction; the elevation ambiguity is resolved with multiple measurements of the target.

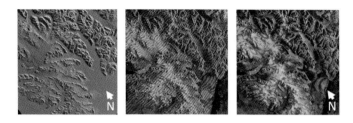

Figure 9.10 SAR images of Comfortlessbreen, Svalbard. Left: Simulated amplitude image from DEM. Center: corresponding raw interferogram. Right: flattened interferogram (one color cycle corresponds to an ambiguity height of 147 m). (Schneevoigt *et al.*, 2012. Reproduced with permission of Cambridge University Press.)

Furthermore, given that the along-track velocity of the aerial platform v is known, the phase difference can be used to estimate the radial velocity u of the target:

$$\phi = \frac{2n\pi u d}{\lambda v} \qquad (9.57)$$

From Figure 9.9(b), it can be seen that the scatterer at zero height (with the same range) contributes an additional term to the measured phase, which is known as the flat-Earth phase ϕ_{flat}, and is given by:

$$\phi_{flat} = \frac{2n\pi d \cos(\theta - \beta)\delta\theta}{\lambda} \qquad (9.58)$$

This flat-Earth phase must be subtracted from the measured phase, so that the resultant phase variations correspond solely to the elevation in the scene with respect to the ground level. The process of removing the flat-Earth phase from the measured phase is called interferogram flattening. Figure 9.10 demonstrates an example of interferogram flattening with the InSAR images taken over Comfortlessbreen in Svalbard.

After flattening the interferogram, it can be seen that the fringes closely follow the topography of the area. Each color cycle in the fringing pattern represents a 2π phase variation, as the phase of the data is sampled within a 2π interval. The elevation difference corresponding to this 2π phase cycle is called the ambiguity height, h_{amb}, and is given by:

$$h_{amb} = \frac{\lambda r \sin\theta}{2d \cos(\theta - \beta)} \qquad (9.59)$$

The sensitivity to the topography is thus proportional to the baseline. To recover the accumulated phase and the true elevation of the scene, phase unwrapping is necessary.

In summary, an InSAR image can be constructed using the following steps:

(i) Formation of two individual SAR images (either with two antennas or repeat-pass).
(ii) Co-registration of the two images.
(iii) Generation of interferogram by differencing the phase of the SAR images.

(iv) Elimination of flat-Earth phase.

(v) Phase unwrapping.

(vi) Height calculation.

(vii) Geometrical correction.

(viii) Geocoding to standard coordination systems.

9.7.3 Practical considerations of InSAR systems

9.7.3.1 Measurement accuracy

In equations 9.55 and 9.56, the accuracy of height estimation depends on the measurement uncertainties in the system parameters. By differentiating the uncertainties with respect to each of the parameters, one could determine the sensitivity of the height estimation as follows (Li and Goldstein, 1990):

Sensitivity to range r : $\qquad \sigma_z^r = \sigma_r \cos\theta$ (9.60a)

Sensitivity to baseline d : $\qquad \sigma_z^d = o_d \dfrac{r}{d} \sin\theta \tan(\theta - \beta)$ (9.60b)

Sensitivity to baseline angle β : $\qquad \sigma_z^\beta = \sigma_\beta r \sin\theta$ (9.60c)

Sensitivity to platform elevation h : $\quad \sigma_z^h = \sigma_h$ (9.60d)

Sensitivity to phase difference φ : $\qquad \sigma_z^\phi = \dfrac{\lambda r \sin\theta}{2\pi d \cos(\theta - \beta)}$ (9.60e)

The height uncertainty is thus directly related to the system geometry via the baseline d, the baseline angle β and the platform elevation h. However, the dependence on the range r is related to system clock timing, sampling clock jitter, atmospheric effects, and so on. The height estimation error due to phase measurement uncertainty can be attributed to four major factors: system noise, pixel misregistration, baseline decorrelation, and motion errors. These are discussed in more detail in the next section.

9.7.3.2 Phase decorrelation

To characterize phase measurement uncertainty quantitatively, we can look at the level of phase decorrelation, or the coherence γ between two SAR images (Askne and Hagberg, 1993),

$$\gamma = \frac{E[s_1 s_2^*]}{\sqrt{E[s_1 s_1^*]E[s_2 s_2^*]}}$$

(9.61)

where $E[s_1 s_2^*]$ denotes the ensemble average of the product of the complex pixel from the first image and the conjugate of the complex pixel from the second image. We can now look at the factors that affect the coherence of InSAR systems.

System noise Thermal noise, atmospheric noise and speckle are the primary noise sources in radar systems. Thermal noise adds uncertainty to the interferometric phase at the pixel level. The phase decorrelation κ resulting from thermal noise can be reduced by increasing the system signal-to-noise ratio (SNR):

$$\kappa = 1 - \gamma = 1 - \frac{1}{1 + \text{SNR} - 1} \qquad (9.62)$$

Atmospheric noise decorrelates the phase because the propagation environment is altered when the atmospheric humidity, temperature and pressure vary over time. Its effect is more obvious to repeat-pass systems. Finally, speckle arises from the fading phenomenon due to random coherent interference among distributed scatterers. Speckle can be efficiently reduced by increasing the number of looks or averages.

The sensitivity of phase standard deviation to phase correlation and number of looks was derived by Zebker and Villasenor (1992). For example, given a phase correlation of 0.9, the phase standard deviation that can be achieved is about 52° with a single look and 21° with four looks.

Pixel misregistration Pixel misregistration is due to an alignment error between two complex SAR images. A general rule of thumb to avoid this error is to register the SAR images to within one-eighth of a pixel. This can be done through two steps – rough registration and precise registration (Gabriel and Goldstein, 1988; Lin *et al.*, 1992; Rodriguez and Martin, 1992; Bamler and Just, 1993). Rough registration identifies the reference points in both images. Transformation and interpolation may be applied to one of the images so that they are registered at the pixel level. Next, precise registration can be performed at the sub-pixel level by aligning the interpolated phase between the two sub-images. By controlling the interpolation level, a one-eighth pixel registration can be achieved. Bamler *et al.* (1993) performed a study on the effect of misregistration on phase standard deviation. They showed that standard deviations of 23° and 42° can be obtained with SNRs of ∞ dB and 10 dB, respectively, given a one-eighth pixel misregistration.

Baseline decorrelation In the previous section, we mentioned that the sensitivity of an InSAR to height variations can be enhanced by increasing the baseline. Nevertheless, the baseline cannot be increased indefinitely for a given range resolution, because the phase would start to decorrelate. The baseline at which the interferogram phase is completely random is called the critical baseline, d_c, which is given by:

$$d_c = \frac{\lambda r}{n \Delta r \cos^2 \theta} \qquad (9.63)$$

where Δr is the range resolution. From the above relation, it is seen that baseline decorrelation is less likely to occur to systems with finer range resolution and longer operating wavelengths. In general, the optimal baseline varies for systems with different purposes. For instance, an InSAR system designed for topographic estimation should have an adequate baseline to avoid reduction in height sensitivity (too short) and decorrelation (too long), while that for surface temporal change estimation should be as short as possible in order to maintain the correlation. More thorough discussions on baseline decorrelation can be found in Zebker and Villasenor (1992) and Rodriguez and Martin (1992). Zebker and Villasenor (1992) also compared the theoretical and observed baseline decorrelation curves with their system.

Other contributions to phase noises, such as temporal decorrelation (due to movement and/or changes in backscattering properties), geometric decorrelation, and volume scattering could also be possible, but their presence would be dependent on the measurement scenarios.

9.7.3.3 Persistent scatterers

One of the challenges in monitoring the temporal change of a scene with InSAR systems, such as routine monitoring of landslides, volcanoes, urban areas and so on, is that temporal decorrelation and atmospheric inhomogeneities degrade the phase correlation over time. Nevertheless, persistent or permanent scatterers that remain coherent over long time intervals can often be found in these scenes. InSAR systems that involve the use of persistent scatterers are termed persistent scatterer interferometry SAR (PSInSAR). The technique was first proposed by Ferretti *et al.* (1999). By taking advantage of these persistent scatterers, the atmospheric phase screen can be estimated and removed from the data, thus enabling accurate and reliable elevation and temporal measurements (Ferretti *et al.*, 2000). Examples of persistent scatterers include objects with large radar cross sections, such as metallic objects, building roofs, prominent natural features, etc. These scatterers could be smaller than a resolution cell of the SAR system, so that geometrical decorrelation is minimal and interferograms with baselines longer than the critical baseline can be used.

9.7.3.4 Implementation considerations

Platform selection InSAR systems can be implemented on spaceborne or airborne platforms. There are several advantages to spaceborne InSAR. First, it is able to provide global coverage and collect data over typically inaccessible areas at a relatively low cost. The effects of motion error in spaceborne platforms are also far less significant than in airborne systems, since airborne platforms are susceptible to vibration and turbulence. Nevertheless, airborne InSAR systems generally give better vertical accuracy and are not affected by ionospheric effects. Airborne platforms are also more flexible in terms of rapid deployment, changes to the survey area, and instrument maintenance and upgrades. Typically, airborne InSAR systems are used in regional applications, such as local mapping, vegetation studies, urban planning, and for militaristic uses.

Modes of operation Whether implemented on spaceborne or airborne platforms, InSAR systems can have three different data collection schemes: single-pass mode, ping-pong mode, and repeat-pass mode. In single-pass mode, the platform carries two antennas separated by a baseline; one antenna is used to transmit and both are used to receive. In ping-pong mode, two antennas are also needed, but both antennas are used to transmit alternatively and receive simultaneously, such that the baseline is effectively doubled. Finally, in repeat-pass mode, the platform flies with a single-antenna over a target area twice to generate two correlated, complex SAR images. The observation interval may vary from seconds to years, depending on the purpose of the system.

Selection of operating wavelength Similar to conventional single SAR systems, there are a number of factors that should be considered when choosing the operating wavelength of an InSAR system. These include the electromagnetic properties of the target of interest, measurement accuracy, mode of operation, availability of the frequency spectrum, and the largest antenna size that can be accommodated. Depending on the application of the InSAR system, different operating wavelengths should be selected to match the scattering characteristics of the target, so that a good phase correlation can be obtained.

The strongest backscattering usually occurs when the dimension of the target is roughly the same as the radar's operating wavelength. For example, in vegetation mapping, a longer wavelength is needed to improve the correlation in densely vegetated areas. Figure 9.11 shows the NASA JPL SIR-C interferograms over Kilauea, Hawaii with two different radar frequency bands (Rosen *et al.*, 1996). The L-band data are much more sensitive to vegetation than the C-band data. The operating mode of an InSAR system also partly determines the radar's wavelength. For example, in repeat-pass, it is critical to maintain the phase correlation between different radar images taken at different times. Since small and light objects, such as leaves, tend to move over time, short wavelength signals should be avoided, if possible, to minimize temporal decorrelation (Rosen *et al.*, 1996).

An excellent and more in-depth discussion on the practical implementation issues of InSAR system can be found in Rosen *et al.* (2000).

9.7.4   Application of InSAR to Cryosphere remote sensing

9.7.4.1   Digital elevation model (DEM) generation

One obvious yet important application of InSAR is the generation of a digital elevation model (DEM) of the Earth. DEM is a digital 3D visualization of the terrain elevation displayed in standard geographic coordinates. DEM has a wide range of scientific and engineering applications, including geology, navigation, urban planning, radio propagation modeling and so on. The first demonstration of DEM generation with InSAR was reported by Graham, in 1974, using an airborne platform (Graham, 1974). After decades of development and the advancement of computer technology, the technique became more mature in the 1990s and early 2000s.

During this time, there were a number spaceborne InSAR systems deployed for topography mapping, including the European Remote Sensing Satellite (ERS-1 in 1991; ERS-2 in 1995) (Shiping, 2000), the Japanese Earth Resources Satellite 1 (JERS-1) in 1992 (Tokunaga, 1998), Radarsat-1 in 1995 (Geudtner *et al.*, 1998), the Shuttle Radar Topography Mission (SRTM) in 2000 (Farr *et al.*, 2007), and Envisat

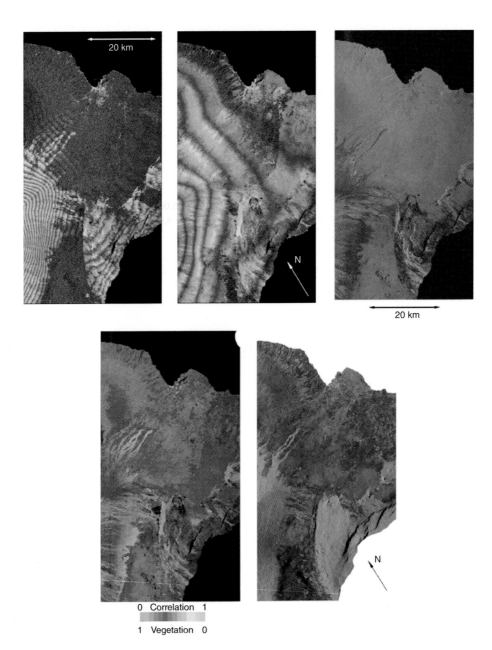

Figure 9.11 NASA JPL SIR-C InSAR data of Kilauea, Hawaii at: (a) C-band interferogram; (b) L-band interferogram; (c) C-band correlation map; (d) L-band correlation map; and (e) NDVI vegetation measure. Bright areas have high correlation and good fringe quality, while dim areas have poor correlation in (a) and (b) (Rosen *et al.*, 1996. Reproduced with permission of John Wiley & Sons Inc).

in 2002. In the SRTM, a near-global scale DEM was produced in eleven days from 60°N latitude to 56°S latitude at one arc sec resolution (≈ 30 m × 30 m), with an absolute vertical accuracy of 9 m (Farr *et al.*, 2007). A more recent spaceborne InSAR system for DEM generation includes ALOS PALSAR in 2006 (Kimura and Ito, 2000), RADARSAT-2 in 2007 (Van der Sande, 2004) and TanDEM-X in 2010 (Krieger *et al.*, 2007).

The DEM data of the Greenland and Antarctic ice sheets can be found at the National Snow and Ice Data Center (NSIDC) (Bamber, 2001, 2009). These data are primarily used for the study of dynamics and properties of ice sheets, glaciers, and ice shelves, which help to facilitate the investigations of ice sheet mass balance as well as to track the long-term evolution of the ice sheets. The Greenland dataset was generated based on the combination of Airborne Topographic Mapper (ATM) data, ERS-1 and Geosat radar altimetry data, and photogrammetric digital height data. It has a grid resolution of 5 km and temporal coverage from 01/01/1970 to 12/01/1970 and from 01/01/1993 to 12/31/1999. The Antarctica DEM data were collected by the ERS-1 radar altimeter from 1994 to 1995 and the Ice, Cloud, and land Elevation satellite (ICEsat) laser altimeter from 2003 to 2008. The dataset is available with a grid resolution of 1 km.

9.7.4.2   Basal ice sheet imaging

The basal terrain of the ice sheets can also be imaged with an InSAR system. Paden *et al.* (2010) and Blake (2010) demonstrated the generation of a 3D basal DEM with KU CReSIS radar depth sounders. It was demonstrated that a 25 m horizontal resolution with an RMS height error of 5 m was achieved over a 10 km² grid around the NEEM site in Greenland. A sample DEM image is shown in Figure 9.12, and more 3D ice bed images can be found at https://www.cresis.ku.edu/research/sensors-development/radar.

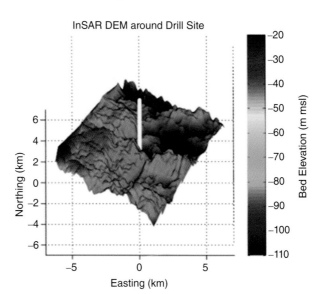

Figure 9.12 InSAR DEM around NEEM drill site in Greenland (Blake 2010. Reproduced with permission).

9.7.4.3 Ice sheet dynamics study

An InSAR system can also be used to study ice dynamics in polar regions (Schneevoigt *et al.*, 2012; Goldstein *et al.*, 1993; Mohr *et al.*, 1998; Joughin *et al.*, 1995, 1999; Rignot *et al.*, 1995, 1997, 2004, 2006, 2011). Understanding the topography and dynamics of the ice sheets is critical for scientists seeking to predict the effects of climate change. The first application of InSAR to measure the ice stream motion was reported by Goldstein *et al.* in 1993, using the SAR images

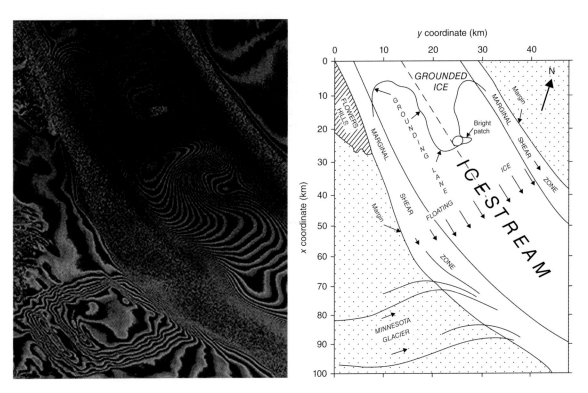

Figure 9.13 Radar interferogram of the Rutford ice stream in Antarctica (one fringe representing 28 mm of range change) (Goldstein *et al.*, 1993. Reproduced with permission of the American Association for the Advancement of Science).

Figure 9.14 Ice velocity superimposed on surface elevation over Ross Ice Shelf, Antarctica (Joughlin *et al.*, 1999. Reproduced with permission of the American Association for the Advancement of Science).

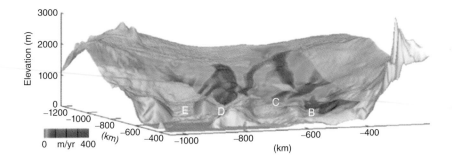

captured by ERS-1 (Goldstein *et al.*, 1993). The achieved detection limit was about 1.5 mm for vertical motions and about 4 mm for horizontal motions. The demonstration was significant, because it implied the possibility of constructing wide-area velocity maps over barely-accessible polar areas. Figure 9.13 shows a radar interferogram of the Rutford ice stream in Antarctica and a diagram of the ice flow adapted from Goldstein *et al.*, 1993.

Figure 9.15 (a) Ice velocity mosaic of Greenland ice sheet assembled from RADARSAT-1 data (color coded on a logarithmic scale from 1 m/yr (brown) to 3 km/yr (purple)) (Rignot and Kanagaratnam 2006. Reproduced with permission of the American Association for the Advancement of Science). (b) Ice velocity map of Antarctica generated by data from ALOS PALSAR, ENVIAT ASAR, RADARSAT-1, RADARSAT-2, ERS-1, and ERS-2 from 1992–2009 (Rignot *et al.*, 2011. Reproduced with permission of the American Asscoaition for the Advancement of Science).

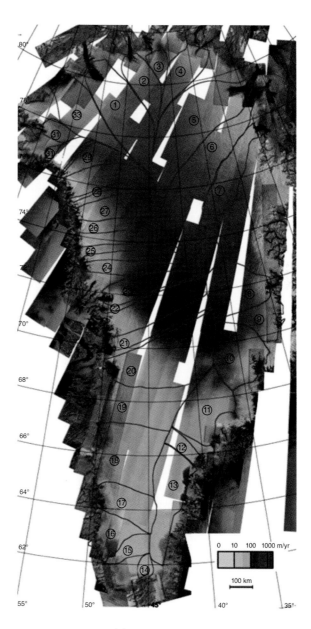

(a)

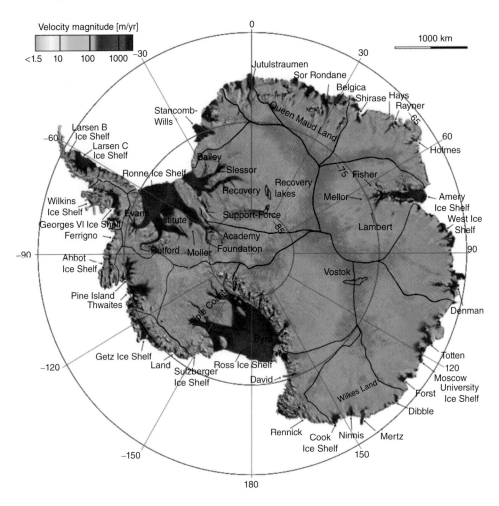

Figure 9.15 (*continued*)

The flow of ice streams can also be tracked with an InSAR system. By using the interferometric data from RADARSAT, Joughin *et al.* (1999) were able to map ice motion in West Antarctic ice streams. Figure 9.14 shows the measured ice speed overlaid on the DEM of the Ross Ice Shelf in Antarctica (Joughin *et al.*, 1999).

In slow-flowing ice streams, where the flow is less than 100 m per annum, direct interferometric measurement is feasible because of the relatively high coherence within the 24-day repeat period of RADARSAT. On the other hand, in faster-flowing areas, tracking techniques, such as intensity tracking and speckle tracking, are needed to improve the data correlation. Another, more recent, example of ice velocity mapping over a wide area was demonstrated by Rignot *et al.* in 2006 and 2011, when the ice velocity was mapped over the entire continents of Greenland and Antarctica (Figure 9.15). The accuracy of glacier

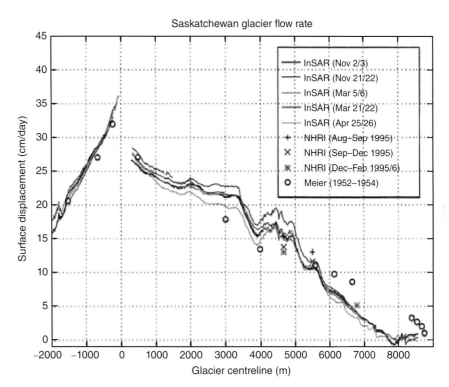

Figure 9.16 A comparison between the measured glacier flow rate over Saskatchewan from InSAR data and NHRI survey data (Mattar *et al.*, 1998. Reproduced with permission of IEEE).

velocity measurements has also been validated experimentally in Mattar *et al.* (1998), and a comparison between InSAR data and National Hydrology Research Institute (NHRI) survey data is given in Figure 9.16.

9.8 Conclusions

An understanding of the operation of modern radar is essential to the design and application of radar to sounding and imaging the polar ice sheets, as well as to the interpretation of radar data. In this chapter, we have provided a brief review of the electrical properties of ice that are important to designing radars for sounding ice and interpreting results. Starting with simple radar equations, we developed expressions for received radar signals from planar and distributed targets, and we showed that radar sensitivity is directly proportional to the power-aperture product. In addition, we noted that high sensitivity radars, required to sound very lossy ice, must have a large power-aperture product. We also presented the trade-offs involved in designing a radar for resolution and signal-to-noise ratio, and discussed the difference between coherent and incoherent integration to improve the signal-to-noise ratio.

Moreover, we discussed the use of Synthetic Aperture Radar (SAR) processing and pulse compression in a radar for improving along-track resolution and obtaining the high sensitivity required to sound ice around ice-sheet margins. We also provided a brief overview of antenna arrays and their use on both short- and long-range aircraft. We presented an example of data collected over the interior of the ice sheet that shows surface and bed echoes, as well as reflections from internal layers. We also provided results over outlet glaciers, in order to illustrate the challenges associated with sounding fast-flowing glaciers like Jakobshavn.

Finally, we provided a brief overview of the principle of SAR and InSAR operation. The chapter included a short discussion on InSAR theory, the steps involved in generating InSAR images, and measurement errors. We have also referenced a few examples to show that InSAR systems were used to collect accurate information on ice sheet topography and ice flow dynamics.

References

ANSYS HFSS. ANSYS. (2012). ANSYS, Inc. Web. <http://www.ansys.com/ProductsSimulation+Technology/Electromagnetics/High-Performance+Electronic+Design/ANSY+HFSS>

Askne, J. & Hagberg, J.O. (1993). *Potential of interferometric SAR for classification of land surfaces*. Proceedings of IEEE 1993 International Geoscience and Remote Sensing Symposium **3**, 985–987.

Balanis, C.A. (2005) *Antenna Theory, Analysis and Design*, 3rd edition. New York, Wiley-Interscience.

Bamler, R. & Just, D. (1993). *Phase statistics and decorrelation in SAR interferograms*. Proceedings of IEEE 1993 International Geoscience and Remote Sensing Symposium **3**, 980–984.

Blake, W.A. (2010). *Interferometric synthetic aperture radar (INSAR) for fine-resolution basal ice sheet imaging*. Ph.D. Dissertation, University of Kansas.

Chuah, T.S. (1997). *Design and Development of a Coherent Radar Depth Sounder for Measurement of Greenland Ice Sheet Thickness*. RSL Technical Report 10470–5.

Cook, C.E. & Paolillo, J. (1964). A pulse compression predistortion function for efficient sidelobe reduction in a high-power radar. *Proceedings of the IEEE* **52**(4), 377–389.

Evans, S. (1963). Dielectric Properties of Ice and Snow – A Review. *Journal of Glaciology* **5**(42), 773–787.

Fahnestock, M., Bindschadler, R., Kwok, R. & Jezek, K. (1993). Greenland ice sheet surface properties and ice dynamics from ERS-1 synthetic aperture radar imagery. *Science* **262**(5139), 1530–1534.

Farr, T.G., Rosen, P.A., Caro, E., Crippen, R., Duren, R., Hensley, S., Kobrick, M., Paller, M., Rodriguez, E., Roth, L., Seal, D., Shaffer, S., Shimada, J., Umland, J., Werner, M., Oskin, M., Burbank, D. & Alsdorf, D. (2007). The shuttle radar topography mission. *Review of Geophysics* **45**(RG2004).

Ferretti, A., Prati, C. & Rocca, F. (1999). *Permanent scatterers in SAR interferometry.* Proceedings of 1999 International Geoscience and Remote Sensing Symposium, 1528–1930.

Ferretti, A., Prati, C. & Rocca, F. (2000). Permanent scatterers in SAR interferometry. *IEEE Transactions on Geoscience and Remote Sensing* **39**(1), 8–20.

Fujita, S., Matsuoka, T., Ishida, T., Matsuoka, K. & Mae, S. (2000). A summary of the complex dielectric permittivity of ice in the megahertz range and its applications for radar sounding of polar ice sheets. In: T. Hondoh (Ed). *Physics of Ice Core Records*, 185–212. Sapporo, Hokkaido University.

Gabriel, A.K. & Goldstein, R.M. (1988). Crossed orbit interferometry: Theory and experimental results from SIR-B. *International Journal of Remote Sensing* **9**(5), 857–872.

Geudtner, D., Vaehon, P.W., Mattar, K.E. & Gray, A.L. (1998). *RADARSAT repeat-pass SAR interferometry.* Proceedings of IEEE 1998 International Geoscience and Remote Sensing Symposium **3**, 1635–1637.

Gogineni, S., Chuah, T., Allen, C., Jezek, K. & Moore, R.K. (1998). An improved coherent radar depth sounder. *Journal of Glaciology* **44**(148), 659–669.

Gogineni, S., Tammana, D., Braaten, D. *et al.* (2001). Coherent Radar Ice Thickness Measurements over the Greenland Ice Sheet. *Journal of Geophysical Research (Climate and Physics of the Atmosphere)* **106**(D24), 33,761–33,772.

Gogineni, P., Li, J., Paden, J. *et al.* (2012). *Sounding and Imaging of Fast Flowing Glaciers and Ice-Sheet Margins*, 23–26. EUSAR, Nuremberg, Germany.

Goldstein, R.M. & Zebker, H.A. (1987). Interferometric Radar Measurements of Ocean Surface Currents. *Nature* **328**(20), 707–709.

Goldstein, R.M., Engelhardt, H., Kamb, B. & Frolich, R.M. (1993). Satellite radar interferometry for monitoring ice sheet motion: Application to an Antarctic ice stream. *Science* **262**, 1525–1530.

Gorham, L.A., Majumder, U.K., Buxa, P., Backues, M.J. & Lindgren, A.C. (2006). *Implementation and analysis of a fast backprojection algorithm.* In Society of Photo-Optical Instrumentation Engineers (SPIE) Conference Series, volume 6237 of Society of Photo-Optical Instrumentation Engineers (SPIE) Conference Series.

Graham, L.C. (1974). Synthetic Interferometric Radar for Topographic Mapping. *Proceedings of the IEEE* **62**(6), 763–768.

Harris, F.J. (1978). On the use of windows for harmonic analysis with the discrete Fourier transform. *Proceedings of the IEEE* **66**(1), 51–83.

Hélière, F., Lin, C.-C., Corr, H. & Vaughan, D. (2007). Radio Echo Sounding of Pine Island Glacier, West Antarctica: Aperture Synthesis Processing and Analysis of Feasibility From Space. *IEEE Transactions on Geoscience and Remote Sensing* **45**(8), 2573–2582.

Jezek, K.C., Farness, K., Carande, R., Wu, X. & Labelle-Hamer, N. (2003). RADARSAT 1 synthetic aperture radar observations of Antarctica: Modified Antarctic Mapping Mission, 2000. *Radio Science* **38**(4), 8067.

Joughin, I.R., Winebrenner, D.P. & Fahnestock, M.A. (1995). Observations of ice-sheet motion in Greenland using satellite radar interferometry. *Geophysical Research Letters* **22**(5), 571–574.

Joughin, I., Gray, L., Bindschadler, R., Price, S., Morse, D., Hulbe, C., Mattar, K. & Werner, C. (1999). Tributaries of west Antarctic ice streams revealed by RADARSAT interferometry. *Science* **286**(5438), 283–286.

Joughin, I., Smith, B., Howat, I.M., Scambos, T. & Moon, T. (2010). Greenland Flow Variability from Ice Sheet-Wide Velocity Mapping. *Journal of Glaciology* **56**(197), 415–430.

Kimura, H. & Ito, N (2000). ALOS/PALSAR: The Japanese second-generation spaceborne SAR and its applications. *Proceedings of SPIE* **4152**, 110–119.

Klauder, J.R., Price, A.C., Darlington, S. & Albersheim, W.J. (1960). Theory and Design of Chirp Radar. *The Bell System Technical Journal* **39**(4), 745–808.

Kowatsch, M. & Stocker, H.R. (1982). Effect of Fresnel ripples on sidelobe suppression in low time-bandwidth product linear FM pulse compression. *IEE Proceedings* **129**(1), 41–44.

Krieger, G., Moreira, A., Fiedler, H., Hajnsek, I., Werner, M., Younis, M. & Zink, M. (2007). TanDEM-X: A satellite formation for high resolution SAR interferometry. *IEEE Transactions on Geoscience and Remote Sensing* **45**(11), 3317–3341.

Kusk, A., & Dall, J. (2010). *SAR focusing of P-band ice sounding data using back-projection.* Proceedings of thr 2010 IEEE International Geoscience and Remote Sensing Symposium, Honolulu, HI, pp. 4071–4074.

Kwok, R. & Fahnestock, M. (1996). Ice Sheet Motion and Topography from Radar Interferometry. *IEEE Transactions on Geoscience and Remote Sensing* **34**(1), 189–200.

Legarsky, J.J., Gogineni, P. & Akins, T.L. (2001). Focused synthetic-aperture radar processing of ice-sounder data collected over the Greenland ice sheet. *IEEE Transactions on Geoscience and Remote Sensing* **39**(10), 2109–2117.

Leuschen, C., Gogineni, P., Allen, C. *et al.* (2010). *The CReSIS Radar Suite for Measurements of the Ice Sheets and Sea Ice during Operation Ice Bridge.* American Geophysical Union Fall Meeting, San Francisco, CA. 13–17.

Li, J. (2009). *Mapping of ice sheet deep layers and fast outlet glaciers with multi-channel-high-sensitivity radar.* Ph.D. Dissertation, University of Kansas.

Li, F.K. & Goldstein, R.M. (1990). Studies of multibaseline spaceborne interferometric Synthetic Aperture Radars. *IEEE Transactions on Geoscience and Remote Sensing* **28**(1), 88–97.

Li, J., Paden, J., Leuschen, C. *et al.* (2012). High-Altitude Radar Measurements of Ice Thickness Over the Antarctic and Greenland Ice Sheets as a Part of Operation IceBridge, *IEEE Transactions on Geoscience and Remote Sensing*, available from 10.1109/TGRS.2012.2203822.

Lin, Q., Vesecky, J.F. & Zebker, H.A. (1992). New approaches in interferometric SAR data processing. *IEEE Transactions on Geoscience and Remote Sensing* **30**(3), 560–567.

Mattar, K.E., Vachon, P.W., Geudtner, D., Laurence, A., Cumming, I.G. & Brugman, M. (1998). Validation of Alpine glacier velocity measurements using

ERS Tandem-Mission SAR data. *IEEE Transactions on Geoscience and Remote Sensing* **36**(3), 974–984.

Misaridis, T. & Jensen, J. (2005). Use of modulated excitation signals in medical ultrasound, part II: Design and performance for medical imaging applications. *IEEE Transactions on Ultrasonics, Ferroelectrics and Frequency Control* **52**(2), 192–207.

Mohr, J.J., Reeh, N. & Madsen, S. (1998). Three-dimensional glacial flow and surface elevation measured with radar interferometry. *Nature* **391**, 273–276.

Paden, J., Akins, T., Dunson, D., Allen, C. & Gogineni, P. (2010). Ice-sheet bed 3-D tomography. *Journal of Glaciology* **56**(195), 3–11.

Peters, E.M., Blankenship, D.D., Carter, S.P., Kempf, S.D., Young, D.A. & Holt, J.W. (2007). Along-Track Focusing of Airborne Radar Sounding Data From West Antarctica for Improving Basal Reflection Analysis and Layer Detection. *IEEE Transactions on Geoscience and Remote Sensing* **45**(9), 2725–2736.

Rignot, E., Jezek, K.C. & Sohn, H.G. (1995). Ice flow dynamics of the Greenland ice sheet from SAR interferometry. *Geophysical Research Letters* **22**(5), 575–578.

Rignot, E., Gogineni, S.P., Krabill, W.B. & Ekholm, S. (1997). North and northeast Greenland ice discharge from satellite radar interferometry. *Science* **276**(5314), 934–937.

Rignot, E., Braaten, D., Gogineni, S.P., Krabill, W.B. & McConnell, J.R. (2004). Rapid ice discharge from southeast Greenland glaciers. *Geophysical Research Letters* **31**(10), doi: 10.1029/2004GL019474.

Rignot, E. & Kanagaratnam, P. (2006). Changes in the Velocity Structure of the Greenland Ice Sheet. *Science* **311**, 986–990, doi: 10.1126/science.1121381.

Rignot, E., Mouginot, J. & Scheuchl, B. (2011). Ice Flow of the Antarctic Ice Sheet. *Science* **333**(6048), 1427–1430.

Rihaczek, A.W. & Hershkowitz, S.J. (2000). *Theory and practice of radar target identification*. Boston: Artech House Radar Library.

Rodriguez, E. & Martin, J. (1992). Theory and design of interferometric SARS. *Proceedings of IEEE* **139**(2), 147–159.

Rogers, A.E.E. & Ingalls, R.P. (1969). Venus: Mapping the surface reflectivity by radar interferometry. *Science* **165**, 797–799.

Rosen, P.A., Hensley, S., Zebker, H.A., Webb, F.H. & Fielding, E.J. (1996). Surface deformation and coherence measurements of Kilauea Volcano, Hawaii, from SIR-C radar interferometry. *Journal of Geophysics Remote Sensing* **268**, 1333–1336.

Rosen, P.A., Hensley, S., Joughin, I.R., Li, F.K., Madsen, S.N., Rodriguez, E. & Goldstein, R.M. (2000). Synthetic aperture radar interferometry. *Proceedings of IEEE* **88**(3), 333–382.

Schneevoigt, N.J., Sund, M., Bogren, W., Kääb, A. & Weydahl, D.J. (2012). Glacier displacement on Comfortlessbreen, Svalbard, using 2-pass differential SAR interferometry (DInSAR) with a digital elevation model. *Polar Record* **48**(244), 17–25.

Shiping, S. (2000). *DEM generation using ERS-1/2 interferometric SAR data*. Proceedings of IEEE 2000 Geoscience and Remote Sensing Symposium **2**, 788–790.

Skolnik, M.I. (1981). *Introduction to Radar Systems*, 3rd edition. New York, McGraw Hill.

Stimson, G.W. (1998). *Introduction to Airborne Radar*, 2nd edition. New Jersey, SciTech Publishing.

Stutzman, W.L. & Thiele, G.A. (1998). *Antenna Theory and Design*, 2nd edition. New York, John Wiley & Sons, Inc.

Tokunaga, K. (1998). *DEM generation using JERS-1 SAR interferometry*. IAPRS Commission IV Symposium on GIS-Between Visions and Applications **32**(4), 625–628.

Van der Sande, J.J. (2004). Anticipated applications potential of RADARSAT-2 data. *Canadian Journal of Remote Sensing* **30**(3), 369–379.

Van Trees, H.L. (2002). Optimum Array Procesing (Detection, Estimation and ModulationTheory, Part IV), Wiley-Interscience; Part IV edition.

Watters, T.R., Leuschen, C.J., Plaut, J.J. *et al.* (2006). MARSIS radar sounder evidence of buried basins in the northern lowlands of Mars. *Nature* **444**(7121), 905–908.

Yan, J.B., Li, J., Rodriguez-Morales, F. *et al.* (2012). Measurements of In-Flight Cross-Track Antenna Patterns of Radar Depth Sounder/Imager. *IEEE Transactions on Antennas and Propagation*, available from 10.1109/TAP.2012.2211327.

Zebker, H.A. & Villasenor, J. (1992). Decorrelation in interferometric radar echoes. *IEEE Transactions on Geoscience and Remote Sensing* **30**(5), 950–959.

Zisk, S.H. (1972). Lunar topography: First radar interferometer measurements of the Alphonsus-Ptolemaeus-Arzachel region. *Science* **178**, 997–980.

Acronyms

CReSIS	Center for Remote Sensing of Ice Sheets
CW	Continuous wave
HF	High Frequency
InSAR	Interferometric Synthetic Aperture Radar
SARs	Synthetic Aperture Radars
SAW	Surface Acoustic Wave
STC	Sensitivity time control
VHF	Very High Frequency

10 Gravimetry measurements from space

Scott B. Luthcke[1], D.D. Rowlands[1], T.J. Sabaka[1], B.D. Loomis[2],
M. Horwath[3] and A.A. Arendt[4]

[1] NASA Goddard Space Flight Center, Greenbelt, USA
[2] SGT Inc., Science Division, Greenbelt, USA
[3] Technische Universitat Munchen, Munich, Germany
[4] University of Alaska, Fairbanks, USA

Summary

Through the measurement of the changes in range and range-rate between two co-orbiting satellites, the GRACE mission has provided unprecedented observations of surface mass variation, revolutionizing our understanding of the Earth's water cycle and, in particular, the mass evolution of land ice. The GRACE observations have made it possible to quantify accurately the decadal, annual and seasonal mass balance of the ice sheets and smaller glacier systems, with temporal resolution on the order of ≈ 15 days and spatial resolution on the order of ≈ 300 km.

Future improvements in GRACE processing and analysis techniques will likely further improve the accuracy and resolution of the surface mass observations. In addition, the GRACE follow-on mission (launch $\approx 2016-17$) and GRACE-II (≈ 2022) will provide the continuity of observations to understand long-term land ice change, sea level rise, and will produce crucial observations for improved modeling and forecasting. The purpose of this chapter is to provide the reader with an understanding of the GRACE mission, its fundamental measurements, and how these measurements are used to observe changes in the Earth's surface mass distribution and, in particular, terrestrial ice mass evolution.

10.1 Introduction

Since its launch in March of 2002, the Gravity Recovery and Climate Experiment (GRACE) mission has acquired ultra-precise inter-satellite K-band range and range-rate (KBRR) measurements taken between two co-orbiting satellites in a

Remote Sensing of the Cryosphere, First Edition. Edited by M. Tedesco.
© 2015 John Wiley & Sons, Ltd. Published 2015 by John Wiley & Sons, Ltd.
Companion Website: www.wiley.com/go/tedesco/cryosphere

450 km altitude polar orbit, approximately 220 km apart (Tapley *et al.*, 2004). These data have vastly improved our knowledge of the Earth's time-variable gravity field. In particular, the GRACE mission has been instrumental in quantifying present day terrestrial ice mass evolution (e.g., Luthcke *et al.*, 2006a, 2008; Velicogna, 2009; Chen *et al.*, 2011; Schrama and Wouters, 2011; Jacob *et al.*, 2012; King *et al.*, 2012; Horwath and Dietrich, 2009). The purpose of this chapter is to provide the reader with an understanding of the GRACE mission, its fundamental measurements and how those measurements are used to observe changes in the Earth's surface mass distribution and, in particular, terrestrial ice mass evolution.

10.2 Observing the Earth's gravity field with inter-satellite ranging

From the beginning of the space age, observations of satellite motions have been used to compute gravity models of the Earth. Over the past decades, there have been several important geopotential model development efforts based on satellite tracking data from a suite of Earth orbiting satellites (e.g., Lemoine *et al.*, 1998). The motion of a satellite is computed by modeling the conservative and non-conservative forces acting on the satellite and integrating the resultant differential equations to obtain a time series of predicted position and velocity (ephemeris). The force models (including the Earth's gravitational potential field) are then refined through parameter estimation, minimizing a measure of the difference between the computed tracking data observations – constructed from the estimated time series of satellite position and velocity – and the observed tracking data. Some examples of modern satellite tracking data for Precision Orbit Determination (POD) and geodetic parameter estimation include: Satellite Laser Ranging (SLR); Global Positioning System (GPS) satellite-satellite ranging through carrier and code phase observations; and Doppler Orbitography and Radiopositioning Integrated by Satellite (DORIS) (Luthcke *et al.*, 2003).

Milo Wolff was the first to introduce the concept of computing the variations in the Earth's gravity field directly from observations of the changing range between two low Earth co-orbiting satellites (Wolff, 1969). The motion of a satellite is dependent on the gravitational potential, which can be understood using conservation of energy. Ignoring non-conservative forces (e.g., drag and radiation pressure), changes in the Earth's potential energy must be compensated by a change in energy from another component. These components include a change in kinetic energy in the three orthogonal velocity directions (e.g., radial, line-of-sight between satellites, and orthogonal to the first two), and a potential energy change due to change in radial position of the satellites. Wolff (1969) showed that the change in kinetic energy in the line of sight direction between the two co-orbiting satellites is by far the dominant mode of energy compensation. Given the simplifying assumption that the two satellites are in a perfectly circular co-orbit, and ignoring non-conservative forces, the line of sight range-rate between the satellites can be expressed as:

$$\dot{\rho}_{12} = \frac{1}{V} U_{12} \tag{10.1}$$

where:

V is the mean orbital velocity given by $\left(\frac{GM}{R} \right)^{1/2}$

U_{12} is the potential difference between satellite position 1 and 2.

The differential potential can be expressed as:

$$U_{12} = U_{12}^{Earth} + U_{12}^{N-body} + U_{12}^{tides} + U_{12}^{oceans} + U_{12}^{atmosphere} + U_{12}^{hydrology} + U_{12}^{land\ ice} + \varepsilon \tag{10.2}$$

Spatial and temporal contributions to the potential include: solid Earth; planetary bodies; ocean tides; solid Earth and pole tides; and mass changes due to the oceans, atmosphere and land ice. Of course simplifying assumptions cannot be used in precisely computing the Earth's time variable gravity field. Therefore, in order to isolate the land ice surface mass changes, it is necessary to compute and model the position and velocity of the satellites, their orientation to get the line-of-sight pointing, the non-conservative forces, and all of the potential contributions noted above, including glacial isostatic adjustment, tides, and ocean and atmosphere mass redistribution.

The GRACE project provides Level 1B (L1B) data which includes:

1 the GPS tracking range and phase observations as well as the position and velocity of the satellites, and timing of observations determined from the GPS tracking;
2 the orientation of the satellites with respect to the inertial frame as a time series of quaternions determined from the onboard star cameras;
3 observations of all non-conservative forces from the onboard accelerometers;
4 inter-satellite biased range, range-rate and range-acceleration observations (along with important light time and geometric corrections) from the onboard K-band ranging system.

For an excellent detailed discussion of the GRACE L1B data product, the reader is directed to the *GRACE Level 1B Data Product User Handbook* (Case *et al.*, 2010).

Time variable gravity is obtained as a time series of delta corrections to an *a priori* geopotential model. The typical time step is on the order of monthly to ten days. The geopotential delta corrections are estimated by minimizing a measure of the difference between computed inter-satellite ranging observations, based on the *a priori* geopotential, and the observed inter-satellite ranging data provided by the

GRACE L1B product. As discussed in detail in the next section, the geopotential corrections can be formulated as either a time series of delta Stokes coefficients or surface mass concentration parameters, expressed in equivalent centimeters of water height for a specified area. The L1B data are processed in a "state of the art" precision orbit determination and geodetic parameter estimation system, such as NASA Goddard Space Flight Center's GEODYN.

These systems provide the necessary reference frame, satellite dynamics, and force modeling infrastructure to compute the inter-satellite ranging residuals (difference between the computed and observed) and the partial derivatives of the delta geopotential parameters with respect to the inter-satellite ranging. The GRACE mission L1B data processing centers (Jet Propulsion Laboratory (JPL), CA; Center for Space Research (CSR), TX; Helmholtz Centre Potsdam GFZ German Research Centre for Geosciences (GFZ), Potsdam, Germany) all use their own sophisticated processing systems but, for the purposes of this chapter, we will use the GEODYN system and processing approach.

The contributions to the computed inter-satellite ranging observations from the Earth's static gravity field, planetary bodies, ocean tides, solid Earth and pole tides are computed within GEODYN. GEODYN is also used to calibrate the accelerometer data, in order to compute the contribution of non-conservative forces (Luthcke *et al.*, 2006b). The contributions from atmospheric and ocean mass variations, and in particular high-frequency variations that would otherwise alias the longer-term (10–30 days) estimated geopotential corrections, must be accounted for during the computation of the inter-satellite ranging residuals and partial derivatives. This is accomplished during the L1B data processing through forward modeling, either using the GRACE mission Atmosphere and Ocean De-aliasing Level 1B data product (AOD1B) (Flechtner, 2007), or a comparable atmosphere and ocean forward model (Luthcke *et al.*, 2006b).

10.3 Surface mass variability from GRACE

The time variable gravitational potential of the Earth can be expressed as a delta geopotential from the static geopotential, and any time-varying gravitational effects one wishes to forward model, as noted in Equation 10.2, and discussed above. For any point on, or above, the surface of the Earth, the time-dependent delta gravitational potential can be expressed in a spherical harmonic expansion, assuming the total mass of the Earth system is time invariant (Hofmann-Wellenhof and Moritz, 2005):

$$\Delta U(r, \lambda, \varphi, t)$$

$$= \frac{GM}{R} \sum_{l=0}^{l_{max}} \sum_{m=0}^{l} \left(\frac{R}{r}\right)^{l+1} \overline{P}_{lm}(\sin \lambda) \left\{ \Delta \overline{C}_{lm}(t) \cos(m\varphi) + \Delta \overline{S}_{lm}(t) \sin(m\varphi) \right\}$$

$$(10.3)$$

where:

$$
\begin{aligned}
\Delta U &= \text{Delta gravitational potential} \\
r, \lambda, \varphi &= \text{Spherical geocentric coordinates of computation point} \\
&\qquad \text{(radius, latitude, longitude)} \\
GM &= \text{The product of the gravitational constant and the mass of} \\
&\qquad \text{the Earth} \\
R &= \text{Mean semi-major axis of the Earth} \\
l, m &= \text{Degree and order of spherical harmonic} \\
\overline{P}_{lm} &= \text{Normalized associated Legendre polynomials} \\
\Delta\overline{C}_{lm}(t), \Delta\overline{S}_{lm}(t) &= \text{Delta Stokes coefficients fully normalized.}
\end{aligned}
$$

As discussed in the previous section, the GRACE data centers process the L1B data and estimate delta gravitational potential fields (gravity fields) from the reduction of the inter-satellite ranging residuals. This Level-2 data product is provided to the public most typically as a time series of spherical harmonic (SH) (Stokes) coefficients. For example, the University of Texas Center for Space Research (CSR) Release-5 Level 2 product is a time series of monthly SH coefficients up to degree and order 60 (Bettadpur *et al.*, 2012). These fields have the atmosphere and oceans mass variations removed, as the AOD1B atmosphere and ocean model product was forward modeled in the processing of the L1B data. The gravity fields are estimated in the Earth's center of mass frame and, therefore, the degree 1 SH coefficients are zero (Hofmann-Wellenhof and Moritz, 2005).

Temporal variations of the gravity field are primarily driven by the global redistribution of water mass that occurs at or near the surface of the Earth. These changes in surface mass are often expressed as changes in density of a thin layer at the Earth's surface. Equation 10.4 relates the delta Stokes coefficients of Equation 10.3 to the change in surface density, $\Delta\sigma$, to a point on the Earth's surface.

$$
\Delta\sigma(\lambda, \varphi, t)
$$

$$
= \frac{M}{4\pi R^2} \sum_{l=0}^{l_{max}} \sum_{m=0}^{l} \overline{P}_{lm}(\sin\lambda)\frac{2l+1}{1+k'_l}\left\{ \Delta\overline{C}_{lm}(t)\cos(m\varphi) + \Delta\overline{S}_{lm}(t)\sin(m\varphi) \right\}
$$

$$(10.4)$$

where k'_l is the Loading Love number of degree l, to account for the Earth's elastic yielding which in general counteracts the additional surface density.

The surface density change, $\Delta\sigma$, has units of kg/m^2, and can be converted to centimeters of equivalent water height by noting that the density of water is 1000 kg/m^3: $\Delta H(t) = \Delta\sigma/10$.

Although Equation 10.4 is rigorously correct, it cannot be used with delta Stokes coefficients estimated from GRACE to compute surface mass density at individual point locations. Two problems with the GRACE SH coefficients prevent this. The first problem is the errors in those coefficients (Wahr *et al.*, 2006). The geographic

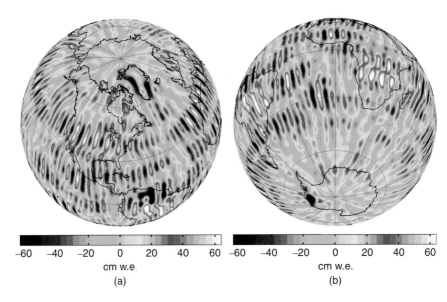

$$-60 \quad -40 \quad -20 \quad 0 \quad 20 \quad 40 \quad 60 \qquad -60 \quad -40 \quad -20 \quad 0 \quad 20 \quad 40 \quad 60$$

cm w.e.   cm w.e.

(a)   (b)

Figure 10.1 Surface mass anomaly from the mean for March 2010, computed directly from CSR RL05 Level-2 monthly spherical harmonics without filtering.

patterns of those errors exhibit significant north-south striping artifacts, as illustrated in Figure 10.1.

To a large part, these meridional striping patterns are due to the aliasing effects of un-modeled short-term temporal variations (Wiehl and Dietrich 2005; Swenson and Wahr 2006; Seo *et al.*, 2008). Over a given GRACE solution period (e.g., one month), the SH coefficient increments that are adjusted with respect to the background model are assumed to be constant, and therefore mass changes within the solution period that are not modeled cannot be captured by this adjustment. In part, they are absorbed by the adjusted spatial variations, thus introducing artifacts into the solution that typically appear as stripes following the orbital sampling pattern. The magnitude of the stripes is reduced toward the high latitudes, due to the fact that the orbit tracks converge at the pole, resulting in the reduction of aliasing due to better spatio-temporal sampling. Improvements in the atmospheric and oceanic models, and strategies to account for short-term variations in the gravity adjustment, are approaches that are being pursued to mitigate the effects of aliasing.

The second problem associated with the direct use of Equation 10.4 is signal leakage, which has multiple causes. The primary cause is the finite expansion used in Equation 10.4 (usually $l_{max} = 60$ for time variable gravity derived from GRACE). Consider the delta Stokes coefficients that represent a change in mass corresponding to an extra 1 cm of water standing over a 1×1 arc degree region. If this set of coefficients is truncated at degree 60, the total mass is correct but it is spread out past the original region boundaries. Signal has leaked out, leaving less in the original area.

For these reasons, GRACE data are most commonly used to quantify the mass changes within a particular region, such as an ice sheet or drainage basin.

Using Equation 10.4 to get an average value over a substantial area diminishes striping and leakage. The process of averaging is further improved with filtering techniques to mitigate striping and leakage. The average surface density change over a region can be obtained by the following expression (Swenson and Wahr, 2002):

$$\overline{\Delta\sigma_{region}}(\lambda, \varphi, t) = \frac{1}{\Omega_{region}} \int \Delta\sigma(\lambda, \varphi, t)\vartheta(\lambda, \varphi)d\Omega \tag{10.5}$$

where $\vartheta(\lambda, \varphi)$ is defined as 1 inside the region and 0 outside, $d\Omega = \cos \lambda d\lambda d\varphi$ is an element of solid angle, and Ω_{region} is the angular area of the region. Combining Equations 10.4 and 10.5, this can be rewritten as (Swenson and Wahr, 2002):

$$\overline{\Delta\sigma_{region}}(\lambda, \varphi, t) = \frac{M}{4\pi R^2 \Omega_{region}} \sum_{l=0}^{l_{max}} \sum_{m=0}^{l} \frac{2l+1}{1+k_l'} \left\{ \vartheta_{lm}^C \Delta\overline{C}_{lm}(t) + \vartheta_{lm}^S \Delta\overline{S}_{lm}(t) \right\} \tag{10.6}$$

where ϑ_{lm}^C and ϑ_{lm}^S are the spherical harmonic coefficients that describe $\vartheta(\lambda, \varphi)$ which are related by:

$$\vartheta(\lambda, \varphi) = \frac{1}{4\pi} \sum_{l=0}^{l_{max}} \sum_{m=0}^{l} \overline{P}_{lm}(\sin \lambda) \left\{ \vartheta_{lm}^C \cos(m\varphi) + \vartheta_{lm}^S \sin(m\varphi) \right\} \tag{10.7}$$

$$\begin{Bmatrix} \vartheta_{lm}^C \\ \vartheta_{lm}^S \end{Bmatrix} = \int \vartheta(\lambda, \varphi)\overline{P}_{lm}(\sin \lambda) \begin{Bmatrix} \cos(m\varphi) \\ \sin(m\varphi) \end{Bmatrix} d\Omega \tag{10.8}$$

As l_{max} approaches infinity, Equations 10.5 and 10.6 yield nearly identical results. For the $l_{max} = 60$ case, Equation 10.6 may reduce signal attenuation and leakage for a particular area of interest. In Equation 10.4, $\Delta\sigma$ truncated to degree 60 has leakage problems and is integrated only over the area of interest. The $\vartheta(\lambda, \varphi)$ function computed in Equation 10.7 also has leakage problems and extends out past the original area of interest. Furthermore, Equation 10.7 gives a form that is convenient for filtering with averaging kernels.

The most common method for extracting localized mass variations from the GRACE project SH coefficients is to apply an averaging kernel, $\overline{W}(\lambda, \varphi)$, in Equation 10.6 in place of $\vartheta(\lambda, \varphi)$ (Swenson and Wahr, 2002). This kernel is computed by convolving the exact kernel, $\vartheta(\lambda, \varphi)$, with a Gaussian filter, resulting in a kernel that smoothly changes from a value of 1 iside the region, to a value of 0 outside the region, over a distance approximately equal to the design smoothing radius. As with the filtering of the global SH coefficients mentioned above, the motivation here is to extract as much signal as possible, while reducing the effect of the errors at the highest degrees.

It is important that this averaging kernel be calibrated to account for the fact that the smoothing function attenuates the signal. The computed calibration factor typically has a value slightly greater than one. It should be noted that the application of the averaging kernel introduces a leakage of signal from outside the region of interest that is caused by the fact that $\overline{W}(\lambda, \varphi)$ is an approximation of $\vartheta(\lambda, \varphi)$.

Typically, the noise covariance of the estimated Stokes coefficients is not taken into account in the post-solution filtering methods, and therefore the filters are not optimal (Klees *et al.*, 2008). Signal attenuation, limited spatio-temporal resolution, and signal leakage in and out of the domain of interest are particular problems when applying the filtering methods. An alternative method is to estimate surface mass concentrations (mascons) directly from the reduction of the GRACE inter-satellite range-rate observation residuals. The mascon technique uses geo-locatable anisotropic constraints to estimate the global mass change directly from GRACE KBRR data, accounting for the full Stokes noise covariance (Luthcke *et al.*, 2006a, 2008; Rowlands *et al.*, 2010; Sabaka *et al.*, 2010; Luthcke *et al.*, 2013). Mascon parameters can be derived using the following relationship, that a change in the gravitational potential caused by adding a small uniform layer of mass over a region at an epoch *t* can be represented as a set of (differential) potential coefficients that can be added to the mean field. The delta coefficients can be computed as in Chao and Au (1987):

$$\Delta \overline{C}_{lm}(t) = \left\{ 10 \left(1 + k_l'\right) R^2 (2l + 1) M \int \overline{P}_{lm}(\sin \lambda) \cos m\varphi d\Omega \right\} \Delta H(t) \qquad (10.9)$$

$$\Delta \overline{S}_{lm}(t) = \left\{ 10 \left(1 + k_l'\right) R^2 (2l + 1) M \int \overline{P}_{lm}(\sin \lambda) \cos m\varphi d\Omega \right\} \Delta H(t) \qquad (10.10)$$

Following Rowlands *et al.* (2005), the estimated mascon parameter, $H_j(t)$, for each mascon area *j*, is a scale factor on the set of differential Stokes coefficients for that mascon area, giving a surface mass change in equivalent centimeters of water. Assembling in matrix notation over *j* mascons, using the above equations provides a set of partial derivatives, L, of the differential Stokes coefficients, Δs, with respect to equivalent water height, Δh, such that

$$\Delta s = L\Delta h \qquad (10.11)$$

The matrix L is pre-computed one time for the defined set of mascons using Equations 10.9 and 10.10 and setting $\Delta H(t)$ to unity.

Standard processing of GRACE L1B KBRR data is usually geared to the estimation of differential Stokes coefficients that represent global change over a chosen time period, such as the GRACE project monthly spherical harmonic products. In orbital software (e.g., GEODYN) that estimate Stokes coefficients from KBRR data, the partial derivative of each KBRR observation, with respect to every estimated Stokes coefficient, is routinely computed. To change the parameterization from global Stokes coefficients to local mascons, all that is required is to compute the partial derivative of each KBRR observation with respect to all of the local mascons. This is accomplished with a change of basis, using the matrix L defined above, to post-multiply the matrix of partial derivatives of the KBRR observations with respect to the Stokes coefficients normally computed in the L1B data processing (see Equation 10.12).

With the change of basis, L, it is then easy to convert the process of estimation of Stokes coefficients to the estimation of mascons. Some researchers (e.g., Jacob *et al.*, 2012), use L to estimate mascons from GRACE level 2 Stokes coefficients, instead of from the original KBRR observations. This is an alternative to the averaging kernel approach described earlier (Equations 10.6–10.8).

The mass distribution described by the matrix L is not a step function of 1 cm water over the area of a mascon set of interest, because it is typically truncated to degree 60, limited by the fundamental resolution of the GRACE observations. This is an advantage when recovering signal leakage out, because L effectively pulls back the signal into the intended mascon region during the estimation process. However, there is now the issue of leakage of signal entering from surrounding regions. As described below, these leakage problems are further mitigated with the application of spatial constraints in the mascon estimation process.

The mascon parameters are non-linear functions of the GRACE inter-satellite K-band range-rate (KBRR) observations, and are therefore estimated using a non-linear least squares Gauss-Newton (GN) method (Seber and Wild, 1989). The application of the GN method to the estimation of the mascon parameters at iteration k, can be expressed following Sabaka *et al.* (2010) and Luthcke *et al.* (2013):

$$\text{GN iteration } k \left\{ \quad \widetilde{\mathbf{h}}_{k+l} = (L^T A^T WAL + \mu P_{hh})^{-1} L^T A^T W(\mathbf{r} + AL\widetilde{\mathbf{h}}_k) \right. \qquad (10.12)$$

where:

$\widetilde{\mathbf{h}}_{k+l}$ = Update of mascon parameters in equivalent cm of water

  L = Partial derivatives of the differential Stokes coefficients with respect to the mascon parameters

  A = Partial derivatives of KBRR observations with respect to the Stokes coefficients

  W = Data weight matrix which accounts for measurement noise and orbital arc parameters

P_{hh} = Mascon regularization matrix

  μ = Regularization matrix damping parameter

  $\mathbf{r}$ = KBRR residuals; difference between observations and model prediction.

The above is an optimal filter, taking into account the noise covariance. A regularization matrix is used, which is constructed from anisotropic constraints that are applied using geophysical boundaries. The constraints are constructed in groups that represent geographical regions so that, if two mascons reside within the same region, the constraint weight is a simple exponential function of the distance and time between the two mascons. If two mascons reside in different regions, the weight is zero, and the constraint vanishes.

The result is a smoothed solution that limits leakage across regional boundaries (e.g., Greenland ice sheet or Gulf of Alaska glacier region). Furthermore, the solution is iterated while minimizing the KBRR observation residuals to recover the full signal and minimizing attenuation (Luthcke *et al.*, 2013). Figure 10.2 shows the monthly surface mass anomaly from a recent iterated global mascon solution (Luthcke *et al.*, 2013). The improvements in signal-to-noise, due to the

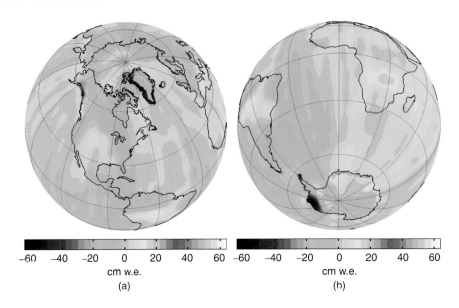

-60 -40 -20 0 20 40 60   -60 -40 -20 0 20 40 60

cm w.e.   cm w.e.

(a)   (b)

Figure 10.2 Surface mass anomaly from the mean for March 2010 computed from a recent global mascon solution with geographic constraints.

regularization and iteration applied in the estimation of the global mascons from the minimization of the GRACE KBRR data, can be seen by comparing with Figure 10.1.

In addition to the mitigation of signal attenuation and leakage, there are additional challenges in extracting ice sheet and glacier region mass balance from GRACE. The GRACE observations measure all contributions to the Earth's temporal and spatial mass variability; therefore, it is necessary to remove the non-glacier signals through independent datasets and models. Models of tidal effects, as well as non-tidal atmospheric and oceanic mass redistributions, are already employed in the gravity field processing, in order to minimize aliasing of the high-frequency mass change signals, thus removing these signals from the gravity field solutions within their model errors (Flechtner, 2007; Bettadpur *et al.*, 2012).

In many terrestrial ice applications, Glacial Isostatic Adjustment (GIA) and terrestrial water storage (TWS) signals are significant in comparison to the ice signals themselves, and therefore must be removed and accounted for in the error analysis (Shepherd *et al.*, 2012; Luthcke *et al.*, 2008). Errors in the atmosphere and ocean forward modeling must also be considered.

10.4   Results

A comprehensive review of the many applications of GRACE data to cryospheric science, over the course of nearly a decade, is beyond the scope of this chapter. However, to illustrate the power of the GRACE observations to observe terrestrial ice mass evolution, we summarize the results from a global mascon solution (Luthcke *et al.*, 2013) that was used in the international Ice Mass Balance Intercomparison Exercise (IMBIE) (Shepherd *et al.*, 2012). It is important to note that, for the duration of the GRACE mission, the processing procedures have been continually enhanced and more accurate forward models have become available, resulting in improved time variable gravity solutions and science products. As the processing methods and models continue to improve, and the time series of data continues to grow, it is acknowledged that the results now presented will also be updated and improved.

Here we summarize the mascon solution presented in (Luthcke *et al.*, 2013). The mascons are estimated globally with ten-day temporal and one-arc-degree equal area spatial sampling. The ice sheet mascons are restricted to grounded ice zones, and are grouped according to major ice regions, applying anisotropic constraints as discussed in the previous section. Mass variations from tides, oceans, atmosphere, and hydrology are forward modeled. For both the Greenland Ice Sheet (GIS) and Gulf of Alaska glaciers (GOA), a GIA model based on the ICE-5G deglaciation history and an incompressible two-layer approximation to the VM2 viscosity profile is used to correct the GRACE mascon solution (Paulson *et al.*, 2007, computed and provided by Geruo A).

For the Antarctic Ice Sheet (AIS), the IJ05_R2 regional GIA model is used (Ivins *et al.*, 2012). The mascon solutions are corrected for the viscous component of post-Little Ice Age (LIA) GIA following the collapse of the Glacier Bay Icefield. As in Luthcke *et al.* (2008), a regional post-LIA GIA model is applied, which is rigidly constrained by approximately 100 precise GPS and Relative Sea Level (RSL) observations of uplift (Larsen *et al.*, 2005). The mascon solutions are further corrected to reflect surface mass variability in the Earth's center of figure frame rather than the center of mass frame in which the solutions are performed. The geocenter correction used in the IMBIE study (Shepherd *et al.*, 2012), derived from the degree 1 Stokes coefficients, determined from Swenson *et al.* (2008), is applied. The geocenter correction accounts for approximately 1% of the GIS and GOA ice trends, and less than 3% for the West AIS and AIS Peninsula ice trends. The largest impact of the geocenter correction is for East AIS at 17% of the trend.

The mascon solution results are summarized in Figures 10.3 and 10.4 for the five most important terrestrial ice regions in terms of contribution to sea level rise: GIS, GOA, East, West and Peninsula regions of the Antarctic Ice Sheet (EAIS, WAIS, and AISP). The figures show mass change results for April 1, 2003 to December 1, 2010, while all reported mass trends are computed over the span of integer number of years, December 1, 2003 to December 1, 2010. It is also important to note that mean annual net balances do not exactly match the mass trends, as they are computed over slightly different time periods (Luthcke *et al.*, 2013). A detailed discussion of these results is provided in Luthcke *et al.* (2013), while here we simply point to a few highlights.

The most important contribution of GRACE to cryospheric sciences is the excellent temporal resolution of mass change observable over the entire areas of the ice sheets and glacier regions. Observations of changes in mass, ranging from ≈15 days to seasonal, annual and decadal, are possible using GRACE. GRACE has

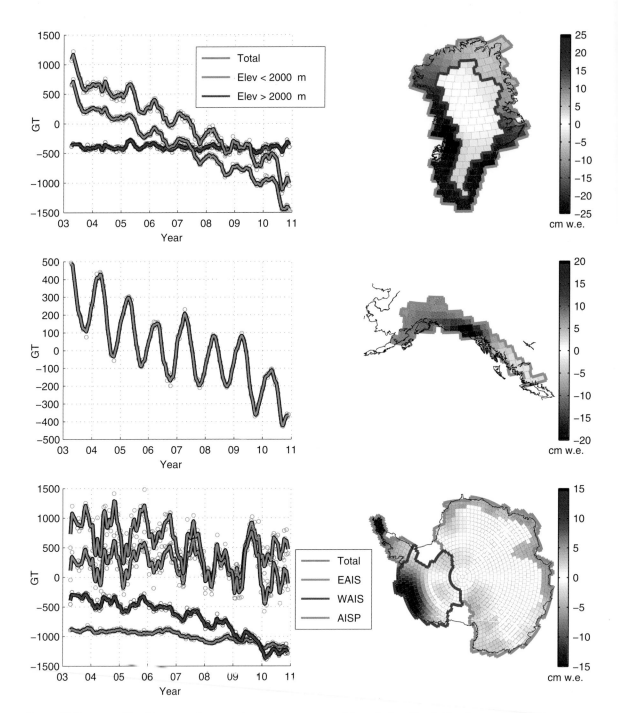

Figure 10.3 Regional land ice mass change and spatial maps of average annual (net) mass balance. The time series of ice mass change includes the ten-day mascon solutions (circles), and the Gaussian smoothed signal with a ten-day smoothing window (lines). The time series colors correspond to the regional outlines defined on each spatial map.

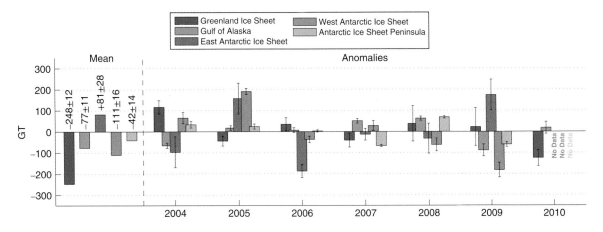

Figure 10.4 Mean and anomalies of annual (net) mass balances by region. The provided error bars define the $1 - \sigma$ uncertainties for each computed balance value.

quantified not only the long-term trends of terrestrial ice, but also the seasonal and inter-annual net balance. Referencing Figures 10.3 and 10.4, we see that the GIS is dominated by long-term negative mass trend from the low elevation margins of the ice sheet. The losses are largest from both the Jakobshavn glacier region on the west coast and the Helheim glacier region on the east coast. The observed mass trend of the GIS over this time period is -230 ± 12 Gt a^{-1}. The error bars include the contributions from the mascon solution itself, forward modeling and correction errors, and signal leakage in and out of the region of interest (Luthcke *et al.*, 2013). Significant annual net balance anomalies are observed for the 2004 balance year (positive) and 2010 balance year (negative), giving rise to an apparent -10 ± 6 Gt a^{-2} acceleration of mass loss.

Unlike the GIS, both the magnitude and uncertainties of the GIA corrections for the AIS are significant with respect to the total trend. The overall AIS trend for the 12/2003 to 12/2010 time period studied is -81 ± 26 Gt a^{-1}. The AIS trend is dominated by significant mass loss from the West AIS (WAIS) with a trend of -106 ± 16 Gt a^{-1}. A trend of -38 ± 14 Gt a^{-1} is found for the AIS Peninsula, while the East AIS trend is 63 ± 28 Gt a^{-1}.

The largest long-term mass losses are found in the WAIS along the Amundsen and Bellingshausen Sea coasts, concentrated along the Pine Island embayment, and at the northern tip of the Peninsula. The WAIS is of particular concern for sea level rise, as this region exhibits the largest accelerated mass loss at -46 ± 6 Gt a^{-2}. While inter-annual variability of the WAIS annual net balances is observed, there is a consistent negative trend in these data (see Figures 10.3 and 10.4), which represents the largest persistent multi-year acceleration of mass loss. The EAIS shows a significant 2009 positive annual net balance concentrated along the Queen Maud Land and Wilkes land coasts, due to a significant accumulation anomaly (Luthcke *et al.*, 2013, Shepherd *et al.*, 2012).

The EAIS, WAIS and Antarctic Peninsula all show very significant annual net balance anomalies relative to the mean annual net balance, with anomalies in some years notably larger than the mean annual net balance. This is in contrast to the GIS, where annual net balance anomalies are only a fraction of the mean annual net balance.

Even with twenty times less ice-covered area than the GIS, the GOA glaciers have lost mass at a significant rate of -69 ± 11 Gt a^{-1} for the time period 12/2003 to 12/2010, and a mean annual net balance of -77 ± 11 Gt a^{-1} for the balance years 2004 to 2010. The largest mass losses occur in the St. Elias and Glacier Bay regions, and the GOA region, as a whole, exhibits significant inter-annual and seasonal variation compared to the long-term trend (Figure 10.4).

What is of particular interest is the rapid response of the GOA region to climate change. Large negative annual mass balance anomalies are observed in 2004, due to record warm temperatures across the region (Truffer *et al.*, 2005), and in 2009, most likely attributed to the March 31, 2009 eruption of Mt. Redoubt. The eruption spread volcanic ash across much of the GOA region (Schaefer *et al.*, 2012), likely enhancing melt rates through a reduction in surface albedo (Arendt *et al.*, 2013). The large positive annual net balance anomaly observed in 2008 is the result of the large 2007–2008 winter accumulation, followed by a cool 2008 summer season.

10.5 Conclusions

Through the measurement of the changes in range and range-rate between two low Earth co-orbiting satellites, the GRACE mission has provided unprecedented observations of surface mass variation. These surface mass change observations have revolutionized our understanding of the Earth's water cycle and, in particular, the mass evolution of land ice. The GRACE observations and their unique processing and analyses have made it possible to accurately quantify the decadal, annual and seasonal mass balance of the ice sheets and smaller glacier systems, with temporal resolution on the order of $\approx$15 days and spatial resolution on the order of $\approx$300 km.

Future improvements in GRACE processing and analysis techniques will likely further improve the accuracy and resolution of the surface mass observations. Finally, the GRACE follow-on mission (launch $\approx$2016–17) and GRACE-II ($\approx$2022) will provide the continuity of observations to understand long-term land ice change, sea level rise, and will provide crucial observations for improved modeling and forecasting.

References

Arendt, A., Luthcke, S., Gardner, A., O'Neel, S., Hill, D., Moholdt, G. & Abdalati, W. (2013). Analysis of a GRACE global mascon Solution for Gulf of Alaska Glaciers. *Journal of Glaciology* **59**(217), doi: 10.3189/2013JoG12J197.

Bettadpur, S. & the CSR Level-2 Team (2012). *Assessment of GRACE mission performance and the RL05 gravity fields.* Paper G31C-02, AGU Fall Meeting, San Francisco, CA.

Case, K., Kruisinga, G. & Wu, S.C. (2010). *GRACE Level 1B Data Product User Handbook.* JPL D-22027, Jet Propulsion Laboratory, CA, USA.

Chao, B.F. & Au, A. (1987). Snow load effect on the Earth's rotation and gravitational field. 1979–1985. *Journal of Geophysical Research* **92**(B9), 9415–9422.

Chen, J.L., Wilson, C.R. & Tapley, B.D. (2011). Interannual variability of Greenland ice losses from satellite gravimetry. *Journal of Geophysical Research* **116**, B07406, 1–11, doi: 10.1029/2010JB007789.

Flechtner (2007). *AOD1B Product Description Document* (2007). GR-GFZ-AOD-0001, GeoForschungszentrum, Potsdam, Germany.

Hofmann-Wellenhof, B. & Moritz, H. (2005). *Physical geodesy.* Vienna, Springer-Verlag.

Horwath, M. & Dietrich, R. (2009). Signal and error in mass change inferences from GRACE: the case of Antarctica. *Geophysical Journal International* **177**(3) 849–864, doi: 10.1111/j.1365-246X.2009.04139.x.

Ivins, E.R., James, T.S., Wahr, J., Schrama, E.J.O., Landerer, F.W. & Simon, K. (2012). Antarctic contribution to sea-level rise observed by GRACE with improved GIA correction, *Journal of Geophysical Research* (In Press) (2012JB009730).

Jacob, T., Wahr, J., Pfeffer, W.T. & Swenson, S. (2012). Recent contributions of glaciers and ice caps to sea level rise. *Nature* **482**(7386) 514–518, doi: 10.1038/nature10847.

King, M.A., Bingham, R.J., Moore, P., Whitehouse, P.L., Bentley, M.J. & Milne, G.A. (2012). Lower satellite gravimetry estimates of Antarctic sea level contribution. *Nature* **49**, 586–589, doi: 10.1038/nature11621.

Klees, R., Revtova, E.A., Gunter, B.C. *et al.* (2008). The design of an optimal filter for monthly GRACE gravity models. *Geophysical Journal International* **175**, 417–432.

Larsen, C.F., Motyka, R.J., Freymueller, J.T., Echelmeyer, K.A. & Ivins, E.R. (2005). Rapid viscoelastic uplift in southeast Alaska caused by post-Little Ice Age glacial retreat. *Earth and Planetary Science Letters* **237**(3–4), 548–560

Lemoine, F.G., Kenyon, S.C., Factor, J.K. *et al.* (1998). *The Development of the Joint NASA GSFC and the National Imagery and Mapping Agency (NIMA) Geopotential Model EGM96.* NASA/TP-1998-206861, NASA Goddard Space Flight Center, Greenbelt, MD 20771.

Luthcke, S.B., Zelensky, N.P., Rowlands, D.D., Lemoine, F.G., Williams, T.A. (2003). The 1-centimeter orbit: Jason-1 precision orbit determination using GPS, SLR, DORIS and altimeter data. *Marine Geodesy* **26**, 399–421, doi: 10.1080/01490410390256727.

Luthcke, S.B., Zwally, H.J., Abdalati, W. *et al.* (2006a). Recent Greenland ice mass loss by drainage system from satellite gravity observations. *Science* **314**(5803), 1286–1289.

Luthcke, S.B., Rowlands, D.D., Lemoine, F.G., Klosko, S.M., Chinn, D. & McCarthy, J.J. (2006b). Monthly spherical harmonic gravity field solutions determined from GRACE inter-satellite range-rate data alone. *Geophysical Research Letters* **33**, L02402, doi: 10.1029/2005GL024846.

Luthcke, S.B., Arendt, A.A., Rowlands, D.D., McCarthy, J.J. & Larsen, C.F. (2008). Recent glacier mass changes in the Gulf of Alaska region from GRACE mascon solutions. *Journal of Glaciology* **54**, 767–777.

Luthcke, S.B., T.J. Sabaka, B.D. Loomis, A.A. Arendt, J.J. McCarthy, J. Camp (2013). Antarctica, Greenland and Gulf of Alaska land ice evolution from an iterated GRACE global mascon solution. *Journal of Glaciology* **59**(216), 613–631, doi: 10.3189/2013jJoG12j147.

Paulson, A., Zhong, S. & Wahr, J. (2007). Inference of mantle viscosity from GRACE and relative sea level data. *Geophysical Journal International* **171**, 497–508.

Rowlands, D.D., Luthcke, S.B., Klosko, S.M. *et al.* (2005). Resolving mass flux at high spatial and temporal resolution using GRACE intersatellite measurements. *Geophysical Research Letters* **32**(L04310), doi: 10.1029/2004GL021908.

Rowlands, D.D., Luthcke, S.B., McCarthy, J.J. *et al.* (2010). Global mass flux solutions from GRACE; a comparison of parameter estimation strategies: mass concentrations versus Stokes coefficients. *Journal of Geophysical Research* **115**, B01403, doi:10.1029/2009JB006546.

Sabaka, T.J., Rowlands, D.D., Luthcke, S.B. & Boy, J.–P. (2010). Improving global mass-flux solutions from GRACE through forward modeling and continuous time-correlation. *Journal of Geophysical Research* **115**, B11403, doi:10.1029/2010JB007533.

Schaefer, J.R., Bull, K., Cameron, C. *et al.* (2012). *The 2009 eruption of Redoubt Volcano, Alaska, report of investigations 2011-5.* State of Alaska Department of Natural Resources.

Schrama, E.J.O. & Wouters, B. (2011). Revisiting Greenland ice sheet mass loss observed by GRACE. *Journal of Geophysical Research* **116**, B02407, doi: 10.1029/2009JB006847.

Seber, G.A.F. & Wild, C.J. (1989). *Nonlinear regression.* New York, Wiley and Sons, 768 pp.

Seo, K.-W., Wilson, C.R., Chen, J. & Waliser, D.E. (2008). GRACE's spatial aliasing error. *Geophysical Journal International* **172**, 41–48, doi: 10.1111/j.1365-246X.2007.03611.x.

Shepherd, A., Erik, R. Ivins, Geruo, A. *et al.* (2012). A reconciled estimate of ice-sheet mass balance. *Science* **338**(1183), doi: 10.1126/science.1228102.

Swenson, S. & Wahr, J. (2002). Methods for inferring regional surface-mass anomalies from gravity recovery and climate experiment (GRACE) measurements of time-variable gravity. *Journal of Geophysical Research* **107**(B9), 2193, doi: 10.1029/2001JB000576.

Swenson, S. & Wahr, J. (2006). Post-processing removal of correlated errors in GRACE data. *Geophysical Research Letters* **33**, L08402, doi: 10.1029/2005GL025285.

Swenson, S., Chambers, D. & Wahr, J. (2008). Estimating geocenter variations from a combination of GRACE and ocean model output. *Journal of Geophysical Research* **113**, B08410, doi: 10.1029/2007JB005338.

Tapley, B.D., Bettadpur, S., Ries, J.C., Thompson, P.F. & Watkins, M.M. (2004). GRACE measurements of mass variability in the earth system. *Science* **305**, 503–505, doi: 10.1126/science.1099192.

Truffer, M., Harrison, W.D. & March, R.S. (2005). Record negative glacier balances and low velocities during the 2004 heat wave in Alaska: implications for the interpretation by Zwally *et al.* in Greenland. *Journal of Glaciology* **51**(175), 663.

Velicogna, I. (2009). Increasing rates of ice mass loss from the Greenland and Antarctic ice sheets revealed from GRACE. *Geophysical Research Letters* **36**, L19503, doi: 10.1029/2009GL040222.

Wahr, J., Swenson, S. & Velicogna, I. (2006). The accuracy of GRACE mass estimates. *Geophysical Research Letters* **33**, L06401, doi: 10.1029/2005GL025305.

Wiehl, M. & Deitrich, R. (2005). Time-variable gravity seen by satellite missions: On its sampling and its parametrization. In: Reighber, C., Lühr, H., Schwintzer, P. & Wickert, J. (eds). *Earth Observation with CHAMP: Results from Three Years in Orbit*, pp.121–126. Berlin, Springer.

Wolff, M. (1969). Direct measurements of the earth's gravitational potential using a satellite pair. *Journal of Geophysical Research* **74**(22), 5295–5300.

Acronyms

GRACE	Gravity Recovery and Climate Experiment
KBRR	K-band range-rate
POD	Precision Orbit Determination
SLR	Satellite Laser Ranging
GPS	Global Positioning System
DORIS	Doppler Orbitography and Radiopositioning Integrated by Satellite
AOD1B	Atmosphere and Ocean De-aliasing Level 1B data product
SH	Spherical harmonic
GN	Gauss-Newton
GIA	Glacial Isostatic Adjustment (GIA)
TWS	terrestrial water storage
IMBIE	Ice Mass Balance Intercomparison Exercise
LIA	Little Ice Age
RSL	Relative Sea Level
AIS	Antarctica Ice Sheet
GIS	Greenland Ice Sheet
WAIS	West Antarctica Ice Sheet
EAIS	East Antarctica Ice Sheet
GOA	Gulf of Alaska

11

Remote sensing of sea ice

Walter N. Meier and Thorsten Markus

NASA Goddard Space Flight Center, Greenbelt, USA

Summary

Sea ice is an important part of the climate system, primarily through its reflection of solar radiation and modification of heat and moisture fluxes. It also affects polar ecosystems and human activities in ice-covered regions. Monitoring of sea ice is difficult, due to its remoteness and extreme climate. Thus, remote sensing represents the best, and often only, option for long-term and large-scale sea ice observations. Fortunately, several sea ice properties can be detected in a wide range of sensors including passive and active microwave, visible, and infrared.

Sea ice concentration is the longest and most complete satellite time series. A nearly-complete daily record of sea ice concentration has been available since late 1978 from a series of multichannel microwave radiometers. This record yields a significant downward trend in Arctic sea ice cover, particularly during summer, which is one of the iconic indicators of climate change over the past three decades. The trend in the Antarctic is slightly positive, but with large regional and interannual variability.

Sea ice thickness is another critical parameter for understanding climate change in the polar regions. Unfortunately, thickness is a more difficult parameter to retrieve remotely. Until the 1990s, other than *in situ* drill holes, the only information on sea ice thickness came from upward-looking sonars on submarines traversing under the Arctic ice cover. In the 1990s, radar altimeters collected some information on ice thickness, but not from the central Arctic.

The first wide-scale coverage of ice thickness did not begin until NASA's ICESat laser altimeter was launched in 2003. Earlier information on changes in ice thickness can be obtained, based on infrared imagery based on thermal heat transfer through the ice, but it is limited to thin ice, unless modeling is included. Proxy ice age data, based on Lagrangian ice motion tracking, can be used to infer thickness because ice tends to thicken as it ages. This method has been used with passive microwave imagery, buoy tracks, and other sources, to produce age fields since the early 1980s. Microwave imagery can also be used to discriminate first-year ice from multi-year ice during non-melt conditions, because of the differing effects of salinity in the ice on the emission/backscatter signal.

Remote Sensing of the Cryosphere, First Edition. Edited by M. Tedesco.
© 2015 John Wiley & Sons, Ltd. Published 2015 by John Wiley & Sons, Ltd.
Companion Website: www.wiley.com/go/tedesco/cryosphere

Several other parameters are retrievable via satellite, including melt onset and freeze-up dates, snow depth over first-year ice, ice temperature, albedo, and melt pond coverage. New sensors, technologies, and methodologies show promise to continue improving our understanding of sea ice processes.

11.1   Introduction

Sea ice forms when ocean waters reach the freezing point. While most sea ice forms in the polar oceans, it can extend toward the mid-latitudes in some locations, particularly in the Northern Hemisphere where sea ice is found as far south as the Bohai Sea, China, at ≈40°N latitude (Figure 11.1). There is a strong seasonal cycle in both hemispheres. In the Arctic, the extent reaches its minimum in September, when sea ice is largely constrained within the boundaries of the Arctic Ocean; it then grows through the autumn and winter, reaching a maximum in late February or March, when ice extends into surrounding seas, including Bering Sea, the Sea of Okhotsk, Hudson Bay, Baffin Bay, and the Labrador Sea.

Sea ice plays an important role in climate and the polar ecosystem. The high albedo of sea ice relative to the ocean results in much less absorption of solar insolation during the summer. The ice also acts as a physical barrier between the ocean

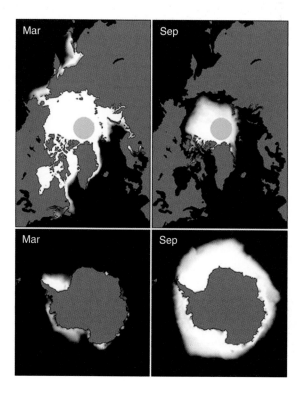

Figure 11.1 Average sea ice coverage in the Northern Hemisphere (top) and Southern Hemisphere (bottom) for March (left) and September (right) for the period 1979–2000. The white shading indicates percent concentration from 15% to 100%. The light gray circle in the Northern Hemisphere images represent a region around the pole that is not imaged by some sensors. (National Snow and Ice Data Center, University of Colorado, Boulder).

and the atmosphere, resulting in a colder, drier winter climate over and near the ice than there would be without ice. When sea ice forms, brine is rejected into the upper layers of the ocean, causing densification of the surface waters. In some locations – particularly in Antarctica and near Greenland – this densification is intense enough for the cold, heavy waters to sink to the bottom; this is an important link in the global ocean thermohaline circulation.

Regions near the sea ice edge can be highly productive biologically because of the mixing of nutrients. Plankton blooms are common during summer near the ice edge, which draws larger animals. The ice is an important platform for birthing, feeding, and transportation among charismatic megafauna (e.g., polar bears, seals, walrus) in the Arctic, and penguins in the Antarctic. Sea ice also plays a role in human activities, particularly in the Arctic, where it is an integral part of indigenous communities (e.g., for hunting and transportation), as well as commercial interests (e.g., shipping, resource extraction, tourism) and national security.

In the Arctic, there has been a significant decline in sea ice over recent decades, and particularly since about 2000, with several record low summer extents (e.g., Cavalieri and Parkinson, 2012). Overall, the area covered by sea ice at the end of summer has been reduced by ≈40% since the early 1980s (Figure 11.2). There are also significant decreasing trends in all other months, and the decrease is essentially pan-Arctic. In the Antarctic, small but statistically significant positive trends have been observed, but with strong year-to-year and regional variability (e.g., Parkinson and Cavalieri, 2012; Stammerjohn *et al.*, 2012). Some regions of the Antarctic show a strong decreasing trend.

These changes in sea ice have been observed using satellite remote sensing platforms. Satellite remote sensing is particularly important for studying the polar

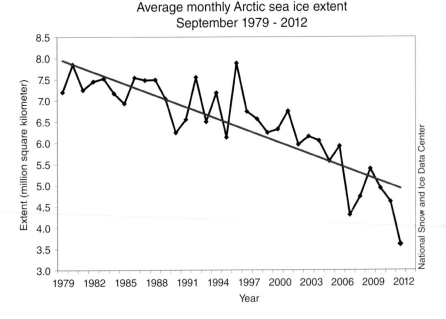

Figure 11.2 Monthly average Arctic September sea ice extent for 1979–2012 (black line and diamonds) and linear trend (blue line). Extent values are estimated from passive microwave radiometer data. (National Snow and Ice Data Center, University of Colorado, Boulder).

regions because of the harsh conditions and limited access at high-latitudes. *In situ* measurements have been collected and are important for satellite validation, process studies, and local climate assessment. However, *in situ* measurements in the polar regions (usually collected during short field camps or from research vessels) are sparse in both space and time.

Autonomous measurements, such as buoys, can provide more continuous observations in sea ice-covered seas, but are still spatially sparse because of the difficult access to deploy the instruments. Also, since sea ice moves with the winds and sea currents, and eventually melts, autonomous instruments have a lifetime of, at most, a few years, assuming that the harsh polar weather does not cause a malfunction sooner – a not uncommon occurrence. The only method to provide a comprehensive, consistent, and continuous record of sea ice conditions is, therefore, through remote sensing observations.

11.2 Sea ice concentration and extent

One key indicator of sea ice conditions is concentration (the fraction area of the ocean covered by ice) and extent (the total surface covered by ice with concentration above a certain threshold, usually the 15% convention used in this chapter). As mentioned above, the decline in Arctic sea ice extent is one of the iconic indicators of climate change in the polar regions. Concentration and extent are also well suited for retrieval by remote sensing instruments, because sea ice is generally well discriminated from the ocean in several regimes of the electromagnetic spectrum. For example, the high albedo of sea ice relative to the ocean is usually easily observable in visible imagery. However, visible and infrared imagery methods are limited, because they are unable to retrieve surface information under clouds, which are often widespread and persistent in the Arctic and Antarctic. In addition, visible imagery is not usable during the long polar nights.

11.2.1 Passive microwave radiometers

Microwave imagery does not have the limitations concerning clouds and solar illumination. Microwave emission and scattering from clouds, and from the atmosphere in general, is small at many frequencies, and the surface emission or scattering from the sea ice in the microwave region is not dependent on solar illumination (Eppler *et al.*, 1992). Thus, the microwave part of the electromagnetic spectrum is particularly useful for observations of sea ice.

Passive microwave radiometers have been in near-continuous operation since 1972, and a consistent record from multichannel radiometers has been collected since late 1978. Passive microwave sensors have relatively wide swaths and are typically flown in polar orbits, resulting in complete coverage (outside of a small circle near the pole) on a daily basis (bi-daily from 1978 to mid-1987). Thus, a consistent and nearly complete daily record of sea ice concentration and extent has been estimated from data collected by spaceborne microwave radiometers, from late 1978 to the present. This represents one of the longest satellite climate records available.

The emitted microwave energy is generally measured by sensors as *brightness temperature*, in units of Kelvin. Essentially, this is the physical temperature of an object, if it were a blackbody at a given frequency and polarization. At microwave frequencies, the Rayleigh-Jeans Law is applicable and the brightness temperature is simply a function of the physical temperature and the emissivity:

$$T_B = \varepsilon \cdot T_S \qquad\qquad\qquad (11.1)$$

where:

T_B is the brightness temperature
T_S is the physical temperature
ε is the emissivity, a measure of how close the emitting body is to a blackbody
   ($\varepsilon = 1$ for blackbody).

The emissivity is a function of the properties of the emitting surface and the microwave frequency and polarization (Eppler *et al.*, 1992). The emissivity of water is highly dependent on the phase state, and the phase state is a much larger factor than physical temperature at many microwave frequencies. This allows microwave radiometers to distinguish between liquid water of the open ocean and frozen sea ice.

The most common radiometer frequencies used for sea ice are at or near 19 GHz and 37 GHz. At these frequencies, the sea ice emission signature clearly contrasts with that from open water (Figure 11.3). The phase state also affects the polarity of the emission. Liquid water is generally more polarized than ice, so differencing two polarization channels at the same frequency provides an even clearer delineation of the ice edge (Figure 11.4)

Salinity is another important factor in the emissivity of sea ice. When sea ice forms, much of the salt is expelled into the upper ocean. However, some salt is initially entrained in highly saline brine pockets within the ice. These pockets slowly drain over time, but the largest change occurs during the summer after ice formation, when meltwater flushes out the brine, significantly lowering the salinity of the ice. Thus, first-year ice (ice that has grown since the end of the previous summer melt season) has a higher salinity than multi-year ice (ice that has survived at least one melt season) in the Arctic (Eppler *et al.*, 1992). The ice salinity differentially affects emission at different frequencies. Differing of frequency channels can be used to discriminate ice age (Figure 11.5). In the Antarctic, there is little multi-year ice, and sea ice formation is complicated by flooding due to the weight of snow on the ice, making ice age discrimination not feasible.

These properties of microwave emission at frequencies and polarizations on satellite microwave radiometers have been used to derive algorithms that use

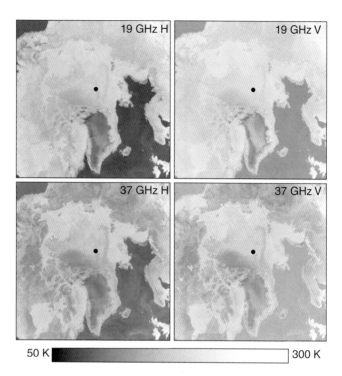

Figure 11.3 Brightness temperature images from March 1, 2008 for 19 GHz (top) and 37 GHz (bottom) at horizontal (left) and vertical (right) polarizations, typical channels used for sea ice. The black circle in the middle represents the region of no data near the North Pole, where the sensor does not image due to the limitations of the polar orbit. (NASA/NSIDC).

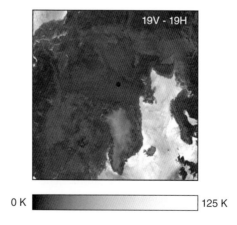

Figure 11.4 Brightness temperature polarization difference at 19 GHz on March 15, 2008. The difference for sea ice is small, while the vertical channel emits much more strongly than the horizontal channel from the ocean. The black circle in the middle represents the region of no data near the north pole, where the sensor does not image due to the limitations of the polar orbit. (NASA/NSIDC).

various combinations of channels, channel differences, and/or channel ratios to determine sea ice concentration in a given grid cell (Steffen *et al.*, 1992). While different algorithms use different approaches, all are essentially empirically derived under the assumption that the measured brightness temperature is due to a mixture of open water emission and emission from sea ice:

$$T_B = (1 - C) \cdot T_{B\text{water}} - C \cdot T_{B\text{ice}} \tag{11.2}$$

Figure 11.5 Brightness temperature vertical polarization difference between 37 and 19 GHz on November 1, 2007. The first-year ice difference is near zero (medium gray), while multi-year ice has a negative difference. The gray circle in the middle represents the region of no data near the north pole, where the sensor does not image due to the limitations of the polar orbit. (NASA/NSIDC).

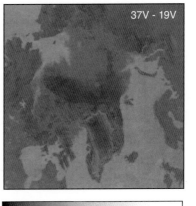

where:

C is the sea ice concentration, ranging between 0 and 1

$T_{B\text{water}}$ and $T_{B\text{ice}}$ are prescribed coefficients, called tiepoints, based on the brightness temperatures over pure (100% ice or 100% water) surface types.

In principle, only one channel is needed to solve equation 11.2 for the concentration. However, nearly all algorithms use a combination of several (three or more) channels. This approach has two advantages. First, with multiple channels, it is feasible (at least in the Arctic) to solve for both total concentration and the concentration of first-year and multi-year ice. Second, multiple channels can better discriminate between ice and water (as seen in Figure 11.4) and can resolve ambiguities in the brightness temperature signature.

Such ambiguities arise due to several factors that limit the capabilities of radiometers to retrieve sea ice concentration. While most radiometers have five or more channels, each channel is not completely independent and, at most, three surface types can be determined (water and two ice types). However, a grid cell may have more than two distinct sea ice signatures. Thin new ice has unique emission properties from thicker first-year and multi-year ice. Snow cover alters the signature, depending on depth, grain size, and liquid water content. Liquid water on the surface during summer melt significantly degrades the ice signature, typically resulting in underestimation of concentration (e.g., Meier, 2005; Andersen *et al.*, 2007). Atmospheric emission, while generally small for channels used in sea ice algorithms, can have a noticeable effect (e.g., Oelke, 1997), particularly over open water.

Some algorithms (e.g., Markus and Cavalieri, 2000) use higher frequency (85–91 GHz) channels that are more influenced by the atmosphere and generally require an atmospheric correction. Wind roughening of the ocean can raise the emission from the open water enough to yield a false non-zero ice concentration, though this effect can largely be removed by using weather filter thresholds.

Sea ice extent is less affected by these issues, because it is simply an ice/no-ice threshold and concentrations errors are important only near the threshold,

Table 11.1 Passive microwave radiometers, platforms, dates of operation, and frequencies. Some sensors have flown on multiple platforms. Operation dates are valid as of June 1, 2014. Frequencies are given to the nearest GHz – small differences occur between sensors for some frequencies. All frequencies have channels for horizontal and vertical polarization, except for 19 GHz on ESMR and 22/23 GHz on all sensors, which are only vertically polarized. SSMIS also has higher frequency sounding channels, which are not used for sea ice.

Sensor	Platform(s)	Dates of operation	Frequencies [GHz]
ESMR	NASA Nimbus-5	Dec 1972 – May 1977	19
SMMR	NASA Nimbus-7	Oct 1978 – Aug 1987	6.63, 10.69, 18.0, 37.0
SSM/I	DMSP F8 – F15	Aug 1987 – present	19.35, 22.235, 37, 85.5
SSMIS	DMSP F16 – F19	Oct 2003 – present	19.35, 22.235, 37, 91.665
AMSR-E	NASA EOS Aqua	Jun 2002 – Oct 2011	6.925, 10.65, 18.7, 23.8, 36.5, 89
AMSR2	JAXA GCOM-W1	Jul 2012 – present	6.925, 7.3, 10.65, 18.7, 23.8, 36.5, 89

a relatively small number of grid cells, usually at the ice edge. However, the precision of extent estimates is limited by the relatively low spatial resolution of passive microwave radiometers. Gridded products over most of the record are limited to 25×25 km grid cells, but the instantaneous field of view (footprint) that defines the effective resolution is on the order of 50–75 km for some channels. Newer sensors such as AMSR-E and AMSR2, have significantly improved resolution (15–25 km), but data have been available only since 2002 (Table 11.1). High-frequency channels (85–91 GHz) have footprints that are about twice the resolution but, as mentioned above, atmospheric correction is needed, and the high-frequency channels need to be combined with lower frequencies to retrieve concentration accurately.

Because of the large sensor footprint, the ice edge can only be determined to $\approx$25–75 km, depending on the character of the ice edge (i.e., diffuse vs. compact) (Partington, 2000; Meier, 2005). Another effect of the large footprint is that near the coast, mixed land-ocean grid cells are common and, in certain conditions, the signal from land and ocean is retrieved as ice by the algorithms, resulting in false ice along the coast. Filters can remove much of this error, but some false ice may remain. While not significant at hemispheric scales, the less precise ice edge limits the application of passive microwave sea ice data for investigation of small-scale processes, such as lead and polynya formation. Due to its low spatial resolution, passive microwave data is of limited utility for navigational support and other operational activities.

11.2.2   Active microwave – scatterometry and radar

Active microwave remote sensing is also applicable for sea ice. Similar to emission, the phase state and salinity strongly affect the backscatter of microwave

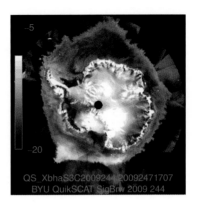

Figure 11.6 QuikScat backscatter for 1 September 2009 from Antarctica. (Brigham Young University Scatterometer Climate Record Pathfinder (http://www.scp.byu.edu/)).

energy emitted from sensors, with sea ice scattering more energy than the ocean (Figure 11.6). Scatterometers, such as the NASA QuikSCAT instrument which operated from June 1999 to November 2009, and the ESA ASCAT instrument, which has been operating since 2009, provide estimates of sea ice extent at a similar spatial scale as passive microwave radiometers (≈ 25 km) (e.g., Remund and Long, 1999), though resolution enhancement techniques can yield substantially higher resolution (e.g., Early and Long, 2001). Backscatter is particularly sensitive to the presence of salt in the ice, and scatterometers are more effective for retrieving first-year and multi-year sea ice types than are passive microwave radiometers (e.g., Nghiem *et al.*, 2007).

Wind roughening of the ocean surface substantially increases the backscatter, resulting in difficulty in discriminating the ice edge in some regions, and a threshold is needed to ameliorate this effect. Scatter from surface melt during summer interferes with the ice backscatter, resulting in higher uncertainty, similar to the effect on passive microwave data; multi-year ice cannot be discriminated during melt conditions.

Synthetic Aperture Radar (SAR) is an imaging sensor that responds to sea ice in a similar manner to scatterometers. However, SAR yields much higher spatial resolution (100–500 m) and is capable of retrieving small-scale processes such as deformation, leads, and polynyas (e.g., Kwok, 2002). Several SAR instruments have flown, including the ESA ERS-1/2 and the Canadian Radarsat-1/2. While providing fine scale details, SAR has limited spatial coverage, so repeat observations are possible only every few days. Also, SAR signatures are complicated, making automated retrieval of ice concentration and extent difficult. Finally, data access is an issue because outside of special agreements, the data is often commercial, and the costs of images are high. Nevertheless, SAR imagery has proven particularly valuable for operational analyses in support of navigation. SAR is also useful for validation of passive microwave estimates (e.g., Kwok, 2002; Heinrichs *et al.*, 2006; Andersen *et al.*, 2007).

11.2.3   *Visible and infrared*

As mentioned above, visible and infrared imagery is of limited use for sea ice because of the prevalence of clouds and (for visible) lack of sunlight. However, where skies are clear, such imagery can often retrieve useful sea ice concentrations. Spatial resolution is much higher than passive microwave (generally 1 – 5 km, but 30 – 250 m for some sensors), thus yielding fine scale details of the ice cover, such as leads and polynyas. In contrast to SAR, the imagery is more easily accessible (many non-commercial sources) and is easier to interpret.

Automated retrieval of concentration is still difficult because of limitations of cloud masking and variation of the surface. For example, while sea ice generally has a much higher reflectance than open water, thin ice is quite dark and is easily missed. Surface temperatures from infrared can be near uniform near the ice edge, particularly during melt and for newly-formed thin ice. Visible and infrared imagery has primarily been used for validation of passive microwave estimates (e.g., Meier, 2005; Cavalieri *et al.*, 2006).

11.2.4   *Operational sea ice analyses*

Analyses of sea ice concentration and extent, along with other parameters of interest (e.g., thickness or stage of development), are regularly produced by operational ice centers, such as the US National Ice Center (NIC) and the Canadian Ice Service (CIS). The primary focus of these operational centers is in support of navigation in ice-infested waters. Because of this, accuracy of the produced ice charts is crucial. The centers thus use the best satellite imagery available when the chart is produced. This generally means SAR, visible/infrared where skies are clear and, as a last resort, the lower-resolution passive microwave or scatterometer sources (Dedrick *et al.*, 2001; Helfrich *et al.*, 2007; Tivy *et al.*, 2011).

The imagery is analyzed and combined by expert human analysts who create the ice charts. The high-quality imagery and expert analysis provides very accurate estimates of the ice edge and concentration, as long as imagery is available. However, the quantity and quality of the input imagery varies and human analysis is, by its nature, subjective. The charts are thus not suitable for climate-scale time series analysis. They can, however, provide useful validation information for satellite products, such as passive microwave estimates (Partington, 2000).

11.3   Sea ice drift

In addition to growth and melt, sea ice drifts with the wind and ocean currents. The drift is modified by gravity (sea surface tilt due to density gradients), the Coriolis effect, and the internal strength of the ice. The drift of the ice can be retrieved by remote sensing imagery through feature tracking algorithms – matching a feature in one image with the same feature in a subsequent image, separated by a period of time. The drift velocity can then be calculated from the displacement distance and the interval between images. Two primary methods have been developed to derive sea ice drift. The first, most commonly-used, method, called cross-correlation, searches for correlation peaks in the second image that match the feature in the first image (Emery *et al.*, 1991). The second method uses wavelet transforms, localized Fourier transforms, to match features between images (Liu and Cavalieri, 1998).

If the spatial resolution is high enough, such as for SAR and some visible imagery, individual ice floes can be tracked, as well as deformation (lead and ridge formation) (Kwok *et al.*, 1999). For lower-resolution sensors, such as passive microwave radiometers, only the average radiometric signature of an amalgamation of features within a grid cell can be tracked (e.g., Kwok *et al.*, 1998). However, passive microwave radiometers have the advantage of complete coverage in the Arctic and Antarctic at daily timescales, even under cloudy skies (Figure 11.7). Thus, passive microwave data has been most commonly used for sea ice drift, though often in conjunction with other data (e.g., buoys) to reduce uncertainties (Meier *et al.*, 2000).

Since features smaller than a grid cell cannot be tracked, the effective velocity resolution limit is the size of a single grid cell over the time interval, although resolution enhancement techniques can yield sub-pixel resolution (Emery *et al.*, 1991; Meier *et al.*, 2000). Because of the low spatial resolution of passive microwave sensors, higher frequency channels (which have a relatively smaller footprint, i.e., 85–91 GHz) are generally used to calculate drift. However, these channels are subject to greater atmospheric interference (clouds), though this can usually be removed by velocity filters. Higher frequencies are also more sensitive to surface meltwater, and ice drift retrieval from passive microwave sensors is limited during summer. Lower frequency channels, such as 37 GHz, are better able to detect motion during melt, but at the expense of lower spatial resolution.

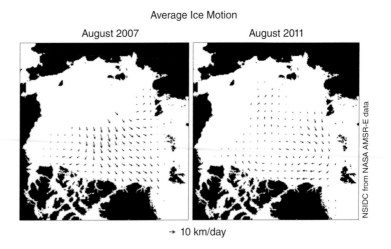

Figure 11.7 Monthly average sea ice drift for August 2007 (left) and August 2011 (right), showing interannual differences in large scale circulation. Drift is estimated from AMSR-E imagery using a cross-correlation feature matching method. (NASA/NSIDC).

Daily motions from passive microwave data are noisy because of the limitation of the grid cell resolution and other errors. However, while the estimates from SSM/I have large RMS errors (6–7 km/day), there is near-zero bias (e.g., Kwok *et al.*, 1998; Meier *et al.*, 2000). This means that the error can be reduced by averaging over a few days and/or performing a spatial interpolation. Optimal interpolation that incorporates higher resolution data (AVHRR and buoys) with the passive microwave data, weighting each source by the number of estimates and their error characteristics, reduces errors to 3–4 km/day (Meier *et al.*, 2000). The higher spatial resolution of the AMSR-E sensor significantly reduces the RMS errors (Meier and Dai, 2006) and improves retrievals during summer, because lower frequency channels (e.g., 19 GHz) can obtain the same spatial resolution as the high frequency SSM/I channels (Kwok, 2008).

Estimating sea ice drift in the Antarctic is more challenging because of greater atmospheric interference, faster ice motion, and rapidly changing surface characteristics (e.g., snow). These factors tend to lower correlation of features over typical time intervals. However, drift has been successfully estimated for Antarctic sea ice (e.g., Emery *et al.*, 1997).

11.4 Sea ice thickness and age, and snow depth

Sea ice thickness represents the third dimension of the ice cover and, thus, is a key element in determining the total sea ice mass or volume. Unlike concentration and extent, remotely sensed thickness estimates are sparse in space and time and are subject to higher uncertainties. The sparseness is due to a lack of sensors capable of estimating thickness in the polar regions. The higher uncertainty is because remote sensing instruments, especially spaceborne instruments, cannot penetrate through the ice to obtain a direct thickness signal and must, instead, infer thickness from surface properties. Despite these limitations, numerous instruments have been used to estimate thickness. The types of measurements fall into two broad categories, altimetric and radiometric, each of which is discussed below. A third approach uses proxy measurements to infer thickness based on ice age classes.

11.4.1 Altimetric thickness estimates

Altimetric types of measurements essentially measure the height of the ice either above or below the surface. Different types of sensors make different physical measurements, as shown in Figure 11.8, from which thickness is derived based on the properties of the ice and the overlying snow cover.

The earliest remotely-sensed sea ice thickness data are from soundings from beneath the ice estimated by upward-looking sonars (ULS), either mounted on submarines or moored on the ocean floor. The ULS instruments measure ice

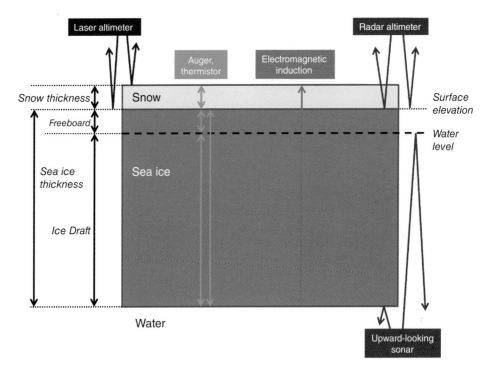

Figure 11.8 Diagram showing definition of sea ice thickness terms (black arrows) and measurement surfaces of different altimetric and sounding sensors. Adapted from Meier and Haas, 2011.

draft – the depth of the ice below the water line. Total thickness is derived from the draft using assumptions of ice and water density. Water density is fairly consistent, but ice density may vary, depending on the salinity of the ice; another factor is the weight of snow on top of the ice. Though sparse in space and time, a climatology of thickness over the central Arctic was produced from submarine records for the 1970s through the mid-1990s (Rothrock *et al.*, 1999; Rothrock *et al.*, 2003).

Satellite altimeters have the potential to provide the most comprehensive thickness estimates. Altimeters measure the freeboard – the height above the water line. As with sonar, density estimates of water, snow, and ice can be used to convert freeboard to total thickness. Again, snow is a source of significant uncertainty – more so for altimeters than for sonars – because freeboard makes up only 10–15% of the total thickness, compared to 85–90% for draft. This means that any errors in the freeboard estimate are magnified by nearly an order of magnitude when converted to total thickness.

Radar altimeters generally penetrate through the snow cover to the snow-ice interface, so only the weight of the snow contributes to the uncertainty. However, the situation is more complicated for laser altimeters, which reflect off the top of the snow surface, retrieving a snow-plus-ice freeboard. Thus, an estimate of snow depth is needed to calculate total ice thickness. Unfortunately, snow thickness

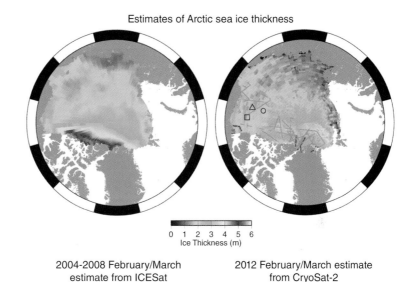

Estimates of Arctic sea ice thickness

0 1 2 3 4 5 6
Ice Thickness (m)

2004-2008 February/March
estimate from ICESat

2012 February/March estimate
from CryoSat-2

measurements are extremely limited in the polar regions. Despite its importance, reliable estimates of snow depth and other snow properties represents a major gap in remote sensing capabilities. Passive microwave sensors have been used to estimate snow depth (Markus and Cavalieri, 1998; Comiso *et al.*, 2003) but, while the spatial distribution is reasonable, there is high uncertainty in the absolute value of depth retrievals. In addition, the method is applicable only over first-year ice.

In addition to snow, there are uncertainties in the altimetry measurement itself, the most important of which is the variation of sea surface height due to the geoid and the ocean dynamics. To minimize this error, the measurements are regularly calibrated (at least every ≈50 km) to open water at the ice edge, or within leads and polynyas. Open water is distinguishable both from being the lowest surface and for its smoothness relative to the signal from sea ice.

Altimetry data in the polar regions has been limited until recently. A radar altimeter on ERS-1/2 provided some Arctic sea ice thickness estimates, but only up to 80°N, thus missing most of the multi-year ice in the central Arctic (Laxon *et al.*, 2003). In January 2003, NASA launched the Ice, Cloud, and land Elevation Satellite (ICESat) with a laser altimeter. It retrieved monthly average freeboard and thickness estimates twice per year (spring and fall) in the Arctic until it ceased operation in February 2010. A time series of ICESat data indicate an accelerating thinning during the 2000s (Kwok *et al.*, 2009), continuing the thinning trend observed by the submarine sonar record (Kwok and Rothrock, 2009). In April, 2010, ESA launched CryoSat-2 with a radar altimeter (CryoSat-1 never orbited, due to a launch failure in 2005) and is producing monthly maps of thickness (Figure 11.9) (Laxon *et al.*, 2013; see also Chapter 15).

The NASA Project IceBridge is a series of airborne campaigns beginning in 2009, collecting altimeter thickness measurements. While there are only a couple of campaigns per year, and the spatial coverage is limited, the data collected provide validation and intercalibration for CryoSat-2 and the planned ICESat-2, as well as filling in the gap after the end of ICESat operations. Both laser and radar altimeter instruments are flown, measuring the height of both the ice surface (by radar) and snow surface (by laser). This allows coincident retrieval of snow depth and ice thickness (Farrell *et al.*, 2012; Kurtz and Farrell, 2011).

Electromagnetic induction (EM) instruments do penetrate through the ice cover and, thus, directly measure thickness. However, the instrument has limited range, so it must either be pulled across the ice surface or flown at low altitude above ice. As such, EM measurements are sparse, but provide more coverage than *in situ* drill holes by ice augers. Both drill holes and EM are important "ground truth" for validation and calibration of the satellite altimeter measurements, and provide information on regional thickness changes (Haas, 2004; Rabenstein *et al.*, 2010).

11.4.2 Radiometric thickness estimates

As mentioned in the first section, electromagnetic emission is sensitive to the presence of sea ice at many frequencies. For some parts of the spectrum, the electromagnetic energy is related to the thickness of ice. This has been most exploited in the thermal infrared. The open ocean radiates thermal energy; as sea ice grows, the emitted energy is attenuated, and the amount of attenuation is dependent on the thickness. There is a threshold at which point the attenuation dominates and any thickness signal is saturated. This threshold varies depending on frequency, but for thermal infrared it is typically at most 1 m (e.g., Yu and Rothrock, 1996). Thus, these methods have been useful only for thin ice.

Nevertheless, such estimates can be a valuable complement to altimetric estimates, which tend to have high errors for thin ice. A recently developed approach that combines satellite derived cloud and surface properties from visible and infrared data with a thermodynamic growth model has been able to obtain thickness estimates up to 2.8 m (Wang *et al.*, 2010), which encompasses all level first-year ice and some multi-year ice regimes.

Passive microwave frequencies used for sea ice concentration detect emission from at or near the surface and, thus, are not sensitive to thickness beyond 10–20 cm (Martin *et al.*, 2004). The ESA Soil Moisture Ocean Salinity (SMOS) sensor, launched in November 2009, has a lower frequency, 1.4 GHz, than previous radiometers. The lower frequency penetrates deeper into the ice, providing thickness estimates up to about 0.5 m.

11.4.3 Sea ice age estimates as a proxy for ice thickness

Sea ice drift algorithms discussed in the previous section can be used to track Lagrangian parcels (i.e., grid cells) of ice. Because the algorithms yield unbiased estimates, the Lagrangian tracking error does not grow significantly over time. This allows parcels of ice to be accurately tracked over several years, allowing an ice age distribution to be built up after an initialization period (Fowler *et al.*, 2004). Ice age is provided simply as an annual age category (i.e., first-year, second-year, etc.), and ice that remains at the end of the summer melt period is aged one year.

Ice age is reasonably well correlated with thickness, because level ice becomes thicker over time, due to thermodynamic growth (Maslanik *et al.*, 2007). The method only tracks one ice age type so, if new ice forms within a grid cell (through divergence or new ice growth after summer melt), it is not accounted for, and the ice age value for a cell is effectively an estimate of the oldest ice in the cell. Because passive microwave data is one of the main sources for the input drift estimates, small-scale motions (e.g., leads) are often missed. Also, dynamic thickening via ridging is not accounted for. Nonetheless, reasonable correlation between thickness and age has been found up to nine year old ice, and the spatial distribution of ice age is consistent with ICESat and CryoSat-2 thickness fields (Maslanik *et al.*, 2007).

A particular advantage of using ice age is that, because it uses passive microwave data, a multi-decadal record over the entire Arctic is possible, in contrast to the sparse submarine estimates and the relatively recent altimetry data. The time series ice indicates a declining trend in first-year ice and near-complete loss of the oldest types (4+ years old) since the early 1980s (Figure 11.10), particularly since 2007, corresponding to the observed thinning by submarine sonar and altimeters.

11.5 Sea ice melt onset and freeze-up, albedo, melt pond fraction and surface temperature

Thermodynamics plays a key role in the seasonal and interannual evolution of the sea ice. In spring, with the onset of melt, albedo begins to drop, resulting in greater absorption of solar energy. In the Arctic, melt ponds (pools of meltwater) form on the ice surface, further lowering albedo. In autumn, as solar illumination decreases, air temperatures fall, cooling the surface until freeze-up begins and ice grows. Many key parameters in these processes are retrievable from visible, infrared, and passive microwave sensors.

11.5.1   Melt onset and freeze-up

The onset of melt of the snow layer overlying the sea ice surface marks the beginning of summer conditions. Because microwave emission is sensitive to the phase state of water, it is optimal for detecting the beginning of melt. The most direct method to determine melt is by a threshold of two brightness temperature channels (19 and 37 GHz) (Drobot and Anderson, 2001). Diurnal effects and

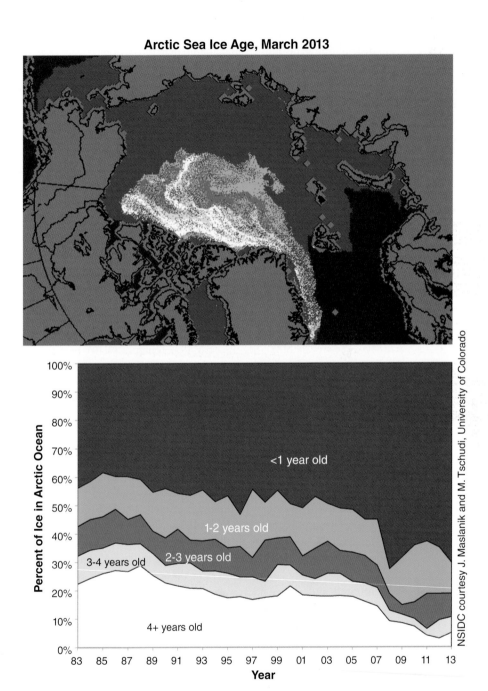

Figure 11.10 Map of sea ice age (top) for March 2013 and time series of age distribution for March from 1983 to 2013. (National Snow and Ice Data Center, image provided by J. Maslanik and M. Tscudi, University of Colorado, Boulder).

refreezing complicate the microwave signature, so a temporal filter is used to remove false melt signals. Other methods incorporate different thresholds and/or ratios to determine melt onset, as well as freeze-up (e.g., Markus *et al.*, 2009; Smith, 1998).

Active microwave (scatterometer and SAR) signatures are also sensitive to melt and thus can obtain melt onset (e.g., Drinkwater and Liu, 2000; Kwok *et al.*, 2003). Differences between passive and active microwave melt detection, and melt indication from surface temperatures (e.g., thermal data, weather stations), arise due to differences in the emitting layer of the melt. Microwave sensors generally penetrate at least partly through the snow, where absorption of solar energy can trigger melt before the surface reaches the melting temperature.

Because of the different characteristic of sea ice and snow in the Antarctic, most melt onset and freeze-up algorithms are only applied to the Arctic. However, methods for melt onset detection in the Arctic have been developed, generally using a combination of sensors (Drinkwater and Liu, 2000).

11.5.2 Sea ice albedo and melt pond fraction

After melt onset occurs, albedo begins to rapidly drop, followed by the ensuing increase in solar energy absorption. Timing of the change in albedo is an important factor in total solar energy absorbed, because melt onset occurs near the summer solstice peak solar insolation (Perovich *et al.*, 2007). Clear-sky broadband albedo is derived from visible reflectance channels by masking clouds, correcting for non-isotropic effects, removing atmospheric interference to obtain a surface albedo from the top-of-the-atmosphere albedo, and finally converting narrow-band albedo to broadband albedo (e.g., Lindsay and Rothrock, 1994). Methods have been developed to correct for the effect of clouds on surface albedo, using a radiative transfer model to create all-sky albedo estimates from AVHRR imagery (Key *et al.*, 2001).

An important contributor to reduced summer albedo in the Arctic is the formation and evolution of melt ponds, small pools of meltwater that form on the surface of the ice. The ponds are generally at a scale too small (usually 10 – 100 m) to be explicitly resolved by satellite remote sensing instruments. However, their spectral signature is unique from the surrounding ice, snow, and ocean. Using characteristic spectral reflectances for each surface type, and a linear mixing approach, the concentration of melt ponds can be determined from visible imagery such as MODIS (Tschudi *et al.*, 2008).

11.5.3 Sea ice surface temperature

Ice surface temperature plays key role in sea ice melt and growth and heat exchange between the ocean and the atmosphere. Passive microwave brightness

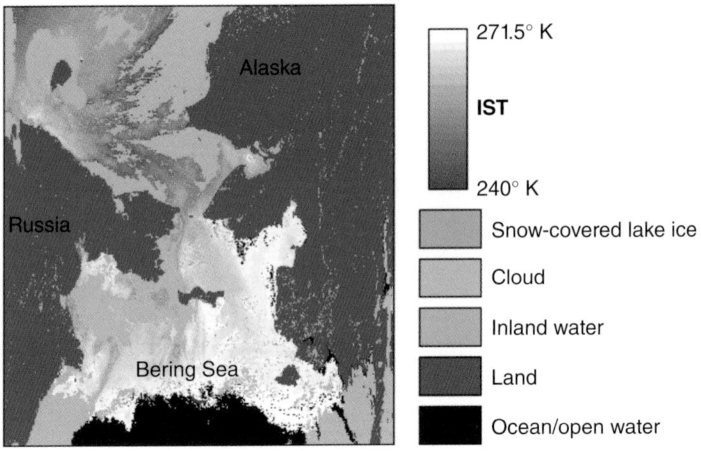

Figure 11.11 Sea ice surface temperature from MODIS. Image from NSIDC MODIS Image Gallery (http://nsidc.org/data/modis/gallery/).

temperatures are a function of the surface temperature. If the emissivity is well-known, deriving surface temperature from brightness temperature is straightforward (equation 11.1). In practice, the variability of the emissivity due to the sea ice properties (snow, salinity, etc.) makes the retrieval of physical temperature impractical at most passive microwave frequencies. Ice temperatures were derived from AMSR-E 6.9 GHz data (Comiso *et al.*, 2003), but the temperature fields showed limited variability – due in part to the fact that the emitting layer is below the surface. Thus, the fields are of less utility than skin temperature.

Thermal infrared imagery is the most useful source for deriving sea ice surface temperature. Temperature algorithms use a split-window technique using two infrared channels, typically 11 and 12 μm. The split-window temperature accounts for atmospheric water vapor and, along with a radiative transfer model, surface skin temperature can be retrieved (Key *et al.*, 1997; Hall *et al.*, 2004). For reliable temperature retrievals, accurate cloud screening is essential, as even thin cirrus clouds can cause significant errors in surface temperature. Cloud detection over sea ice is difficult, because clouds often have similar radiometric properties (albedo, temperature) to the underlying sea ice. To prevent such errors, a fairly conservative cloud mask needs to be applied, but that limits the area where temperature is retrieved and clear-sky regions are likely to be masked out (Hall *et al.*, 2004; see Figure 11.11).

Visible and infrared data can be combined with radiative transfer models to estimate cloud properties and radiative fluxes, complementing albedo and temperature to obtain a suite of thermodynamic parameters over sea ice (e.g., Maslanik *et al.*, 1998). Albedo, temperature, cloud, and radiative flux fields have been produced as part of the Extended AVHRR Polar Pathfinder (APP-x) project since the first AVHRR sensor was launched in 1982 (http://stratus.ssec.wisc.edu/products/appx/appx.html).

11.6 Summary, challenges and the road ahead

Remote sensing is an essential element for characterizing sea ice. Because of its remote location, harsh environment, and ephemeral nature (drift, melt/growth), *in situ* observations from the ice are expensive, difficult, and limited in space and time. Only satellite-borne remote sensing instruments can provide comprehensive coverage of the sea ice-covered oceans over a long period of time.

New and planned future missions will increase our capabilities. Since July 2012, AMSR2 has been providing enhanced passive microwave fields. ESA CryoSat-2 and SMOS are yielding improved thickness estimates. ESA and Eumetsat MetOp instruments, and the recently launched US NPP VIIRS instrument, will carry on the legacy of previous visible and infrared sensors, such as AVHRR and MODIS. The ESA Sentinel-1 C-band SAR, which was launched in April 2014, will continue this key tool for sea ice characterization, and future Sentinel missions will provide useful sea ice data. The planned NASA ICESat-2, with a planned launch in 2017 will continue altimetric estimates of sea ice thickness. Future JAXA GCOM planned missions are planned to have follow-on MODIS and AMSR instruments.

However, there is a risk of gaps in the future. There are no concrete plans for future passive microwave radiometers beyond the DMSP SSMIS series and the JAXA GCOM missions, which have a planned end of missions around 2020. Thus, by the end of the current decade, one of the most iconic and longest satellite climate records will be at risk unless new missions are developed. Unless CryoSat-2 significantly outlasts its planned three-year lifetime, there will be a gap in altimetry missions before NASA's ICESat-2 mission, as there was between ICESat-1 and CryoSat-2.

In terms of sea ice parameters, the possible lack of overlap between CryoSat-2 and ICESat-2 is particularly unfortunate, because the combination of radar and laser altimeters is perhaps the best method to determine snow depth on sea ice (Farrell *et al.*, 2012). Snow is an essential element of heat and radiative fluxes, as well as biogeochemical cycles in sea ice, but currently there are no reliable basin-scale estimates of snow depth. Snow also contributes to the uncertainty of passive microwave sea ice retrievals and, especially, altimeter estimates of sea ice thickness.

Continuity of sensors across multiple satellite missions is critical to our understanding of the state and fate of sea ice. This includes redundancy and overlap of sensor missions to hedge against potential satellite failures, and to provide intercalibration between missions, which are necessary for continuous and consistent parameters.

So, there are challenges. Nonetheless, the legacy to date of sea ice remote sensing has been one of great success. The passive microwave record of sea ice concentration and extent is one of the longest satellite-derived climate records and it provides one of the most iconic indicators of climate change. New methods and new sensors are providing detailed information on sea ice thickness, which are essential for assessing the state and evolution of sea ice mass balance. Visible and infrared imagery provide a long record of albedo, temperature, and radiative fluxes. In short, over the last several decades, remote sensing has tremendously increased our knowledge of sea ice and the role of sea ice in the climate system, and will continue to do so into the future.

References

Andersen, S., Tonboe, R., Kaleschke, L., Heygster, G. & Pedersen, L.T. (2007). Intercomparison of passive microwave sea ice concentration retrievals over the high-concentration Arctic sea ice. *Journal of Geophysical Research* **112**, C08004, doi: 10.1029/2006JC003543.

Cavalieri, D.J., Markus, T., Hall, D.K., Gasiewski, A.J., Klein, M. & Ivanoff, A. (2006). Assessment of EOS Aqua AMSR-E Arctic Sea Ice Concentrations Using LANDSAT-7 and Airborne Microwave Imagery. *IEEE Transactions on Geoscience and Remote Sensing* **44**, 11, 3057–3069, doi: 10.1109/TGRS.2006.878445.

Cavalieri, D.J. & Parkinson, C.L. (2012). Arctic sea ice variability and trends, 1979–2010. *The Cryosphere* **6**, 881–889, doi: 10.5194/tc-6-881-2012.

Comiso, J.C., Cavalieri, D.J. & Markus, T. (2003). Sea ice concentration, ice temperature, and snow depth using AMSR-E data. *IEEE Transactions on Geoscience and Remote Sensing* **41**(2), 243–252, doi: 10.1109/TGRS.2002.808317.

Dedrick, K., Partington, K., Van Woert, M., Bertoia, C. & Benner, D. (2001). US National/Naval Ice Center digital sea ice data and climatology. *Canadian Journal of Remote Sensing* **27**(5), 457–475.

Drinkwater, M.R. & Liu, X. (2000). Seasonal to interannual variability in Antarctic sea–ice surface melt. *IEEE Transactions on Geoscience and Remote Sensing* **38**, 1827–1842, doi: 10.1109/36.851767.

Drobot, S. & Anderson, M. (2001). Comparison of interannual snowmelt onset dates with atmospheric conditions. *Annals of Glaciology* **33**, 79–84.

Early, D.S. & Long, D.G. (2001). Image Reconstruction and Enhanced Resolution Imaging from Irregular Samples. *IEEE Transactions on Geoscience and Remote Sensing* **39**, 2, 291–302.

Emery, W.J., Fowler, C.W., Hawkins, J. & Preller, R.H. (1991). Fram Strait satellite image-derived ice motion. *Journal of Geophysical Research* **96**(C3), 4751–4768 (Correction: *Journal of Geophysical Research* **96**(C5), 8917–8920).

Emery, W.J., Fowler, C.W. & Maslanik, J.A. (1997). Satellite-derived maps of Arctic and Antarctic sea ice motion: 1988 to 1994. *Geophysical Research Letters* **24**(8), 897–900, doi: 10.1029/97GL00755.

Eppler, D.T., *et al.* (1992). Passive microwave signatures of sea ice. In: Carsey, F.D. (ed). *Microwave Remote Sensing of Sea Ice*, pp. 47–71. AGU Geophysical Monograph 68.

Farrell, S.L., Kurtz, N., Connor, L.N., *et al.* (2012). A first assessment of IceBridge snow and ice thickness data over Arctic sea ice. *IEEE Transactions on Geoscience and Remote Sensing* **50**(6), 2098–2111, doi: 10.1109/tgrs.2011.2170843.

Fowler, C., Emery, W.J. & Maslanik, J. (2004). Satellite-derived evolution of Arctic sea ice age: October 1978 to March 2003. *IEEE Geoscience and Remote Sensing Letters* **1**(2), 71–74, doi: 10.1109/LGRS.2004.824741.

Hall, D.R., Key, J.R., Casey, K.A., Riggs, G.A. & Cavalieri, D.J. (2004). Sea ice surface temperature product from MODIS. *IEEE Transactions on Geoscience and Remote Sensing* **42**(5), 1076–1087.

Haas, C. (2004). Late-summer sea ice thickness variability in the Arctic Transpolar Drift 1991–2001 derived from ground-based electromagnetic sounding. *Geophysical Research Letters* **31**, L09402, doi: 10.1029/2003GL019394.

Heinrichs, J.F., Cavalieri, D.J. & Markus, T. (2006). Assessment of the AMSR-E Sea Ice-Concentration Product at the Ice Edge Using RADARSAT-1 and MODIS Imagery. *IEEE Transactions on Geoscience and Remote Sensing* **44**(11), 3070–3080, doi: 10.1109/TGRS.2006.880622.

Helfrich, S.R., McNamara, D., Ramsay, B.H., Baldwin, T. & Kasheta, T. (2007). Enhancements to and Forthcoming Developments To the Interactive Multi-sensor Snow and Ice Mapping System (IMS). *Hydrological Processes* **21**(12), 1576–1586.

Kaleschke, L., Tian-Kunze, X., Maaß, N., Mäkynen, M. & Drusch, M. (2012). Sea ice thickness retrieval from SMOS brightness temperatures during the Arctic freeze-up period. *Geophysical Research Letters* **39**, L05501, doi: 10.1029/2012GL050916, 2012.

Key, J., Collins, J., Fowler, C. & Stone, R. (1997). High-latitude surface temperature estimates from thermal satellite data. *Remote Sensing of Environment* **61**, 302–309.

Key, J.R., Wang, X., Stoeve, J.C. & Fowler, C. (2001). Estimating the cloudy-sky albedo of sea ice and snow from space. *Journal of Geophysical Research* **106**(D12), 12489–12497, doi: 10.1029/2001JD900069.

Kurtz, N.T. & Farrell, S.L. (2011). Large-Scale Surveys of Snow Depth on Arctic Sea Ice from Operation Icebridge. *Geophysical Research Letters* **38**(L20), L20505, doi: 10.1029/2011GL049216.

Kwok, R. (2002). Sea ice concentration from passive microwave radiometry and openings from SAR ice motion. *Geophysical Research Letters* **29**(10), 10.1029/2002GL014787.

Kwok, R. (2008). Summer sea ice motion from the 18 GHz channel of AMSR-E and the exchange of sea ice between the Pacific and Atlantic Sectors. *Geophysical Research Letters* **35**, L03504, doi: 10.1029/2007GL032692.

Kwok, R. & Rothrock, D.A. (2009). Decline in Arctic sea ice thickness from submarine and ICESat records: 1958 2008. *Geophysical Research Letters* **36**, L15501, doi: 10.1029/2009GL039035.

Kwok, R., Schweiger, A., Rothrock, D.A., Pang, S. & Kottmeier, C. (1998). Sea ice motion from satellite passive microwave imagery assessed with ERS SAR and buoy motions, *Journal of Geophysical Research* **103**(C4), 8191–8214.

Kwok, R., Cunningham, G.F., LaBelle-Hamer, N., Holt, B. & Rothrock, D.A. (1999). Ice thickness derived from high-resolution SAR imagery. *Eos, Transactions American Geophysical Union* **80**(42), 495–497.

Kwok, R., Cunningham, G.F. & Nghiem, S.V. (2003). A study of melt onset in RADARSAT SAR imagery. *Journal of Geophysical Research* **108**(C11), 3363, doi: 10.1029/2002JC001363.

Kwok, R., Cunningham, G.F., Wensnahan, M., Rigor, I., Zwally, H.J. & Yi, D. (2009). Thinning and volume loss of Arctic sea ice: 2003–2008. *Journal of Geophysical Research*, doi: 10.1029/2009JC005312.

Laxon, S.W., Peacock, N. & Smith, D. (2003). High interannual variability of sea ice thickness in the Arctic region. *Nature* **425**, 947–950, doi: 10.1038/nature02050.

Laxon, S.W., Giles, K.A., Ridout, A.L., *et al.* (2013). CryoSat-2 estimates of Arctic sea ice thickness and volume. *Geophysical Research Letters* **40**, 732–737, doi: 10.1002/grl.50193.

Lindsay, R.W. & Rothrock, D.A. (1994). Arctic sea ice albedo from AVHRR. *Journal of Climate* **7**, 1737–1749, doi: 10.1175/1520-0442(1994)007<1737:ASIAFA> 2.0.CO;2.

Liu, A.K. & Cavalieri, D.J. (1998). On Sea Ice Drift from the Wavelet Analysis of the Defense Meteorological Satellite Program (DMSP) Special Sensor Microwave Imager (SSM/I) Data. *International Journal of Remote Sensing* **19**(7), 1415–1423, doi: 10.1080/014311698215522.

Markus, T. & Cavalieri, D.J. (1998). Snow depth distribution over sea ice in the southern ocean from satellite passive microwave data. In: Jeffries, M. (ed). *Antarctic Sea Ice: Processes, Interactions, and Variability*. AGU Antarctic Research Series, vol. 74, pp. 183–187.

Markus, T. & Cavalieri, D.J. (2000). An enhancement of the NASA Team sea ice algorithm. *IEEE Transactions on Geoscience and Remote Sensing* **38**(3), 1387–1398, doi: 10.1109/36.843033.

Markus, T., Stroeve, J.C. & Miller, J. (2009). Recent changes in Arctic sea ice melt onset, freeze-up, and melt season length. *Journal of Geophysical Research* **114**, C12024, doi:10.1029/2009JC005436.

Martin, S., Drucker, R., Kwok, R. & Holt, B. (2004). Estimation of the thin ice thickness and heat flux for the Chukchi Sea Alaskan coast polynya from Special Sensor Microwave/Imager data, 1990–2001. *Journal of Geophysical Research* **109**, C10012, doi: 10.1029/2004JC002428.

Maslanik, J., Fowler, C., Key, J., Scambos, T., Hutchinson, T. & Emery, W. (1998). AVHRR-based Polar Pathfinder products for modeling applications. *Annals of Glaciology* **25**, 388–392.

Maslanik, J.A., Fowler, C., Stroeve, J., *et al.* (2007). A younger, thinner Arctic ice cover: Increased potential for rapid, extensive sea-ice loss. *Geophysical Research Letters* **34**, L24501, doi: 10.1029/2007GL032043.

Meier, W.N. (2005). Comparison of passive microwave ice concentration algorithm retrievals with AVHRR imagery in arctic peripheral seas Geoscience and Remote Sensing. *IEEE Transactions on Geoscience and Remote Sensing* **43**(6), 1324–1337, doi: 10.1109/TGRS.2005.846151.

Meier, W.N. & Dai, M. (2006). High-resolution sea-ice motion from AMSR-E imagery. *Annals of Glaciology* **44**, 352–356.

Meier, W.N. & Haas, C. (2011). Changes in the physical state of sea ice. In: *Snow, Water, Ice and Permafrost in the Arctic (SWIPA): Climate Change in the Cryosphere*, Chapter 9, Sea Ice, Arctic Monitoring and Assessment Programme (AMAP), Oslo, Norway, xii+538 pp., ISBN:978-82-7971-071-4.

Meier, W.N., Maslanik, J.A. & Fowler, C.W. (2000). Error analysis and assimilation of remotely sensed ice motion within an Arctic sea ice model. *Journal of Geophysical Research* **105**(C2): 3339–3356.

Nghiem, S.V., Rigor, I.G., Perovich, D.K., Clemente-Colon, P., Weatherly, J.W. & Neumann, G. (2007). Rapid reduction of Arctic perennial sea ice. *Geophysical Research Letters* **34**, L19504, doi: 10.1029/2007GL031138.

Oelke, C. (1997). Atmospheric signatures in sea-ice concentration estimates from passive microwaves: Modeled and observed. *International Journal of Remote Sensing* **185**, 1113–1136.

Parkinson, C.L. & Cavalieri, D.J. (2012). Antarctic sea ice variability and trends, 1979–2010 *The Cryosphere* **6**, 871–880, doi: 10.5194/tc-6-871-2012.

Partington, K.C. (2000). A data fusion algorithm for mapping sea-ice concentrations from Special Sensor Microwave/Imager data. *IEEE Transactions on Geoscience and Remote Sensing* **38**(4), 1947–1958, doi: 10.1109/36.851776.

Perovich, D.K., Nghiem, S.V., Markus, T. & Schweiger, A. (2007). Seasonal evolution and interannual variability of the local solar energy absorbed by the Arctic sea ice–ocean system. *Journal of Geophysical Research* **112**, C03005, doi: 10.1029/2006JC003558.

Rabenstein, L., Hendricks, S., Martin, T., Pfaffhuber, A. & Haas, C. (2010). Thickness and surface-properties of different sea-ice regimes within the Arctic Trans Polar Drift: Data from summers 2001, 2004 and 2007. *Journal of Geophysical Research* **115**, C12059, doi: 10.1029/2009JC005846.

Remund, Q.P. & Long, D.G. (1999). Sea Ice Extent Mapping Using Ku-Band Scatterometer Data. *Journal of Geophysical Research* **104**(C5), 11515–11527.

Rothrock, D.A., Yu, Y. & Maykut, G.A. (1999). Thinning of the arctic sea-ice cover. *Geophysical Research Letters* **26**, 3469–3472.

Rothrock, D.A., Zhang, J. & Yu, Y. (2003). The arctic ice thickness anomaly of the 1990s: A consistent view from observations and models. *Journal of Geophysical Research* **108**(C3), 3083, doi: 10.1029/2001JC001208.

Smith, D.M. (1998). Observation of perennial Arctic sea ice melt and freeze-up using passive microwave data. *Journal of Geophysical Research* **103**(27), 753–27, 769, doi: 10.1029/98JC02416.

Stammerjohn, S., Massom, R., Rind, D. & Martinson, D. (2012). Regions of rapid sea ice change: An inter-hemispheric seasonal comparison. *Geophysical Research Letters* **39**, L06501, doi: 10.1029/2012GL050874.

Steffen, K., *et al.* (1992). The estimation of geophysical parameters using passive microwave algorithms. In: Carsey, F.D. (ed). *Microwave Remote Sensing of Sea Ice*. AGU Geophysical Monograph **68**, pp. 47–71.

Tivy, A., Howell, S.E.L., Alt, B., *et al.* (2011). Trends and variability in summer sea ice cover in the Canadian Arctic based on the Canadian Ice Service Digital Archive, 1960–2008 and 1968–2008. *Journal of Geophysical Research* **116**, C03007, doi: 10.1029/2009JC005855.

Tschudi, M.A., Maslanik, J.A. & Perovich, D.K. (2008). Derivation of melt pond coverage on Arctic sea ice using MODIS observations. *Remote Sensing of Environment* **112**(5), 2605–2614, doi: 10.1016/j.rse.2007.12.009.

Wang, X., Key, J.R. & Liu, Y. (2010). A thermodynamic model for estimating sea and lake ice thickness with optical satellite data. *Journal of Geophysical Research* **115**, C12035, doi: 10.1029/2009JC005857.

Yu, Y. & Rothrock, D.A. (1996). Thin ice thickness from satellite thermal imagery. *Journal of Geophysical Research* **101**(C11), 25,753–25,766, doi: 10.1029/96JC02242.

Acronyms

AMSR2	Advanced Microwave Scanning Radiometer 2
AMSR-E	Advanced Microwave Scanning Radiometer - Earth Observing System
APP-x	Extended AVHRR Polar Pathfinder
AVHRR	Advanced Very High Resolution Radiometer
CIS	Canadian Ice Service
EM	Electromagnetic induction instruments
ESA ASCAT	European Space Agency Advanced Scatterometer
ESA ERS-1/2	European Space Agency European Remote Sensing Satellite-1/2
ICESat	Ice, Cloud, and Land Elevation Satellite
NIC	US National Ice Center
SAR	Synthetic Aperture Radar
SMOS	Soil Moisture Ocean Salinity sensor
SSM/I	Special Sensor Microwave Imager
ULS	Upward-looking sonars

Website cited

http://stratus.ssec.wisc.edu/products/appx/appx.html

12 Remote sensing of lake and river ice

Claude R. Duguay[1], Monique Bernier[2], Yves Gauthier[2] and
Alexei Kouraev[3]

[1] *University of Waterloo, Waterloo, Ontario, Canada*
[2] *Centre Eau, Terre, Environnement, Québec, Canada*
[3] *Laboratoire d'Etudes en Géophysique et Océanographie Spatiales (LEGOS), Toulouse, France*

Summary

Lake and river ice play an important role in the biological, chemical, and physical processes
of cold region freshwater. The presence of freshwater ice also has many economic implications, ranging from hydroelectricity and transportation (e.g., duration of ice-road and
open-water shipping seasons) to the occurrence and severity of ice-jam flooding that can
cause serious damage to infrastructure and property. In addition to its significant influence
on bio-physical and socio-economic systems, freshwater ice is also a sensitive indicator of
climate variability and change.

Documented trends and variability in freshwater ice conditions have largely been related
to air temperature changes. Long-term trends observable from ground-based records reveal
increasingly later freeze-up and earlier break-up dates, closely corresponding to increasing
air temperature trends, but with greater sensitivity at the more temperate latitudes. Broad
spatial patterns in these trends are also related to major atmospheric circulation patterns
originating from the Pacific and Atlantic oceans, such as the El Niño-La Niña/Southern
Oscillation, the Pacific North American pattern, the Pacific Decadal Oscillation, and the
North Atlantic Oscillation/Arctic Oscillation.

Despite the wide-ranging influences of freshwater ice and its robustness as an indicator of climate change, a dramatic reduction in ground-based observational recordings has
occurred globally since the 1980s. Consequently, satellite remote sensing has assumed a
greater role in observing lake ice and river ice. This chapter provides an overview of the
recent progress on remote sensing of lake and river ice. For lake ice, topics reviewed comprise the determination of:

1 ice cover concentration, extent and phenology;
2 ice types;
3 ice thickness and snow on ice;
4 snow/ice surface temperature; and
5 grounded and floating ice covers on shallow Arctic and sub-Arctic lakes.

Remote Sensing of the Cryosphere, First Edition. Edited by M. Tedesco.
© 2015 John Wiley & Sons, Ltd. Published 2015 by John Wiley & Sons, Ltd.
Companion Website: www.wiley.com/go/tedesco/cryosphere

Regarding remote sensing of river ice, topics covered include:

1 the determination of ice extent, ice phenology, ice types, ice jams, flooded areas, ice thickness, and surface flow velocities; and
2 the incorporation of SAR-derived ice information into a GIS-based system for river-flow modeling and flood forecasting.

The chapter concludes with an outlook on anticipated developments, in light of recent and upcoming satellite missions.

12.1 Introduction

Lake and river ice play an important role in the biological, chemical, and physical processes of cold region freshwater. The presence of freshwater ice also has many economic implications, ranging from hydroelectricity and transportation (e.g., duration of ice-road and open-water shipping seasons) to the occurrence and severity of ice-jam flooding, which can cause serious damage to infrastructure and property (Prowse *et al.*, 2011c). Freshwater ice is estimated to cover a total area of 1.7×10^6 km^2 over the Northern Hemisphere (determined at peak thickness, north of the January 0°C isotherm, excluding the Greenland ice sheet) and a volume of 1.6×10^3 km^3 (Brooks *et al.*, 2013; Figure 12.1). The estimated area of freshwater ice is nearly equal to that of the Greenland ice sheet, and its volume to that of snow on land.

Lakes that form a seasonal ice cover are a major component of the terrestrial landscape. They cover approximately 2% of the Earth's land surface, with the majority of them located in the Northern Hemisphere (Brown and Duguay, 2010). Estimates of their areal coverage can reach up to 40–50% in some regions of the Arctic and sub-Arctic. Lakes have the highest evaporation rates of any high latitude terrestrial surface (Rouse *et al.*, 2008a). Their frequency and size greatly influence the magnitude and timing of landscape-scale evaporative and sensible heat inputs to the atmosphere, and they are important to regional climatic and meteorological processes. Shallow lakes warm quickly in spring and have very high evaporation rates until they freeze in autumn. Large lakes take a substantially longer period to warm, but stay ice-free (or partly ice-free) into early winter, and their total evaporation amounts are significantly greater (Rouse *et al.*, 2005).

The duration of lake ice, in particular, controls the seasonal heat budget of lake systems, thus determining the magnitude and timing of evaporation (Rouse *et al.*, 2008b). The presence (or absence) of ice cover on lakes during the winter months also has an effect on both regional climate and weather events (e.g., thermal moderation and lake-effect snow) (Brown and Duguay, 2010). Monitoring of lake ice is therefore critical to our skill at forecasting high-latitude weather, climate, and river runoff. Recent investigations, indeed, emphasize the importance of considering lake ice cover for modeling the energy and water balance of high-latitude river basins, for regional climate modeling, and for improving numerical weather prediction in regions where lakes occupy a significant fraction of the landscape

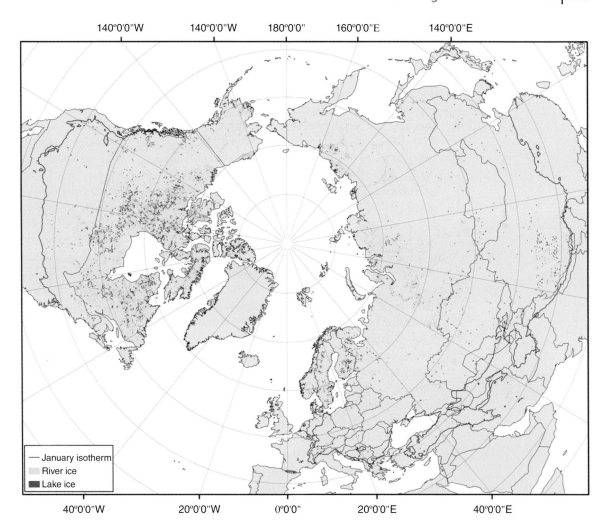

140°0'0"W   140°0'0"W   180°0'0"   160°0'0"E   140°0'0"E

January isotherm
River ice
Lake ice

40°0'0"W   20°0'0"W   0°0'0"   20°0'0"E   40°0'0"E

Figure 12.1 Approximate areal extent of freshwater ice in the Northern Hemisphere at peak thickness, north of the January 0°C isotherm, excluding areas of perennial land ice (from Brooks *et al.*, 2013). The water body mask used is from the Global Lake and Wetlands Database (GLWD). (Lehner and Doll, 2004. Reproduced with permission of Elsevier).

(Lindström *et al.*, 2010; Kheyrollah Pour *et al.*, 2012; Martynov *et al.*, 2012; Zhao *et al.*, 2012).

River ice also affects an extensive portion of the global hydrologic system, particularly in the Northern Hemisphere, where major ice covers develop on 29% of the total river length and seasonal ice affects 58% (Prowse *et al.*, 2007). For large rivers in cold continental regions, such as the Lena and lower Mackenzie, or at higher latitudes, such as the Yukon, ice conditions can persist for more than half the year over the entire river length (Prowse *et al.*, 2011a). In contrast, rivers with more temperate headwaters only have sections (e.g., 73% of Ob River length) that experience such long-term ice effects. For these and other rivers of the Northern Hemisphere, ice duration and break-up exert a significant control on the timing and magnitude of extreme hydrologic events such as low flows and floods (Prowse *et al.*, 2007).

River geomorphology, vegetation regimes, and nutrient/sediment fluxes that support aquatic ecosystems are especially sensitive to changes in the severity and timing of river ice break-up (Prowse *et al.*, 2011c). Given the broad ecological and socio-economic significance of river ice, scientific concern has been expressed about how future changes in climate might affect river-ice regimes (IGOS, 2007).

In addition to its significant influence on bio-physical and socio-economic systems, freshwater ice is also a sensitive indicator of climate variability and change. Documented trends and variability in freshwater ice conditions have largely been related to air temperature changes. Long-term trends observable from ground-based records reveal increasingly later freeze-up and earlier break-up dates, closely corresponding to increasing air temperature trends, but with greater sensitivity at the more temperate latitudes (Brown and Duguay, 2010; Prowse *et al.*, 2011b). Broad spatial patterns in these trends are also related to major atmospheric circulation patterns originating from the Pacific and Atlantic oceans, such as the El Niño-La Niña/Southern Oscillation, the Pacific North American pattern, the Pacific Decadal Oscillation, and the North Atlantic Oscillation/Arctic Oscillation (see Bonsal *et al.*, 2006; Prowse *et al.*, 2011b for details).

Despite the wide-ranging influences of freshwater ice and its robustness as an indicator of climate change, a dramatic reduction in ground-based observational recordings has occurred globally since the 1980s (Lenormand *et al.*, 2002; Duguay *et al.*, 2006; IGOS, 2007; Prowse *et al.*, 2011a; Jeffries *et al.*, 2012). Satellite remote sensing provides the necessary means to increase the spatial coverage and temporal frequency of ground-based observations. This chapter provides an overview of the recent literature on remote sensing of lake and river ice, and builds on a previous review by Jeffries *et al.* (2005). For a comprehensive review of lake ice and river ice characteristics, properties and processes, the reader is invited to consult Jeffries *et al.* (2012). Remote sensing of lake ice is reviewed first followed by river ice. The chapter concludes with an outlook on anticipated developments in light of recent and upcoming satellite missions.

12.2   Remote sensing of lake ice

Remote sensing of freshwater ice is a topic that has received little attention when compared to other elements of the cryosphere. In this section, we review the main developments that have recently taken place in satellite remote sensing of lake ice, more specifically:

1 ice cover concentration, extent and phenology;
2 ice types;
3 ice thickness and snow on ice;
4 snow/ice surface temperature; and
5 grounded and floating ice covers on shallow Arctic and sub-Arctic lakes.

12.2.1   Ice concentration, extent and phenology

Ice concentration is the fraction of the water surface that is covered by ice. It is typically reported as a percentage (0 to 100% ice), a fraction (0 to 1), or in tenths (0/10 to 10/10). Ice concentration on lakes is not usually determined at the satellite pixel-scale, as is done operationally for sea ice from passive microwave-based algorithms. Rather, it is estimated over various areas of a lake, as is done for the production of ice charts by the National Oceanic and Atmospheric Administration (NOAA) for the Great Lakes of southern Canada/northern United States, or over the entire lake surface, as done by the Canadian Ice Service (CIS) (Duguay et al., 2011).

Ice extent defines a section of a water body as either ice-covered or ice-free. From remote sensing data, each pixel is usually attributed a single class value for ice (i.e., not discriminating for ice types) and one for open water, as done in NOAA's Interactive Multisensor Snow and Ice Mapping System (IMS) 4 km resolution grid daily product (Helfrich et al., 2007) and the National Aeronautics and Space Administration's (NASA) 500 m Moderate Resolution Imaging Spectroradiometer (MODIS) snow products (Hall et al., 2002). Extent is frequently reported in terms of area (in km^2) covered by ice.

"Ice phenology" is the term used to describe the seasonal cycle of lake ice cover. It encompasses the freeze-up and break-up periods and, by extent, ice cover duration. "Freeze-up" and "break-up" are often used interchangeably with "ice on" and "ice off". Although freeze-up and break-up are often reported in the literature to occur on specific calendar dates, they rather represent processes or sequences of processes. As stated by Jeffries et al. (2012):

> "... freeze-up can be viewed as the period of time between initial ice formation and establishment of a complete ice cover, and break-up as the period of time between the exposure of bare ice (once all snow has melted) and the complete clearance of the ice cover."

Table 12.1 provides definitions for ice phenology variables from a remote sensing perspective at the pixel level, and for entire lakes or lake sections. The terminology presented in this table could be adopted as a basis for remote sensing of ice phenology. This would remove some of the existing confusion regarding the meaning of freeze-up and break-up, particularly when comparing remote sensing-derived dates against ground-based observations and numerical lake ice model output.

12.2.1.1   Optical remote sensing

To date, operational methods to determine ice concentration and extent have largely relied on the visual (or semi-automated) interpretation of optical and, to a lesser extent, microwave imagery. Such is the case for the creation of the IMS and Canadian Ice Service (CIS) ice products (see Duguay et al., 2011 for details), and NOAA ice charts (or ice analysis product) of the Great Lakes (Figure 12.2).

Table 12.1   Definition of ice phenology variables at per pixel level and for entire lake or lake section (Kang *et al.*, 2012. Reproduced under Creative Commons Attribution Licence 3.0).

	Pixel level	Entire lake or lake section
Freeze-up Period	**Freeze onset (FO)**: First day of the year on which the presence of ice is detected in a pixel and remains until ice-on. **Ice-on**: Day of the year on which a pixel becomes totally ice-covered. **Freeze duration (FD)**: number of days between freeze-onset and ice-on dates.	Complete freeze over (CFO): Day of the year when all pixels become totally ice-covered.
Break-up period	**Melt onset (MO)**: First day of the year on which generalized spring melt begins in a pixel. **Ice-off**: Day of the year on which a pixel becomes totally ice-free. **Melt duration (MD)**: numbers of days between melt-onset and ice-off dates.	Water clear of ice (WCI): Day of the year when all pixels become totally ice-free.
Ice season	**Ice cover duration (ICDp)**: number of days between ice-on and ice-off dates.	**Ice cover duration (ICDe)**: number of days between CFO and WCI.

The three products are generated using imagery obtained from a range of satellite sensors operating at different frequencies. As an example, the 4 km grid resolution IMS product incorporates a wide variety of satellite imagery (e.g., AVHRR, GOES, SSM/I), as well as derived mapped products (e.g., USAF Snow/Ice Analysis, AMSU) and surface observations (see Helfrich *et al.*, 2007 for details). Ice phenology (freeze-up, break-up and ice cover duration) anomalies have recently been derived from the IMS product for the largest lakes of the Arctic (Duguay *et al.*, 2011, 2012, 2013).

Few algorithms have been developed to automatically map lake ice cover from optical (visible to mid-infrared) satellite data. One algorithm, known as Snowmap, has been devised to operationally generate snow data products from the MODIS sensor onboard of NASA's Aqua and Terra satellites. The products provide information on snow-covered land and ice on inland water at spatial resolutions of 500 m and 0.05° at daily, eight-day, and monthly temporal resolutions (monthly products are only generated at 0.05°). The Snowmap algorithm may experience difficulties in detecting the presence of lake ice in areas where snow cover is patchy due to wind redistribution, a situation often encountered during the early winter period on lakes found at northernmost latitudes. However, the performance of the Snowmap algorithm over lake ice is a topic that has yet to be fully investigated.

MODIS snow data products with a spatial resolution of 500 m have been utilized recently as part of climate-related investigations. Brown and Duguay (2012) assessed the utility of the MODIS daily snow product (Collection 5) from Aqua and Terra for determining lake ice phenology at the sub-grid cell level of the

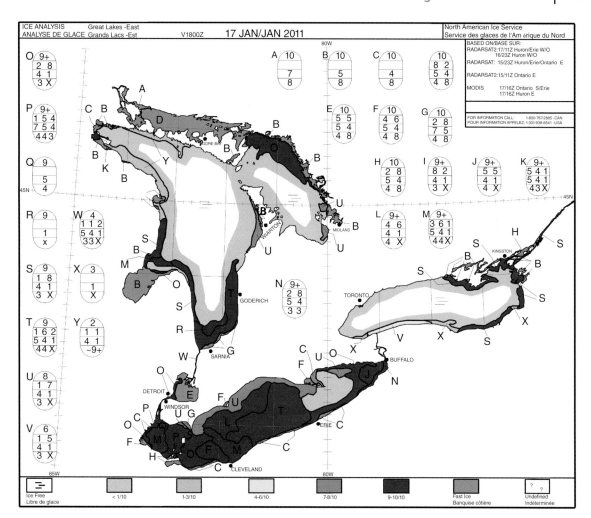

Figure 12.2 Ice analysis (chart) product of the eastern Great Lakes (Lake Ontario, Lake Erie, and Lake Huron) on 17 January 2011, constructed from the interpretation of RADARSAT-1/2 and MODIS imagery by ice analysts of the US National Ice Center (NIC). (US National Ice Center http://www.natice.noaa.gov/products/great_lakes.html).

North American Regional Reanalysis (NARR) 32-km grid product. The NARR product provided atmospheric forcing variables for lake ice model simulations with the Canadian Lake Ice Model (CLIMo; Duguay *et al.*, 2003). Both IMS 4 km and MODIS 500 m products were found to be useful for detecting ice-off dates when compared to CLIMo output. However, the MODIS product was advantageous for detecting ice-on, mainly due to the finer resolution and resulting spatial detail (Brown and Duguay, 2012).

Kropáček *et al.* (2013) used eight-day MODIS to analyze the ice phenology of 59 large lakes on the Tibetan Plateau for the period 2001–2010. The authors report that ice cover duration shows high variability due to both climatic and local factors, and that freeze-up dates appear to be more thermally determined than break-up for the studied lakes.

An algorithm based on image thresholding of reflectance has been presented for extracting lake ice phenology events from a historical satellite record, Advanced Very High Resolution Radiometer (AVHRR; 1.1 km spatial resolution) imagery (Latifovic and Pouliot, 2007). The development of the algorithm was motivated by interest in extending existing *in situ* observational records of 36 Canadian lakes (Duguay *et al.*, 2006) and to develop records for six additional lakes in Canada's far north for the period 1985–2004. A strong agreement (i.e., similar ice trends over the overlapping period) between freeze-up and break-up dates obtained from the AVHRR record and *in situ* observations was observed, suggesting that the two data sources are a useful complement to each other. The algorithm presented by Latifovic and Pouliot (2007) is not sensor-specific and is, therefore, claimed to be applicable to data from other optical sensors such as MetOp/AVHRR, MODIS, MERIS (MEdium Resolution Imaging Spectrometer), and SPOT/VGT (Système Pour l'Observation de la Terre/VEGETATION).

While imagery from optical sensors is highly desirable for monitoring lake ice, the presence of cloud cover, which can be extensive during some periods of the year, as well as late fall/early winter darkness (or low sun angle) experienced at high latitudes, limit its usefulness during the freeze-up period, particularly when applying automated algorithms to satellite data (e.g., MODIS Snowmap). Given these limitations, investigations have also focused on the development of approaches to monitor ice cover from microwave scatterometry, radiometry, and altimetry, and from synthetic aperture radar (SAR).

12.2.1.2 *Microwave remote sensing*

Microwave remote sensing provides the capability to obtain ice phenology parameters under cloudy and polar darkness conditions. Algorithms that rely on the temporal evolution of backscatter (scatterometry and altimetry) and brightness temperature, or a combination of both, have recently been developed to monitor ice cover conditions on large lakes. High temporal sampling over large areas (at times twice daily or better) is possible with some of the satellite sensors operating at microwave frequencies (e.g., passive radiometers and scatterometers). This is, however, to the detriment of spatial resolution (several km to tens of km), which is much higher for SAR systems (1 m to about 100 m).

Using data from the SeaWinds scatterometer on board the NASA's QuikSCAT satellite (operational from June 1999 until November 2009), Howell *et al.* (2009) developed an algorithm to map the spatial distribution of ice phenology parameters on Great Bear Lake (GBL) and Great Slave Lake (GSL), Northwest Territories, Canada. Ku-band backscatter coefficients $\sigma°$ (VV) from QuikSCAT (enhanced at 4 km spatial resolution) can be used to detect melt onset, ice-off and freeze onset (ice-on) dates. In winter, $\sigma°$ exhibits relatively high returns, whereas melt onset is marked by a strong decrease in $\sigma°$.

Following the first significant downturn in the QuikSCAT $\sigma°$ temporal evolution, the $\sigma°$ begins a series of up and downturn oscillations. The first oscillation (i.e., up-turn and down-turn) is related to freeze-thaw processes. The sharp drop in QuikSCAT $\sigma°$ after this period marks the date when the lake becomes clear of ice (ice-off). The QuikSCAT $\sigma°$ then increases sharply from the relatively low open-water $\sigma°$ values to higher $\sigma°$ values, to indicate freeze onset (ice-on). Observed changes in $\sigma°$ above or below certain threshold values form the basis of the algorithm developed by Howell *et al.* (2009). Results from the application of the algorithm to GBL and GSL have revealed contrasting patterns in ice phenology parameters within and between the two lakes (2000–2006 period), with ice cover duration lasting five weeks longer on GBL, and the ice regime of GSL being significantly influenced by water inflow from Slave River to the south.

More recently, Kang *et al.* (2012) developed an approach similar to that of Howell *et al.* (2009), but using daily time series of brightness temperature from the Advanced Microwave Scanning Radiometer – Earth Observing System (AMSR-E). The study shows that 18.7 GHz H-pol is the most suitable channel for detecting ice phonological events (freeze-onset/melt-onset and ice-on/ice-off dates as well as ice cover duration; see Table 12.1) on GBL and GSL. The algorithm proposed tracks changes in brightness temperature above or below predefined threshold values to detect ice dates.

A comparison of dates for several ice phenology parameters derived from other satellite remote sensing products (e.g., NOAA IMS, QuikSCAT from the study of Howell *et al.* (2009)) show that AMSR-E 18.7 GHz H-pol (or sensors operating onboard of other satellite platforms at a similar frequency) provides a promising means for routinely monitoring ice phenology on large northern lakes. The same algorithm concept could be applied to historical records of the Scanning Multi-channel Microwave Radiometer (SMMR) and Special Sensor Microwave/Imager (SSM/I), to reconstruct ice conditions on GBL and GSL since 1979.

In addition to examining the value of scatterometry and radiometry individually, potential also exists for monitoring ice cover on northern lakes, by exploiting the synergy of more than 20 years of radar altimetric and passive microwave data to improve spatial and temporal coverage. A dedicated methodology for ice discrimination has been developed and tested for the Caspian and Aral seas, as well as for the lakes Ladoga, Onega and Baikal (Kouraev *et al.*, 2003, 2007a, 2007b, 2008, and in press). The approach uses the simultaneous nadir-looking active and passive data from several altimetric missions (TOPEX/Poseidon, Jason-1, Envisat and GFO), complemented by passive microwave SSM/I data to discriminate ice from open water (Figure 12.3).

Combination of the two types of satellite observations is beneficial, due to the wide spatial coverage and good temporal resolution offered by SSM/I, and the high radiometric sensitivity and along-track spatial resolution of the altimeters. By analyzing this data, it is possible to define specific dates of ice events (the first appearance of ice, the formation of a stable ice cover, the first appearance of open water and the complete disappearance of ice) for the water bodies of interest.

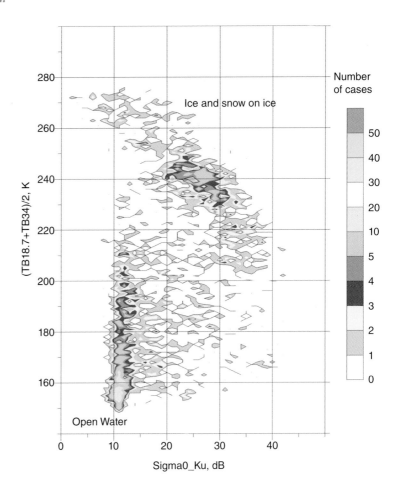

Figure 12.3 Frequency distribution (number of cases; total number of observations (*n*) = 3,130) of Envisat radar altimeter backscatter observations (Sigma0_Ku in decibel, dB) versus passive (TB18.7 + TB34)/2 microwave observations (degrees Kelvin, K) for Cycles 11 to 30 (5 November 2002 to 4 October 2004) over Lake Baikal, Russia. Two well-defined clusters are easily identifiable, making it possible to discriminate open water from ice cover.

In most cases, it is possible to define ice event dates using these maps with a five-day temporal resolution, and an uncertainty of ± 2.5 days. Such an approach enhances the potential of microwave measurements for ice studies, and can be used reliably to extend the existing time series available for coastal stations, as well as to create new time series for regions that were not previously covered by continuous *in situ* observations.

Co-polarized (HH and VV) SAR data from the ERS-1/2 (C-band), JERS-1 (L-band), RADARSAT-1/2 (C-band) and Envisat ASAR (C-band) sensors have also been analyzed, mostly through visual interpretation, to map changes in lake ice cover in response to climate over the last decade or so at Arctic locations. For example, Mueller *et al.* (2009) used SAR data acquired in late summer to show that five lakes located at the northern tip of Ellesmere Island (Nunavut, Canada) experienced significant reductions in summer ice cover from 1992 to 2007, with some lakes transitioning from perennially to annually ice-covered conditions following the warm El Niño year of 1998.

In a separate study, Cook and Bradley (2010) analyzed RADARSAT (HH-pol) and LANDSAT Thematic Mapper (TM)/Enhanced Thematic Mapper Plus (ETM+) imagery (115 SAR and 19 LANDSAT images) to evaluate recent (1997–2007) changes in the ice cover of Upper and Lower Murray Lakes (81°20′N, 69°30′W), also situated on Ellesmere Island, during the melt period. Similar to the results reported in Mueller *et al.* (2009), the authors suggest that the observed ice melt at Upper and Lower Murray Lakes, due to recent warming in the high Arctic, has forced the lakes near a threshold from a state characterized by perennial ice cover to the current state, which includes seasonal melting of lake ice.

The potential of multi-polarized (co-, cross-, and quad-polarization) SAR data has recently been examined to automatically map lake ice cover during the break-up (melt) period (Geldsetzer *et al.*, 2010; Sobiech and Dierking, 2013). Geldsetzer *et al.* (2010) analyzed RADARSAT-2 SAR imagery for monitoring ice cover during spring melt on lakes located in the Old Crow Flats, Yukon, Canada. The authors were successful at identifying initial break-up with a simple threshold applied to HH backscatter data (> 81% accuracy) and the main break-up period with a threshold on cross-polarized (HV) data (66–97%).

Sobiech and Dierking (2013) evaluated the performance of the unsupervised k-means classification in a binary classification of ice cover and open water on lakes and river channels of the central Lena Delta, northern Siberia. K-means is an unsupervised image classification approach frequently used for mapping river ice types (see section 12.3.2). Using six TerraSAR-X (X-band, single HH co-pol) and three RADARSAT-2 images (C-band, quad-pol HH, VV, HV, and VH simultaneously) obtained during spring, 2011, Sobiech and Dierking found that the performance of the k-means classification is comparable to that of a fixed-threshold approach. Application of a low-pass filter prior to the classification of river channels, and a closing filter on the classification results of lakes, strongly improved the overall k-means classification results.

12.2.2   Ice types

Unlike ice cover concentration and extent, classification of lake ice types has been the object of very few investigations. This is in striking contrast to river ice, for which classification of ice types has been the topic of several publications (see section 12.3.2). This situation likely stems from the fact that knowledge of river ice types is important not only for public safety and navigation, as is also the case for lake ice, but most notably for the prediction and mitigation of river ice jams (Bérubé *et al.*, 2009).

SAR data is useful for obtaining some information about the internal structure of ice (texture) due to the ability of the microwave signal to penetrate ice (Hall *et al.*, 1994) under dry snow conditions. The most comprehensive data set of coincident ice type and radar observations over lake ice was collected during the 1997 Great LAkes Winter Experiment (GLAWEX'97; Nghiem and Leshkevich, 2007).

As part of this experiment, a C-band polarimetric scatterometer operated by the Jet Propulsion Laboratory (JPL) was installed on separate occasions on two ice-breaker vessels for periods of two weeks each (February and March, 1997) in order to gather backscatter signatures of various ice types (new ice, pancake ice, consolidated ice, stratified ice, brash ice, and lake ice with crusted snow) and open water on Lake Superior at incidence angles from 0–60°. The scatterometer measurements included incidence angles and polarizations of spaceborne SAR instruments on ERS, RADARSAT, and Envisat satellites.

The ultimate goal of GLAWEX'97 was to compile a library of radar signatures, together with *in situ* observations, of ice types that could be used with the interpretation of spaceborne SAR data, acquired concurrently with Earth Resource Satellite 2 (ERS-2; VV) and RADARSAT-1 ScanSAR (HH) imagery, for ice classification and mapping (Nghiem and Leshkevich, 2007). Using the library of backscatter signatures, Leshkevich and Nghiem (2007) were able successfully to identify and map different ice types through a supervised classification technique. However, wind speed and direction was found to confound the discrimination between open water and ice, since RADARSAT-1 and ERS-2 provide single frequency, single polarization data. Cross-polarization data (e.g., RADARSAT-2 and TerraSAR-X) are less sensitive to wind effects than co-polarized data and, therefore, could be seen as optimal for discriminating between open water and ice cover on windy days. However, Geldsetzer *et al.* (2010) recently reported that the signal from open water surfaces (as well as that from lake ice) can be at or below the noise floor, making cross-polarized data of limited use. Sobiech and Dierking (2013) conclude that HH-polarized images are best suited for separation of ice and water surfaces.

12.2.3 Ice thickness and snow on ice

Lake ice has been shown to respond to changing climate conditions, particularly changes in air temperature and snow accumulation (Brown and Duguay, 2011). Trends and variability in ice phenology are typically associated with variations in air temperatures, while trends in ice thickness tend to be associated more with changes in snow cover (Brown and Duguay, 2010). During the ice growth season, the dominant factors that control the thickening of lake ice are temperature and snowfall. Once ice has formed, snow accumulation on the ice surface slows the growth of ice below due to the insulating properties as a result of the lower thermal conductivity (thermal conductivity of snow, 0.08–0.54 $Wm^{-1}K^{-1}$ vs 2.24 $Wm^{-1}K^{-1}$ for ice; Sturm *et al.*, 1997).

Snow mass can also change the composition of the ice by promoting snow ice development, and hence influence the thickness of the ice cover (Brown and Duguay, 2010). Observations of ice thickness and snow on ice (depth, mass) are required for both climate monitoring, and for evaluating and improving models of lake-ice growth. Microwave sensors provide the capabilities for estimating ice thickness and snow on ice over broad areas and with high temporal coverage. However, few studies have explored this potential.

Until very recently, no approach had been developed to estimate ice thickness on large lakes from satellite remote sensing. The potential of passive microwave data for ice thickness determination had, however, been previously demonstrated in an airborne study by Hall *et al.* (1981), using a limited number of field observations. Other studies have shown that ERS or RADARSAT C-band SAR could be used to determine ice thickness in northern shallow lakes that freeze to bed (see section 12.2.5), when combined with optical data that can provide estimates of lake depth (e.g., Duguay and Lafleur, 2003).

Lately, Kang *et al.* (2010) have shown that the temporal evolution of brightness temperature measurements from the AMSR-E 10.7 GHz and 18.7 GHz frequency channels during the ice growth season on Great Bear Lake (GBL) and Great Slave Lake (GSL), Canada, is strongly correlated with ice thickness as estimated with a numerical lake ice model (Duguay *et al.*, 2003). Using AMSR-E data from 2002–2007, the authors showed that over 90% of the variations in brightness temperature on GBL and GSL could be explained by the seasonal evolution of ice thickness on these lakes in winter. The strong relationship between brightness temperature at 18.7 GHz V-pol and ice growth from the lake ice model has since been explored for the development of regression-based ice thickness (ICT) retrieval algorithms (Kang *et al.*, 2014). Simple linear regression equations ($ICT_{GBL} = 3.53 \times T_B - 737.929$ for GBL $ICT_{GSL} = 2.83 \times T_B - 586.305$ for GSL) allow for the estimation of ice thickness (in cm) on a monthly basis from January to April (Figure 12.4). The transferability of the regression equations remains to be examined for other large lakes of the Northern Hemisphere (e.g., Lake Baikal, Lake Ladoga, and Lake Onega in Russia).

While some progress has been made in estimating ice thickness from coarse-resolution passive microwave data, no method has yet been developed to our knowledge for estimating snow depth or snow water equivalent (SWE) on lake ice. Duguay *et al.* (2005) have shown that traditional passive microwave algorithms used to estimate SWE over land (difference between 37 GHz and 19 GHz frequency channels) do not work when applied over lake ice. Derksen *et al.* (2009) further illustrate that 19 and 37 GHz frequencies have penetration depths (≈ 2 at 19 GHz and 0.75 m at 37 GHz) that are strongly influenced by water beneath the ice for part of the season, but are also influenced by the ice and overlying snowpack by the end of the season.

The estimation of snow depth (and SWE) on lake ice is a challenging problem that needs to be further investigated using both passive and active microwave data acquired at Ka-/Ku-band frequencies. Kouraev *et al.* (2007a) report that snow accumulation, ice ageing (growth), and decay induce a decrease in Ku-band radar altimetry backscatter values and changes in passive microwave brightness temperature measurements (see Figure 12.3).

12.2.4   *Snow/ice surface temperature*

It is now well recognized that satellite observations of lake ice extent/concentration and lake surface temperature are valuable for improving numerical weather

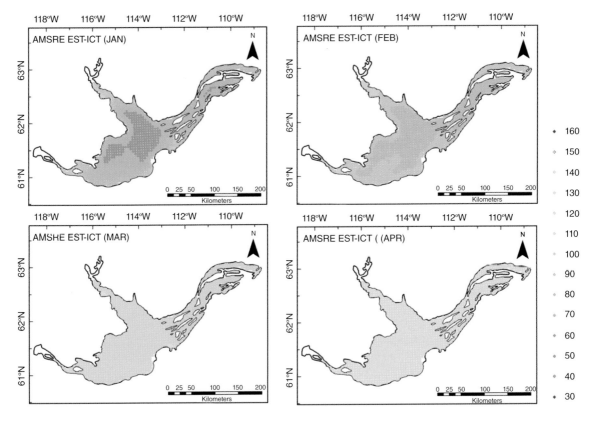

Figure 12.4 Ice thickness maps of Great Slave Lake for the months of January, February, March and April (2002–2009 average) derived from AMSR-E 18.7 GHz V-pol brightness temperature data (Kang *et al.*, in press. Reproduced with permission of Elsevier).

predictions in regions where lakes occupy a significant fraction of the landscape (e.g., Kheyrollah Pour *et al.*, 2012; Zhao *et al.*, 2012). Furthermore, it has been shown that lake surface temperature (LST) products derived from satellite sensors such as MODIS are useful for evaluating lake snow/ice surface temperature output simulated with one-dimensional lake models. For example, Kheyrollah Pour *et al.* (2012) showed that MODIS daily LST is helpful at identifying current limitations of the Freshwater Lake model (FLake), one of the most common lake models used as a lake parameterization scheme in numerical weather prediction (NWP) and regional climate models (RMCs), particularly regarding its lack of proper representation of snow on ice.

Compared to FLake, the Canadian Lake Ice Model (CLIMO; Duguay *et al.*, 2003), which simulates the seasonal evolution of snow cover on the lake ice surface, generates snow/ice surface temperatures that are closer to those observed with MODIS (Figure 12.5). Statistics of model performance given by the index of agreement (Ia), the root mean square error (RMSE), and mean bias error (MBE), reveal that CLIMO outperforms FLake, even if both models slightly overestimate the LSTs

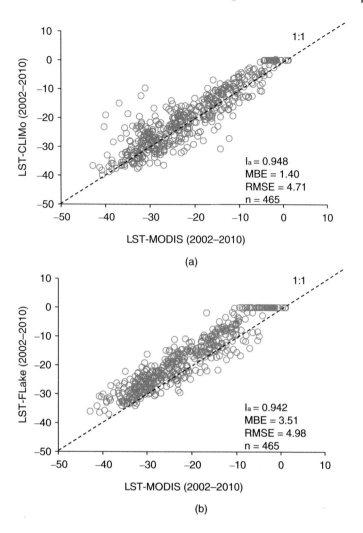

Figure 12.5 Comparison of modeled LST from (a) CLIMo and (b) FLake models with MODIS-derived LST (°C) data for ice cover seasons 2002–2010 in Back Bay (Great Slave Lake), Canada. Modified from Kheyrollah Pour *et al.*, 2012.

from MODIS. However, land/lake surface temperatures from MODIS products have been reported to suffer from a 1–2°C cold bias (ignoring the possible effect of cloud contamination) in other studies (e.g., Soliman *et al.*, 2012).

12.2.5 Floating and grounded ice: the special case of shallow Arctic/sub-Arctic lakes

Shallow lakes (less than ≈4 m deep) are a ubiquitous feature of the Arctic coastal plains of Siberia, northern Alaska and Canada. In Canada, for example, they are particularly prevalent in the Hudson Bay Lowlands and the Mackenzie River Delta region. Within the Arctic, shallow water bodies (lakes and ponds) are estimated to occupy between 15% and 50% of total land area (Duguay and Pietroniro, 2005).

There has been interest for many years in monitoring/mapping the seasonal evolution of floating and grounded ice (i.e., ice that is frozen to the bottom of the lake) of shallow lakes from remote sensing. Knowing when (i.e., the timing) and where the ice becomes grounded or remains afloat on shallow lakes during the ice growth season is relevant for climate monitoring (e.g., Arp *et al.*, 2012; Surdu *et al.*, 2014; Figure 12.6), the determination of water availability (e.g., Jeffries *et al.*, 1996; White *et al.*, 2008; Grunblatt and Atwood, 2013), and mapping of fish overwintering habitat (e.g., Hirose *et al.*, 2008; Brown *et al.*, 2010).

With Arctic climate warming, shallow tundra lakes are expected to develop thinner ice covers, likely resulting in a smaller fraction of lakes that freeze to their bed in winter (Surdu *et al.*, 2014). A shift from a grounded-ice to a floating-ice regime can initiate talik development and could potentially release large stocks of carbon previously frozen in permafrost in the form of methane (Arp *et al.*, 2012).

In SAR imagery, floating ice that contains bubbles (a typical situation for shallow tundra lakes) shows strikingly different backscatter intensities than that of ground ice. The change from high (floating ice) to low (grounded ice) backscatter during the ice growth season has been documented and explained in several investigations using C-band SAR data from ERS-1/2, RADARSAT-1, Envisat Advanced SAR alternating polarization and wide swath modes (Jeffries *et al.*, 1994; Morris *et al.*, 1995; Duguay *et al.*, 2002; Duguay and Lafleur, 2003; Brown *et al.*, 2010; Arp *et al.*, 2011, 2012; Engram *et al.*, 2012; Surdu *et al.*, 2014) and,

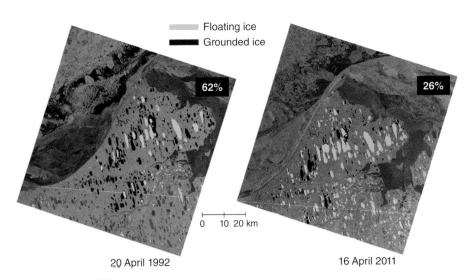

Figure 12.6 Image segmentation results (grounded ice and floating ice) of ERS-1/2 SAR acquisitions (20 April 1992 and 16 April 2011), obtained with the Iterative Region Growing with Semantics (IRGS) algorithm, as implemented in the MAp-Guided Ice Classification System (MAGIC) software (Clausi *et al.*, 2010). Ice season 1991–1992 was colder than 2010–2011, resulting in a larger number of lakes freezing to their bed (grounded) in April 1992 (62% grounded ice), compared with the same period in April 2011 (26% grounded ice).

recently, with L-band data (23.6 cm) acquired by the Japan Aerospace eXploration Agency's (JAXA) Phased-Array type L-band SAR (PALSAR) instrument aboard the Earth on the Advanced Land Observing Satellite (ALOS) (Engram *et al.*, 2012).

12.3 Remote sensing of river ice

In many northern rivers, the development of ice covers leads to important critical issues: ice jamming and, therefore, flooding of large areas; reduction of hydroelectric power at generating stations; impediment to navigation; and damage to human structures. The dynamic ice and flooding that accompany dynamic freeze-up and break-up, in particular, pose a significant risk to human life. In this section, we review advances in remote sensing of river ice, specifically:

1 the determination of ice extent, ice phenology, ice types, ice jams, flooded areas, ice thickness, and surface flow velocities and
2 the incorporation of SAR-derived ice information into a GIS-based system for river-flow modeling and flood forecasting.

12.3.1 Ice extent and phenology

Ice extent defines a section of a river as either ice-covered or ice-free, without consideration for ice types. Ice phenological parameters are determined by monitoring the evolution of ice extent along rivers from freeze onset, until water becomes clear of ice. Both optical and radar altimeter data have been used for this purpose on larger rivers. Pavelsky and Smith (2004) used daily time series of MODIS (250 m) and AVHRR (1.1 km) satellite images to monitor the spatial and temporal patterns of ice break-up along 1600–3300 km segments of the Lena, Ob', Yenisey, and Mackenzie rivers. The authors were able to visually identify the first day of predominantly ice-free water for ten years (1992–1993, 1995–1998, and 2000–2003), with a mean accuracy of 1.75 days when compared to ground-based observations. Large ice jams were also observable (see section 12.3.4 for more on this topic), particularly at confluences. However, smaller ice jams could not be detected, due to the limited spatial resolution of the imagery used. As shown in Figure 12.7, MODIS imagery can detect ice jams and flooding from the largest rivers.

In a more recent study, Chaouch *et al.* (2012) developed an automated approach that incorporates a threshold-based decision-tree image classification algorithm to determine ice extent from MODIS (visible and near-infrared bands at 250 m) on the Terra satellite. The algorithm, which generates three ice extent products (i.e., daily ice maps, weekly composite ice maps, and running cloud-free composite ice maps), was evaluated over nine ice seasons (2002–2010). Evaluation of the MODIS derived products for the Susquehanna River, one of the longest (≈ 715 km) and widest (≈ 1609 m at Harrisburg, PA) rivers in the northeastern USA, reveals

Figure 12.7 Ice jam on the Yukon River and flooding of the small town of Galena, Alaska, from the Moderate Resolution Imaging Spectroradiometer (MODIS) on NASA's Terra satellite, captured on May 28, 2013. In this color composite image, river water appears navy blue; ice appears teal; and vegetation is bright green. Clouds are white to pale blue-green, and cast shadows black. (NASA).

a good agreement with aerial photographs, *in situ* observations-based ice charts, and LANDSAT imagery (91% probability of ice detection).

Satellite radar altimetry is another tool that has recently been explored for monitoring the ice regime of large rivers. Troitskaya *et al.* (in press) propose a methodology for ice discrimination from altimetric satellite missions based on the analysis of two parameters:

    a) backscatter values (ICE-2 retracker) from Ku-band (18 Hz waveform) Envisat radar altimeter data; and

    b) brightness temperature differences ($dT_B = T_{B36.5} - T_{B23.8}$) from passive microwave radiometer data.

The approach has been developed and tested for the Ob' River in Russia (Troitskaya *et al.*, in press). For each crossover of the altimetric track over the main river channel, data is processed and grouped into 11 zones around selected

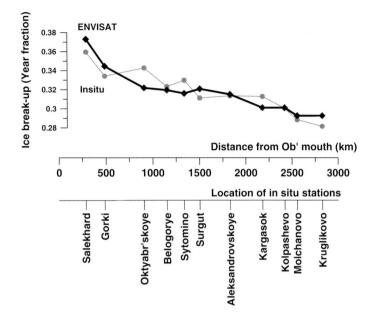

Figure 12.8 Average *in situ* (grey line, 2001–2011) and altimeter-derived (black line, 2002–2011) values of ice break-up along the main Ob' river channel between Salekhard and Novosibirsk, Russia.

ground-based stations, in order to increase the temporal resolution. The spatial variability of altimeter-derived dates of ice formation and break-up is in good agreement with ground-based observations for each zone (Figure 12.8). The algorithm and chosen threshold perform best for ice break-up detection, showing the dates of full clearance of ice, with an average bias equal to 0 and maximum bias ranging from −5 to 8 days.

12.3.2   Ice types, ice jams and flooded areas

Information on river ice types is needed for accurate hydrological forecasts to predict break-up of ice jams, and to issue timely flood warnings (Chaouch *et al.*, 2012). SAR is the preferred tool for the classification of river ice types, due to its penetrating capabilities and sensitivity to surface roughness and ice texture, particularly the size and density of volume scatterers (e.g., air bubbles) (Gherboudj *et al.*, 2007, 2010). C-band SAR at high spatial resolution (ca. 5–30 m) has been used successfully to map various ice types on small to medium-size rivers, or small stretches of large rivers.

Weber *et al.* (2003) used fine beam-mode RADARSAT-1 (HH) data acquired in winter and spring, 2000 and 2001, to map major ice cover types on the Peace River (northern British Columbia and Alberta, Canada). Video footage of the ice conditions on the Peace River was obtained from aerial ice observations conducted simultaneously with SAR image acquisitions. Image analysis was accomplished through visual interpretation, and by performing an unsupervised Fuzzy k-means classification applied to the RADARSAT-1 data. The unsupervised classification produced seven classes that represented the major ice cover types observed on the Peace River. The spatial distribution of ice cover types, as generated by the ice classification, coincided generally well with airborne observations and the visual interpretation of the original RADARSAT-1 images. Weber *et al.* (2003) indicated that the location of the boundaries between the ice types appeared to be accurate, but not necessarily precise.

The work of Weber *et al.* (2003) has lead to several follow-up investigations to automatically map, via image segmentation techniques, ice cover types on large rivers. The peer-reviewed literature on the subject reveals that the development of improved methods or the adoption and validation of existing ones has mainly, if not all, taken place in Canada. For example, Töyrä *et al.* (2005) further validated the approach developed by Weber *et al.* (2003) for the lower Peace River, Alberta. Using RADARSAT-1 standard beam mode images acquired during a break-up ice jam event on April 29 and May 1, 2003, Töyrä *et al.* (2005) evaluated its use for rapid-response mapping, using minimal ground verification. After a visual comparison with aircraft observations, it was noted that ice jams, deteriorated ice and solid ice, as well as the open water leads and reaches, were mapped very well.

The RADARSAT-1 images were also used to generate ice-jam flood maps using the unsupervised Fuzzy k-means classifier. It was found that spring ice-jam flood extents could easily be delineated. Results of the study of Töyrä *et al.* (2005) showed that RADARSAT-1 images were useful for the rapid mapping of river ice break-up conditions and spring flood extent. Unterschultz *et al.* (2009) also investigated the viability of using RADARSAT-1 satellite images to characterize river ice during the winter ice growth and break-up periods (Athabasca River, Alberta, Canada). They concluded that SAR shows excellent potential for identifying ice jams, intact ice, and open water during break-up.

Further advances in the determination of river ice types from SAR show that a two-step process involving image texture and unsupervised classification with the Fuzzy k-means algorithm is the method that provides the best results to date (Gauthier *et al.*, 2007, 2008). The procedure, known as the Ice Mapping Automated Procedure from Radar data (IceMap-R; Gauthier *et al.*, 2010), allows for the fully automated segmentation of up to nine ice classes and open water areas from RADARSAT HH data. The classified ice types are then merged into the main ice types (i.e., thermal ice, juxtaposed (or agglomerated) frazil ice and consolidated ice, along with open water; Figure 12.9) and can also be converted into roughness classes (e.g., water and floating pans, smooth ice and rough ice) needed as input in hydraulic flood routing models such as River1D (She and Hicks, 2006).

The global accuracy of IceMap-R in correctly classifying the main ice types with RADARSAT HH data is consistently around 70–80% (Gauthier *et al.*, 2012). Known limitations of the automated procedure applied to co-polarization HH data include the discrimination between smooth thermal ice and water, the presence of frazil pans, the presence of melting snow or water over ice, and the presence of rapids. Jasek *et al.* (2013) showed that the accuracy obtainable with IceMap-R can be significantly increased by exploring the use of both co- and cross-polarized (HH and HV) data available from RADARSAT-2. Using a set

Ice type	Legend	Roughness
Open water		Low
Thermal ice		Low
Frazil pans		Low to light
Agglomerated frazilice		Light
Consolidated ice (packedice, blocks)		Medium to high

Peace River, Dec.25, 2008   Koksoak River, Dec.01, 2008   Chaudière River, Jan.10, 2005

Figure 12.9 River ice types and related roughness classes derived from RADARSAT-1 HH imagery using IceMap for the Peace, Koksoak, and Chaudière rivers, Canada.

of five RADARSAT-2 images acquired for a section of the Peace River (British Columbia, Canada), the authors showed that HV is most efficient in correctly classifying water and juxtaposed ice, while HH is better for discriminating thermal ice and consolidated ice. The global accuracy achieved with this combination of polarizations is in the order of 90%. This is a significant improvement over the use of RADARSAT HH data alone (70–80%).

12.3.3 Ice thickness

Knowledge of ice thickness is important for river-ice hydraulic models. Together with ice roughness, it is one of the key ice parameters for flood routing models (She and Hicks, 2006). As part of an investigation on the evaluation of RADARSAT-1 data to characterize river ice during winter and breakup on the Athabasca River at Fort McMurray, Canada, Unterschultz *et al.* (2009) also assessed the possibility of estimating river ice thickness with C-band SAR. To determine whether a correlation did exist between ice thickness and C-band HH radar backscatter, observations of ice thickness, snow depth, and ice structure/type (e.g., thermal, juxtaposed, hummocky) were made during a field measurement campaign (February 17–19, 2004) along nine transects on the river, and compared to backscatter values extracted from fine beam mode RADARSAT-1 images acquired around the same time period (February 7 and March 2). The authors reported that a relationship did exist between ice thickness (transect average values) and mean backscatter, but that the sample size was too small for this to be considered a conclusive demonstration of any ability to determine average ice thickness based on radar backscatter alone. Results from the study suggested that knowledge of ice structure and inclusions (i.e., bubbles) needs to be considered for developing a model for ice thickness determination using SAR images. Unterschultz *et al.* (2009) concluded that further investigations are needed before ice thickness can be determined from SAR backscatter data.

In a more recent study, Mermoz *et al.* (2013) developed a method for the retrieval of river ice thickness from RADARSAT-2 polarimetric C-band SAR data. The authors first performed a Wishart classification to derive river ice types (Mermoz *et al.*, 2009b), followed by the masking of bubble-free thermal ice and consolidated ice. These two ice types were determined to represent surface areas ranging from 29.9% to 60.4% of a given river and excluded from further analysis because their radar backscatter signature was found to be almost insensitive to thickness variations. The relationship between radar backscatter and river ice thickness measurements was then examined over the unmasked areas of the three rivers (at 70 locations in total).

For the other ice types, polarimetric entropy (a measure that captures variability in terms of scattering mechanisms) was used to obtain ice thickness estimates from RADARSAT-2 data (the fitted model explained up to 85% of the observed variance in ice thickness). A leave-one-out cross-validation was then applied to assess the accuracy of the river ice thickness estimates. The RMSE was found to be 9.2 cm, and the effective RMSE 16.6 %. However, Mermoz *et al.* (2013) suggest that the robustness of this empirical model remains to be assessed. Interferometric SAR data also provides a promising means for the retrieval of pure thermal ice thickness (Wegmuller *et al.*, 2010).

Thermal infrared imagery is another technology that shows potential for estimating thin river ice. In a recent study, airborne images from a FLIR A40M thermal camera were acquired and successfully used to estimate thickness of thin ice (< 20 cm) on the St. Lawrence River, between Montreal and Quebec City, Canada. A simplified one-dimensional heat diffusion equation was solved using a finite element method to obtain ice thickness from airborne ice temperature measurements. The thermal images were also shown to provide significant details about ice floe characteristics, ice concentration, shape, size, structure, and number of floes per hectare, in addition to surface temperature and ice thickness. The approach developed in this study could be explored further to estimate thin ice of large rivers and lakes from thermal infrared data acquired by MODIS on the Terra and Aqua satellites, for example.

12.3.4 *Surface flow velocities*

Quantification of river surface velocity is important for understanding a wide range of biological, chemical and physical processes in northern rivers, and for the evaluation of river hydraulic models. Kääb and Prowse (2011) recently developed a novel approach that uses river-ice debris as an index of surface water velocity. The river-ice debris are tracked over the typical time lapse of two or more image acquisitions by spaceborne optical sensors that form a data set for stereo mapping. Using a satellite stereo image pair from the Advanced Spaceborne Thermal Emission and Reflection Radiometer (ASTER) onboard the NASA Terra satellite (15 m spatial resolution and 55.3 s time lapse), a triplet stereo scene from the Panchromatic Remote Sensing Instrument for Stereo Mapping (PRISM) aboard the Japanese ALOS satellite (2.5 m spatial resolution and 45 s or 90 s lapse rate, respectively), and an IKONOS satellite stereo pair (1 m spatial resolution and 53 s time lapse), ice debris are tracked between two images. The resulting displacements are then converted to velocity, using the time interval between the stereo images.

With this approach, Kääb and Prowse were able to measure and visualize, for the first time, the almost complete surface velocity field along approximately 80 km and 40 km long reaches of the St. Lawrence and Mackenzie rivers (Canada), respectively. Surface flow velocities derived from the ALOS PRISM satellite stereo triplet for a section of the Mackenzie River are shown in Figure 12.10. The approach proposed by the authors is, indeed, very promising, but it needs to be tested for more stereo-image data sets and validated with coincident *in situ* measurements of surface flow velocity and observations of river-ice debris with shore-based or airborne cameras.

12.3.5 Incorporating SAR-derived ice information into a GIS-based system in support of river-flow modeling and flood forecasting

While satellite imagery can help monitor the evolution of ice jams, fronts, and flooded areas, there is a need to improve models to help forecast flooding and become early warning systems (e.g., Puestow *et al.*, 2004). The FRAZIL system, developed by Gauthier *et al.* (2007, 2008), is a GIS-based system designed to support winter river-flow modeling and ice-related flood forecasting. It has been tested and improved since its inception in 2005 on rivers in the province of Quebec and, more recently, on the Athabasca River, northern Alberta, Canada.

The FRAZIL system includes two major components. The first of these consists of a set of GIS tools developed in the Python programming language, for use with ArcGIS (ESRI), and the second one incorporates a series of image processing routines (IceMap-R), developed in the image-processing software Geomatica (PCI Geomatics). FRAZIL is used to help define the morphological characteristics of a river channel before hydraulic modeling, as well as to characterize the ice cover state (ice coverage, roughness and ice-jam length) and evolution (ice front progression, freeze-up completion, signs of breakup) from SAR-derived ice maps. Some of the data generated by the FRAZIL system can then be used as input into hydraulic routing models or breakup forecasting models.

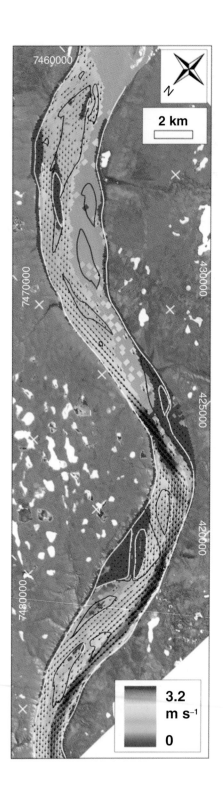

Figure 12.10 Surface flow velocities and vectors on the Mackenzie River, Canada, derived from an ALOS PRISM satellite stereo triplet of 21 May 2008 acquired at around 20:30 UTC. The red outlines indicate sand bars visible at low water level, the white lines vegetated islands and river margins assumed to be not or only slightly flooded during high water. Grey data voids in the river indicate open water without ice debris tracked. Image center latitude and longitude are approximately 67.33°N/130.70°W. Coordinate grid UTM zone 9. (Kääb, A. and Prowse, T. 2011. Reproduced with permission of John Wiley & Sons Inc).

12.4 Conclusions and outlook

Freshwater ice is one of, if not the least, studied components of the cryosphere from remote sensing. With the increasing recognition of the importance of freshwater ice as a sensitive indicator of climate variations, and its impact on ecological and human systems in a changing climate, there is a rising demand for timely information on lake and river ice conditions (e.g., timing of phenological events, ice thickness, ice jams and flooding). However, as alluded to earlier in this chapter and in recent studies, this happens at a time when ground-based observational networks have "hit bottom" in many countries of the Northern Hemisphere, to the point where they can no longer form the primary basis of observations.

In fact, the lamentable state of ground-based freshwater ice observation networks is not a recent phenomenon in countries such as Canada; this is something that began 20–30 years ago (e.g., Lenormand *et al.*, 2002). Unfortunately, this occurred at a time when satellite remote sensing technology was not fully ready to complement or replace ground-based observations.

As shown throughout this chapter, significant progress has taken place in remote sensing of lake and river ice over the last decade, but many of the approaches (algorithms) developed to date have not reached maturity to the extent required by the environmental modeling community (e.g., numerical weather forecasting, climate and hydrological modeling) or by public policy- and decision-makers (e.g., transportation on ice roads, prediction and mitigation of ice jams and flood warning). However, there are some signs that we are progressing in the right direction, as demonstrated in pilot studies (e.g., the assimilation of lake surface temperature and ice cover in NWP models, and the integration of SAR-derived river ice products in a GIS-based system for flood forecasting). Upcoming satellite missions, such as ESA's Sentinels (the first Sentinel was planned for launch in 2013) and Canada's RADARSAT constellation (planned for 2018), will further encourage the development and greater operational usage of remote-sensing freshwater ice products. Exciting times do lie ahead for satellite remote sensing of lake and river ice.

Acknowledgments

The work of M. Bernier and C. Duguay was supported by European Space Agency (ESA-ESRIN) Contract No. 4000101296/10/I-LG (Support to Science Element, North Hydrology Project) and Discovery Grants from the Natural Sciences and Engineering Research Council of Canada (NSERC). The early development of FRAZIL was also supported by the GEOIDE Network of Excellence, the Canadian Space Agency (CSA), and Hydro-Québec. The development of IceMAP-R was supported by the Quebec Public Safety Department, BC Hydro, and the CSA SOAR program. The research of A. Kouraev was completed within the framework of the Russian-French cooperation GDRI "CAR-WET-SIB", PICS "BaLaLaICA", Project 13-05-91051-RFBR-a "Lakes Baikal and Ladoga – joint complex studies", French ANR "CLASSIQUE", CNES TOSCA "Lakes" and "SWOT" Projects, Russian FZP 1.5 "Kadry", and EU FP7 "MONARCH-A" projects. Finally, we wish to thank Kyung-Kuk (Kevin) Kang, Homa Kheyrollah Pour and Cristina Surdu for the production of Figures 12.4, 12.5 and 12.6, respectively.

References

Arp, C.D., Jones, B. ., Urban, F, and Grosse, G. (2011). Hydrogeomorphic processes of thermokarst lakes with grounded-ice and floating-ice regimes on the Arctic coastal plain. *Alaska, Hydrological Processes* **25**, 2422–2438.

Arp, C.D., Jones, B.M., Lu, Z., and Whitman, M.S. (2012). Shifting balance of thermokarst lake ice regimes across the Arctic Coastal Plain of northern Alaska. *Geophysical Research Letters* **39**, L16503, doi: 10.1029/2012GL052518.

Bérubé, F., Bergeron, N., Gauthier, Y., and Choquette, Y. (2009). *Investigation of the use of GPR for characterizing river ice types.* Proceedings of the 15th Workshop on the Hydraulics of Ice Covered Rivers, St. John's, Canada, June 15–17, 2009, 55–64.

Bonsal, B.R., Prowse, T.D., Duguay, C.R., and Lacroix, M.P. (2006). Impacts of large-scale teleconnections on freshwater-ice duration over Canada. *Journal of Hydrology* **330**, 340–353.

Brooks, R., Prowse, T.D., and O'Connell, I.J. (2013). Quantifying Northern Hemisphere freshwater ice. *Geophysical Research Letters* **40**, 1128–1131, doi: 10.1002/grl.50238.

Brown, L.C. and Duguay, C.R. (2010). The response and role of ice cover in lake-climate interactions. *Progress in Physical Geography* **34**, 671–704, doi: 10.1177/0309133310375653.

Brown, L.C. and Duguay, C.R. (2011). A comparison of simulated and measured lake ice thickness using a Shallow Water Ice Profiler. *Hydrological Processes* **25**, 2932–2941, doi: 10.1002/hyp.8087.

Brown, L.C. and Duguay, C.R. (2012). Modelling lake ice phenology with an examination of satellite detected sub-grid cell variability. *Advances in Meteorology* **2012**, Article ID 529064, 19 pages, doi: 10.1155/2012/529064.

Brown, R.S., Duguay, C.R., Mueller, R.P., Moulton, *et al.* (2010). Use of synthetic aperture radar to identify and characterize overwintering areas of fish in ice-covered arctic rivers: a demonstration with broad whitefish and their habitats in the Sagavanirktok River, Alaska. *Transactions of the American Fisheries Society* **139**, 1711–1722, doi: 10.1577/T09–176.1.

Chaouch, N., Temimi, M., Romanov, P., Cabrera, R., *et al.* (2012). An automated algorithm for river ice monitoring over the Susquehanna River using the MODIS data. *Hydrological Processes*, doi: 10.1002/hyp.9548.

Clausi, D., Qin, A., Chowdhury, M., Yu, P., and Maillard, P. (2010). MAGIC: Map-guided ice classification system. *Canadian Journal of Remote Sensing* **36**, S13–S25.

Cook, T.L. and Bradley, R.S. (2010). An analysis of past and future changes in the ice cover of two High-Arctic lakes based on synthetic aperture radar (SAR) and LANDSAT imagery. *Arctic, Antarctic, and Alpine Research* **42**, 9–18, doi: 10.1657/1938-4246-42.1.9.

Derksen, C., Silis, A., Sturm, M., *et al.* (2009). Northwest Territories and Nunavut snow characteristics from a subarctic traverse: Implications for passive microwave remote sensing. *Journal of Hydrometeorology* **10**, 448–463, doi: 10.1175/2008JHM1074.1.

Duguay, C., Brown, L., Kang, K.-K., and Kheyrollah Pour, H. (2011). Lake ice. In: Richter-Menge, J., Jeffries, M.O. & Overland, J.E. (eds). *Arctic Report Card 2011*. http://www.arctic.noaa.gov/reportcard.

Duguay, C., Brown, L., Kang, K.-K., and Kheyrollah Pour, H. (2012). [Arctic]. Lake ice. [In State of the Climate in 2011]. *Bulletin of the American Meteorological Society* **93**, S156–S158.

Duguay, C., Brown, L., Kang, K.-K., and Kheyrollah Pour, H. (2013). [Arctic]. Lake ice. [In State of the Climate in 2012]. *Bulletin of the American Meteorological Society* **94**, S152–S154.

Duguay, C.R. and Lafleur, P.M. (2003). Estimating depth and ice thickness of shallow subarctic lakes using spaceborne optical and SAR data. *International Journal of Remote Sensing* **24**, 475–489.

Duguay, C.R. and Pietroniro, A. (2005). *Remote Sensing in Northern Hydrology: Measuring Environmental Change*. Geophysical Monograph 163, American Geophysical Union, Washington, DC, 160 pp., doi: 10.1029/GM163.

Duguay, C.R., Pultz, T.J., Lafleur, P.M., and Drai, D. (2002). RADARSAT backscatter characteristics of ice growing on shallow sub-arctic lakes, Churchill, Manitoba, Canada. *Hydrological Processes* **16**, 1631–1644.

Duguay, C.R., Flato, G.M., Jeffries, M.O., Ménard, P., *et al.* (2003). Ice cover variability on shallow lakes at high latitudes: Model simulations and observations. *Hydrological Processes* **17**, 3465–3483.

Duguay, C.R., Green, J., Derksen, C., English, *et al.* (2005). *Preliminary assessment of the impact of lakes on passive microwave snow retrieval algorithms in the Arctic*. Proceedings of 62nd Eastern Snow Conference, Waterloo, Ontario, Canada, June 7–10, 2005, 223–228.

Duguay, C.R., Prowse, T.D., Bonsal, B.R., Brown, R.D., *et al.* (2006). Recent trends in Canadian lake ice cover. *Hydrological Processes* **20**, 781–801.

Engram, M., Walter Anthony, K., Meyer, F., and Grosse, G. (2012). Synthetic aperture radar (SAR) backscatter response from methane ebullition bubbles trapped by thermokarst lake ice. *Canadian Journal of Remote Sensing* **38**, 667–682.

Gauthier, Y., Paquet, L.-M., Gonzalez, A., Hicks, F., *et al.* (2007). *Using the FRAZIL system in support of winter river flow modeling*. Proceedings of the 14th Workshop on the Hydraulics of Ice Covered Rivers, Quebec City, Canada, June 19–22, 2007, 17pp.

Gauthier, Y., Paquet, L.-M., Gonzalez, A., and Bernier, M. (2008). Using radar and GIS to support ice related flood forecasting *Geomatica* **62**, 273–285.

Gauthier, Y., Tremblay, M., Bernier, M. and Furgal, C. (2010). Adaptation of a radar-based river ice mapping technology to the Nunavik context. *Canadian Journal of Remote Sensing* **36**(S1), 168–185, doi: 10.5589/m10-018.

Gauthier, Y., Poulin, J. Bernier, M. (2012). *Rapport d'évaluation de l'algorithme ICEMAP-R (v3.0) pour la cartographie radar de la glace de rivière*. Rapport de recherche remis à la Direction de la sécurité civile, Ministère de la Sécurité publique du Québec.

Geldsetzer, T., van der Sanden, J., and Brisco, B. (2010). Monitoring lake ice during spring melt using RADARSAT-2 SAR. *Canadian Journal of Remote Sensing* **36**(S2), S391–S400.

Gherboudj, I., Bernier, M., Hicks, F., and Leconte, R. (2007). Physical characterization of air inclusions in river ice. *Cold Regions Science and Technology* **49**, 179–194.

Gherboudj, I., Bernier, M., and Leconte, R. (2010). A backscatter modelling for river ice cover: Analysis and numerical results. *IEEE Transactions on Geoscience and Remote Sensing* **48**, 1788–1798.

Grunblatt, J. and Atwood, D. (2013). Mapping lakes for winter liquid water availability using SAR on the North Slope of Alaska. *International Journal of Applied Earth Observation and Geoinformation*, 10.1016/j.jag.2013.05.006.

Hall, D.K., Foster, J.L., Chang, A.T.C., and Rango, A. (1981). Freshwater ice thickness observations using passive microwave sensors. *IEEE Transactions on Geoscience and Remote Sensing* **GE-19**, 189–193.

Hall, D.K., Fagre, D.B., Klasner, F., Linebaugh, G., *et al.* (1994). Analysis of ERS–1 synthetic aperture radar data of frozen lakes in northern Montana and implications for climate studies. *Journal of Geophysical Research* **99**, 22,473–22,482.

Hall, D.K., Riggs, G.A., Salomonson, V.V., DiGiromamo, N., *et al.* (2002). MODIS snow-cover products. *Remote Sensing of Environment* **83**, 181–194.

Helfrich, S.R., McNamara, D., Ramsay, B.H., Baldwin, T., *et al.* (2007). Enhancements to, and forthcoming developments in the Interactive Multisensor Snow and Ice Mapping System (IMS). *Hydrological Processes* **21**, 1576–1586.

Hirose, T., Kapfer, M., Bennett, J., Cott, P., *et al.* (2008). Bottomfast ice mapping and the measurement of ice thickness on tundra lakes using C-band synthetic aperture radar remote sensing. *Journal of the American Water Resources Association* **44**, 285–292.

Howell, S.E.L., Brown, L.C., Kang, K.-K., and Duguay, C.R. (2009). Variability in ice phenology on Great Bear Lake and Great Slave Lake, Northwest Territories, Canada, from SeaWinds/QuikSCAT: 2000–2006. *Remote Sensing of Environment* **113**, 816–834.

IGOS (2007). *Integrated Global Observing Strategy Cryosphere Theme Report – For the Monitoring of our Environment from Space and from Earth*. Geneva: World Meteorological Organization. WMO/TD-No. 1405. 100 p.

Jasek, M., Gauthier, Y., Poulin, J., and Bernier, M. (2013). *Monitoring of freeze-up on the Peace River at the Vermilion rapids using RADARSAT-2 SAR data.* Proceedings of the 17th Workshop on River Ice, Edmonton, Alberta, July 21–24, 2013, 29 pp.

Jeffries, M.O., Morris, K., Weeks, W.F., and Wakabayashi, H. (1994). Structural and stratigraphic features and ERS 1 synthetic aperture radar backscatter characteristics of ice growing on shallow lakes in NW Alaska, winter 1991–1992. *Journal of Geophysical Research* **99**, 22459–22471.

Jeffries, M.O., Morris, K., and Liston, G.E. (1996). A method to determine lake depth and water availability on the north slope of Alaska with spaceborne radar and numerical ice growth modeling. *Arctic* **49**, 367–374.

Jeffries, M.O., Morris, K., and Kozlenko, N. (2005). Ice characteristics and processes, and remote sensing of frozen river and lakes, In: Duguay, C.R. and Pietroniero, A. (eds). *Remote Sensing in Northern Hydrology.* AGU Monograph **163**, 63–90.

Jeffries, M.O., Morris, K., and Duguay, C.R. (2012). Floating ice: lake ice and river ice. In: Williams, R.S., Jr. and Ferrigno, J.G. (eds). *Satellite Image Atlas of Glaciers of the World – State of the Earth's Cryosphere at the Beginning of the 21st Century: Glaciers, Global Snow Cover, Floating Ice, and Permafrost and Periglacial Environments.* US Geological Survey Professional Paper 1386-A, A381–A424.

Kääb, A. and Prowse, T. (2011). Cold regions river flow observed from space. *Geophysical Research Letters* **38**, L08403, doi: 10.1029/2011GL047022.

Kang, K.-K., Duguay, C.R., Howell, S.E., Derksen, C.P., *et al.* (2010). Sensitivity of AMSR-E brightness temperatures to the seasonal evolution of lake ice thickness. *IEEE Geoscience and Remote Sensing Letters* **7**, 751–755, doi: 10.1109/LGRS.2010.2044742.

Kang, K.-K., Duguay, C.R., and Howell, S.E. (2012). Estimating ice phenology on large northern lakes from AMSR-E: Algorithm development and application to Great Bear Lake and Great Slave Lake, Canada. *The Cryosphere* **6**, 235–254, doi: 10.5194/tc-6-235-2012.

Kang, K.-K., Duguay, C.R., Lemmetyinen, J., and Gel, Y. (2014). Estimation of ice thickness on large northern lakes from AMSR-E brightness temperature measurements. *Remote Sensing of Environment* **150**, 1–19, doi: 10.1016/j.rse.2014.04.016.

Kheyrollah Pour, H., Duguay, C.R., Martynov, A., and Brown, L.C. (2012). Simulation of surface temperature and ice cover of large northern lakes with 1-D models: A comparison with MODIS satellite data and *in situ* measurements. *Tellus Series A: Dynamic Meteorology and Oceanography* **64**, 17614, doi: 10.3402/tellusa.v64i0.17614.

Kouraev A.V., Papa, F., Buharizin P.I., Cazenave, A., *et al.* (2003). Ice cover variability in the Caspian and Aral seas from active and passive satellite microwave data. *Polar Research* **22**, 43–50.

Kouraev, A.V., Semovski, S.V., Shimaraev, M.N., Mognard, N.M., *et al.* (2007a). Observations of lake Baikal ice from satellite altimetry and radiometry. *Remote Sensing of Environment* **108**, 240–253.

Kouraev, A.V., Semovski, S.V., Shimaraev, M.N., Mognard, N.M., *et al.* (2007b). Ice regime of lake Baikal from historical and satellite data: Influence of thermal and dynamic factors. *Limnology and Oceanography* **52**, 1268–1286.

Kouraev A.V., Shimaraev, M.N., Buharizin, P.I., Naumenko, M.A., *et al.* (2008). Ice and snow cover of continental water bodies from simultaneous radar altimetry and radiometry observations. *Survey in Geophysics*, doi 10.1007/s10712-008-9042-2.

Kouraev A.V., Cretaux, J.F., Zakharova, E., Mercier, F., *et al.* (in press). Seasonally-frozen lakes in boreal regions. In: Benveniste, J., Vignudelli, S., and Kostianoy, A. (eds). *Inland water altimetry*. Springer.

Kropáček, J., Maussion, F., Chen, F., Hoerz, S., *et al.* (2013). Analysis of ice phenology of lakes on the Tibetan Plateau from MODIS data. *The Cryosphere* **7**, 287–301.

Latifovic, R. and Pouliot, D. (2007). Analysis of climate change impacts on lake ice phenology in Canada using the historical satellite data record. *Remote Sensing of Environment* **16**, 492–507.

Lehner, B. and Doll, P. (2004). Development and validation of global database of lakes, reservoirs and wetlands. *Journal of Hydrology* **296**, 1–22, doi: 10.1016/j.jhydrol.2004.03.028.

Lenormand, F., Duguay, C.R., and Gauthier, R. (2002). Development of a historical ice database for the study of climate change in Canada. *Hydrological Processes* **16**, 3707–3722.

Leshkevish, G.A. and Nghiem, S.V. (2007). Satellite SAR remote sensing of Great Lakes ice cover. Part 2. Ice classification and mapping. *Journal of Great Lakes Research* **33**, 736–750.

Lindström, G., Pers, C.P., Rosberg, R., Strömqvist, J., *et al.* (2010). Development and test of the HYPE (Hydrological Predictions for the Environment) model – a water quality model for different spatial scales. *Hydrology Research* **41**, 295–319.

Martynov, A., Sushama, L., Laprise, R., Winger, K., *et al.* (2012). Interactive lakes in the Canadian Regional Climate Model, version 5: the role of lakes in the regional climate of North America. *Tellus Series A: Dynamic Meteorology and Oceanography* **64**, 16226, doi: 10.3402/tellusa.v64i0.16226.

Mermoz, S., Allain, S., Bernier, M., Pottier, E., *et al.* (2009a). Classification of river ice using polarimetric data. *Canadian Journal of Remote Sensing* **35**, 460–473.

Mermoz, S., Dribault, Y., Bernier, M., Allain, S., *et al.* (2009b). *Investigation of RADARSAT-2 and Terrasar-X data for river ice characterization from remote sensing*. Proceedings of the 15th Workshop on the Hydraulics of Ice Covered Rivers, St. John's, Canada, June 15 – 17, 2009, 389–400.

Mermoz, S., Allain-Bailhache, S., Bernier, M., Pottier, E., *et al.* (2013). Retrieval of river ice thickness from C-band PolSAR data. *IEEE Transactions on Geoscience and Remote Sensing* **99**, 1–11, doi: 10.1109/TGRS.2013.2269014.

Morris, K., Jeffries, M.O., and Weeks, W.F. (1995). Ice processes and growth history on Arctic and sub-Arctic lakes using ERS-1 SAR data. *Polar Record* **31**, 115–128.

Mueller, D.R., Van Hove, P., Antoniades, D., Jeffries, M.O., *et al.* (2009). High Arctic lakes as sentinel ecosystems: Cascading regime shifts in climate, ice cover, and mixing. *Limnology and Oceanography* **54**, 2371–2385.

Nghiem, S.V. and Leshkevich, G.A. (2007). Satellite SAR remote sensing of Great Lakes ice cover, Part 1. Ice backscatter signatures at C Band. *Journal of Great Lakes Research* **33**, 722–735.

Pavelsky, T. and Smith, L.C. (2004). Spatial and temporal patterns in Arctic river ice breakup observed with MODIS and AVHRR time series. *Remote Sensing of Environment* **93**, 328–338.

Prowse, T.D., Bonsal, B.R., Duguay, C.R., and Lacroix, M.P. (2007). River-ice break-up/freeze-up: A review of climatic drivers, historical trends, and future predictions. *Annals of Glaciology* **46**, 443–451.

Prowse, T., Alfredsen, K., Beltaos, S., Bonsal, B., *et al.* (2011a). Arctic freshwater ice and its climatic role. *Ambio* **40**(S1), 46–52, doi 10.1007/s13280-011-0214-9.

Prowse, T., Alfredsen, K., Beltaos, S., Bonsal, B., *et al.* (2011b). Past and future changes in lake and river ice. *Ambio* **40**(S1), 53–62, doi 10.1007/s13280-011-0216-7.

Prowse, T., Alfredsen, K., Beltaos, S., Bonsal, B., *et al.* (2011c). Effects of changes in Arctic lake and river ice. *Ambio* **40**(S1), 63–74, doi 10.1007/s13280-011-0217-6.

Puestow, T.M., Randell, C.J., Rollings, K.W., Khan, A.A., *et al.* (2004). *Near real-time monitoring of river ice in support of flood forecasting in eastern Canada: Towards the Integration of Earth observation technology in flood hazard mitigation.* Proceedings of the IEEE International Geoscience and Remote Sensing Symposium, Vol. 4, Anchorage, USA, 20–24 September 2004, 2268–2271, doi: 10.1109/IGARSS.2004.1369736.

Rouse, W.R., Binyamin, J., Blanken, P.D., Bussières, N., *et al.* (2005). Role of northern lakes in a regional energy balance. *Journal of Hydrometeorology* **6**, 291–305.

Rouse, W.R., Blanken, P.D., Duguay, C.R., Oswald, C.J., *et al.* (2008a). The Influence of lakes on the regional heat and water balance of the central Mackenzie River Basin. Chapter 18 in *Cold Region Atmospheric and Hydrologic Studies: The Mackenzie GEWEX Experience, Volume 1: Atmospheric Dynamics*, Springer-Verlag, 309–325.

Rouse, W.R., Blanken, P.D., Duguay, C.R., Oswald, C.J., *et al.* (2008b). Climate-lake interactions. Chapter 8 in *Cold Region Atmospheric and Hydrologic Studies: The Mackenzie GEWEX Experience, Volume 2: Hydrologic Processes*, Springer-Verlag, 139–160.

She, Y. and Hicks, F.E. (2006). Modeling ice jam release waves with consideration for ice effects. *Journal of Cold Regions Science and Technology* **45**, 137–147.

Sobiech, J. and Dierking, W. (2013). Observing lake- and river-ice decay with SAR: advantages and limitations of the unsupervised k-means classification approach. *Annals of Glaciology* **54**, 65–72, doi: 10.3189/2013AoG62A037.

Soliman, A., Duguay, C., Saunders, W., and Hachem, S. (2012). Pan-Arctic land surface temperature from MODIS and AATSR: Product development and inter-comparison. *Remote Sensing* **4**, 3833–3856; doi: 10.3390/rs4123833.

Sturm, M., Holmgren, J., König, M., and Morris, K. (1997). The thermal conductivity of seasonal snow. *Journal of Glaciology* **43**, 26–41.

Surdu, C., Duguay, C.R., Brown, L.C., and Fernández Prieto, D. (2014). Response of ice cover on shallow lakes of the North Slope of Alaska to contemporary climate conditions (1950–2011): radar remote sensing and numerical modeling data analysis. *The Cryosphere* **8**, 167–180, doi: 10.5194/tc-8-167-2014.

Töyrä, J., Pietroniro, A., Carter, T., and Beltaos, S. (2005). *RADARSAT-1 SAR classification of breakup river ice condition and ice jam flooding.* Proceedings of the 26th Canadian Symposium on Remote Sensing, Wolfville, Canada, June 14–16, 2005, 97–98.

Troitskaja Y., Lebedev, S.A., Kostianoy, A.G., Kouraev, A.V., *et al.* (in press). Rivers and reservoirs in boreal zone of Eurasia. In: Benveniste, J., Vignudelli, S., and Kostianoy, A. (eds). *Inland water altimetry.* Springer.

Unterschultz, K.D., van der Sanden, J., and Hicks, F.E. (2009). Potential of RADARSAT-1 for the monitoring of river ice: Results of a case study on the Athabasca River at Fort McMurray, Canada. *Cold Regions Science and Technology* **55**, 238–248.

Weber, F., Nixon, D., and Hurley, J. (2003). Semi-automated classification of river ice types on the Peace River using RADARSAT-1 synthetic aperture radar (SAR) imagery. *Canadian Journal of Civil Engineering* **30**, 11–27.

Wegmuller, U., Santoro, M., Werner, C., Strozzi, T., *et al.* (2010). *Estimation of ice thickness of tundra lakes using ERS-ENVISAT cross-interferometry.* Proceedings of the IEEE International Geoscience and Remote Sensing Symposium, Honolulu, USA, 25–30 July 2010, 316–319, doi: 10.1109/IGARSS.2010.5649026.

White, D.M., Prokein, P., Chambers, M.K., Lilly, M.R., *et al.* (2008). Use of synthetic aperture radar for selecting Alaskan lakes for winter water use. *Journal of the American Water Resources Association* **44**, 276–284.

Zhao, L., Jin, J., Wang, S.-Y., and Ek, M.B. (2012). Integration of remote-sensing data with WRF to improve lake-effect precipitation simulations over the Great Lakes region. *Journal of Geophysical Research* **117**, D09102, doi: 10.1029/2011JD016979.

Acronyms

ALOS	Advanced Land Observing Satellite
AMSR-E	Advanced Microwave Scanning Radiometer – Earth Observing System
AMSU	Advanced Microwave Sounding Unit
ASAR	Advanced Synthetic Aperture Radar

ASTER	Advanced Spaceborne Thermal Emission and Reflection Radiometer
AVHRR	Advanced Very High Resolution Radiometer
CIS	Canadian Ice Service
CLIMo	Canadian Lake Ice Model
Envisat	Environmental Satellite
ERS-1/2	European Remote Sensing satellite-1/2
ETM+	Enhanced Thematic Mapper Plus
FLake	Freshwater Lake model
GBL	Great Bear Lake
GFO	Geosat Follow-On satellite
GLAWEX'97	Great LAkes Winter Experiment
GOES	Geostationary Operational Environmental Satellites
GSL	Great Slave Lake
HH	Co-polarization (horizontal send – horizontal receive)
HV	Cross-polarization (horizontal send – vertical receive)
IGOS	Integrated Global Observing Strategy
IMS	Interactive Multisensor Snow and Ice Mapping System
IRGS	Iterative Region Growing with Semantics algorithm
JAXA	Japan Aerospace eXploration Agency
JERS-1	Japanese Earth Resources Satellite 1
JPL	Jet Propulsion Laboratory
LST	Lake surface temperature
MAGIC	Map-Guided Ice Classification System software
MBE	Mean Bias Error
MERIS	MEdium Resolution Imaging Spectrometer
MetOp	Meteorological Operational satellite program
MODIS	Moderate Resolution Imaging Spectroradiometer
NARR	North American Regional Reanalysis
NASA	National Aeronautics and Space Administration
NIC	National Ice Service, U.S.
NOAA	National Oceanic and Atmospheric Administration
NWP	Numerical Weather Prediction
PALSAR	Phased-Array type L-band SAR
PRISM	Panchromatic Remote Sensing Instrument for Stereo Mapping
QuikSCAT	Quick Scatterometer
RADARSAT	Radar Satellite of the Canadian Space Agency
RCM	Regional Climate Model
RMSE	Root Mean Square Error
SAR	Synthetic Aperture Radar
SMMR	Scanning Multi-channel Microwave Radiometer
SPOT-5	Système Pour l'Observation de la Terre
SSM/I	Special Sensor Microwave/Imager
SWE	Snow Water Equivalent

TB	Brightness temperature
TM	Thematic Mapper
TOPEX	TOPography EXperiment
USAF	United States Air Force
VGT VEGETATION	Sensor on SPOT satellite
VV	Co-polarization (vertical send – vertical receive)

Websites cited

http://www.natice.noaa.gov/products/great_lakes.html
http://earthobservatory.nasa.gov/IOTD/view.php?id=81227
http://earthobservatory.nasa.gov/IOTD/view.php?id=81227

13 Remote sensing of permafrost and frozen ground

Sebastian Westermann[1], Claude R. Duguay[2], Guido Grosse[3] and Andreas Kääb[1]

[1] *University of Oslo, Oslo, Norway*
[2] *University of Waterloo, Waterloo, Ontario, Canada*
[3] *Alfred Wegener Institute, Helmholtz Centre for Polar and Marine Research, Potsdam, Germany*

Summary

Permafrost and frozen grounds are key elements of the terrestrial cryosphere that will be strongly affected by a warming climate. With widespread permafrost degradation likely to occur in this century, remote sensing of permafrost is seeking to unveil the processes and causal connections governing this development, from the monitoring of variables related to the permafrost state to the mapping of the impacts of degradation and potential natural hazards on the ground.

However, remote sensing of permafrost is challenging. The physical subsurface variables which characterize its thermal state – ground temperature, ice content and thaw depth – are not directly measurable through current remote sensing technologies. Instead, there is a large diversity of target characteristics for remote sensing from which the permafrost state can be indirectly derived. Mountain permafrost environments are characterized by a strong heterogeneity on small spatial scales, so that high-resolution remote sensors are generally required. Permafrost landforms and surface features, such as rock glaciers, thermokarst lakes and push moraines, can be identified by image classification techniques in a variety of remote sensing products. Permafrost-related vertical and horizontal surface deformations can be identified by repeat digital elevation models, or radar interferometry.

Similar techniques are applied in the Arctic lowland permafrost areas to identify and map indicators of thawing ice-rich permafrost, such as thermokarst, thaw slumps, or coastal erosion. In some areas, the presence of permafrost correlates with certain vegetation types or surface covers, which can be mapped from satellite sensors. Furthermore, in the vast lowland permafrost areas, physical variables directly or indirectly related to thermal subsurface conditions are accessible through more coarsely resolved remote sensing techniques. These include the land surface temperature, the freeze-thaw state of the surface and subsurface, and the gravimetric signal from the ground.

Remote Sensing of the Cryosphere, First Edition. Edited by M. Tedesco.
© 2015 John Wiley & Sons, Ltd. Published 2015 by John Wiley & Sons, Ltd.
Companion Website: www.wiley.com/go/tedesco/cryosphere

Recently, there is progress towards quantitative monitoring of ground temperatures and thaw depths by employing remote sensing data in conjunction with thermal subsurface modeling. By exploiting the cumulative information content of several remote sensing data sets through data fusion strategies, the best possible estimate for the ground thermal state can be achieved.

13.1 Permafrost – an essential climate variable of the "Global Climate Observing System"

Permafrost is defined as sub-surface earth materials that remain continuously at or below the freezing temperature of water for at least two consecutive years (Harris *et al.*, 1988). A distinction is made between the perennially frozen permafrost and the active layer on top of the permafrost, which is subject to annual thawing and freezing (Harris *et al.*, 1988). The active layer is of great importance, as biological processes, including the formation and decomposition of organic material, are restricted to this zone.

Due to the distribution of the land masses on Earth, permafrost is strongly concentrated in the Northern Hemisphere, where approximately a quarter of the land area belongs to the permafrost zone (Brown *et al.*, 1997). The largest extent of permafrost is found in Siberia, followed by Arctic Canada and Alaska. Furthermore, permafrost is found at high elevations of many mountain regions, e.g., the Central Asian Mountains, the European Alps, the Rocky Mountains of North America, and the Andes (Gorbunov, 1978; Haeberli *et al.*, 1993).

In addition to permafrost, seasonally frozen ground in permafrost-free areas is an important transient element of the cryosphere. In winter, approximately 55% of the exposed land area of the Northern Hemisphere is in frozen state (Zhang *et al.*, 2003b), which has strong impacts on hydrology, biological processes, land surface energy budget, agriculture, and infrastructure.

While the occurrence of permafrost is largely driven by hemispheric-scale climate zonation, local surface and subsurface properties play an important role in determining the thermal regime and the thickness of the active layer. As a result, areas both with and without permafrost can coexist in some regions, which gives rise to the classification of permafrost areas by its subsurface extent in continuous (more than 90% of the area underlain by permafrost), discontinuous (50–90% of the area), sporadic (less than 50% of the area) and isolated (only small patches) zones (Brown *et al.*, 1997; Figure 13.1). Permafrost formation and degradation occur on timescales of decades up to millennia, so that the present permafrost distribution must be regarded as a mixed result of past and current climate conditions.

In the last decade, a number of studies have gathered convincing observational and modeling evidence for a sustained warming in the Arctic and many mountain regions (e.g., Serreze *et al.*, 2000; Comiso and Parkinson, 2004; ACIA, 2005; Overland *et al.*, 2008). The ongoing warming trend has an impact on permafrost, with increasing permafrost temperatures being reported for most permafrost regions

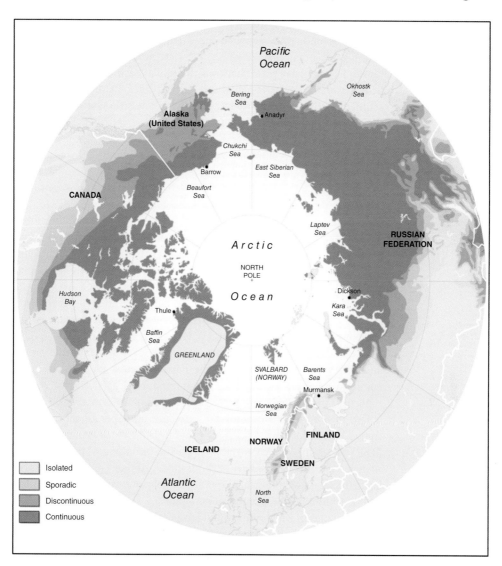

Figure 13.1 Circum-polar permafrost distribution with classification in continuous, discontinuous, sporadic, and isolated permafrost zones. (Map by Philippe Rekacewicz, UNEP/GRID-Arendal; data from International Permafrost Association, 1998).

(Romanovsky *et al.*, 2010, 2012). General circulation models (GCMs) predict an accelerated future warming trend for the Arctic, with increases of the mean air temperature over land surfaces of up to 8°C until the end of the century (IPCC, 2007). Accordingly, a sizable reduction of the permafrost area is projected until 2100 (e.g., Delisle, 2007; Lawrence *et al.*, 2008), which will have strong impacts on the ecology (Jorgenson *et al.*, 2001), hydrology (Smith *et al.*, 2005), infrastructure (Parker, 2001; Larsen *et al.*, 2007), and economy (Prowse *et al.*, 2009) of these regions.

In mountain areas, thawing permafrost can affect the stability of slopes and rock walls and, thus, can become a trigger for natural hazards (Kääb *et al.*, 2005). Widespread thawing of organic-rich permafrost, largely found in northern lowlands, may have a feedback on the global carbon cycle (Zimov *et al.*, 2006; Schaefer *et al.*, 2011). The total soil organic carbon stored in permafrost areas is estimated to be about half of the worldwide soil carbon stock, or more than twice the current atmospheric carbon pool (Schuur *et al.*, 2008; Tarnocai *et al.*, 2009). Degradation of permafrost and an increase in average thickness and temperature of the active layer will make large amounts of previously frozen organic material that has been accumulated over millennia available for microbial decomposition (Walter *et al.*, 2006; Schuur *et al.*, 2009). The activation of portions of this carbon stock would convert northern permafrost regions from net sinks to net sources of greenhouse gases, potentially resulting in an amplification of the global warming trend. In a future warming Arctic, an additional increase in frequency and magnitude of natural pulse disturbances such as wildfires, droughts, thermokarst, and erosion may result in further increase of permafrost degradation and carbon release not yet considered in current numerical models (Grosse *et al.*, 2011).

Due to its dominant role for Arctic land ecosystems, permafrost has been included in the Global Climate Observing System (GCOS) as an Essential Climate Variable for terrestrial systems (WMO, 2010). The ambition of GCOS is "to provide continuous, reliable, comprehensive data and information on the state and behavior" (WMO, 2010) of the Earth system with the goal of compiling a common database for national and international agencies, for example for the reports of the Intergovernmental Panel on Climate Change (IPCC).

Two *in situ* monitoring programs are currently operational – the Circumpolar Active Layer Monitoring (CALM) program, and the Thermal State of Permafrost (TSP) program – which together form the backbone of permafrost monitoring required for GCOS. The CALM program coordinates measurements of the active layer thickness at more than 150 sites (Nelson *et al.*, 2008; Shiklomanov *et al.*, 2008), while the TSP program (Romanovsky *et al.*, 2010) comprises more than 500 boreholes in which the ground temperature profile is measured. However, the distribution of the monitoring sites is limited to a few regions with supporting infrastructure, while vast regions are not covered at all. It is thus desirable to better exploit the potential of remotely sensed data sets to complement the *in situ* monitoring programs.

However, remote sensing of permafrost is challenging. The physical variables ground temperature, water and ice content and thaw depth, which characterize its thermal state, are not directly measurable through current remote sensing technologies. Instead, there is a very large diversity of target characteristics for remote sensing from which the permafrost state can be derived. As a result, remote sensing of permafrost is highly influenced by the spatial and temporal scales over which the processes and target variables under investigation vary.

In this chapter, we distinguish high (spatial resolution <5 m), medium (5–100 m), low (100–1000 m) and very low (>1000 m) resolution remote sensors. In many permafrost regions, the surface cover can vary dramatically over lateral distances of tens of meters which can only be captured by high resolution sensors. On the other hand, rapid changes, such as the onset of freezing or thawing or the diurnal temperature cycle, require data acquisition at high temporal resolution which, in general, is only achieved by low and very low resolution sensors. In such applications, evaluating and understanding subpixel variability is a major challenge to make better use of remote sensing products in permafrost investigations.

We loosely distinguish two approaches for remote sensing of permafrost:

a) Remote identification and mapping of surface features and objects typical for permafrost and related processes, such as rock glaciers, thermokarst lakes, pingos, or ice wedge polygons, and their changes over time.

b) Remote sensing of physical variables that are directly or indirectly related to thermal subsurface conditions, such as land surface temperature (LST) or freeze-thaw state of the surface.

Some remote sensing systems can be employed for both applications, for example if surface features are mapped based on remotely sensed physical variables. In many cases, the changes in physical variables can be linked to permafrost evolution (e.g., increasing LST), while the impacts of such changes are reflected in the development of surface features (e.g., thaw slumps). With widespread permafrost degradation likely to occur in this century, remote sensing of permafrost is seeking to unveil the processes and causal connections governing this development, from the monitoring of variables related to the permafrost state, to the mapping of the impacts of degradation and potential natural hazards on the ground.

A number of reviews of remote sensing of permafrost and related phenomena have been published (Hall, 1982; Zhang *et al.*, 2004; Duguay *et al.*, 2005; Kääb *et al.*, 2005; Kääb, 2008). In this chapter, we update the overview on remote sensing methods for permafrost studies with current sensor technologies. The following sections are dedicated to remote sensing methods in two permafrost environments that differ substantially in their character: mountain permafrost and lowland permafrost. For remote sensing of permafrost in mountain areas (Section 13.2), evaluating surface features is the dominant remote sensing application. Here, the topographic setting imposes a strong spatial variability (Harris, 1988), so that only high-resolution sensors play a major role. Remote sensing techniques employed in the vast arctic and subarctic lowland areas, however, include a wide range of sensors that allow both mapping of surface features (Section 13.3) and monitoring of physical surface variables that can be used to assess the permafrost state (Section 13.4). We conclude with an outlook (Section 13.5) on thermal permafrost modeling using remotely sensed data sets as input.

13.2   Mountain permafrost

In contrast to lowland permafrost, the main factor influencing mountain permafrost distribution is its topographic setting, which includes elevation, slope and aspect. Mountain permafrost occurrence is, therefore, intimately connected to slope stability in a range of spatial and temporal scales. In this section, we cover remote sensing techniques for:

a) detection and mapping of surface features related to mountain permafrost either as indicator or important boundary condition;
b) creating digital elevation models (DEMs), a crucial prerequisite of most mountain permafrost studies; and
c) detecting terrain elevation change and displacement.

A number of methods introduced here are also useful for applications in lowland permafrost areas.

13.2.1   Remote sensing of surface features and permafrost landforms

Among typical surface features that indicate the presence of permafrost in mountain areas are rock glaciers, thermokarst lakes and push moraines (Etzelmüller *et al.*, 2001; Kääb *et al.*, 2005). Furthermore, surface roughness, snow cover and topographic parameters, such as elevation, aspect and slope are factors for the presence and properties of permafrost (Etzelmüller *et al.*, 2001). Similarly, objects and processes that can be closely connected to mountain permafrost can be investigated through spectral remote sensing analysis, including, for example, glaciers (Wessels *et al.*, 2002; Kääb *et al.*, 2003) or glacierized rock walls (Salzmann *et al.*, 2004; Fischer *et al.*, 2006, 2011).

Manual or automatic image classification techniques range from classifications based on mono-spectral (i.e., grayscale), multispectral, hyperspectral, radar backscatter intensity or polarization data, to spatio-spectral analyses utilizing not only the spectral information of the image pixels, but also their spatial context (object-oriented classification). In the microwave region of electromagnetic spectrum, analysis of the backscatter, the coherence of the interferometric phase of Synthetic Aperture Radar (SAR) signals, and the signal polarization enable delineation and characterization of terrain and its dynamics, such as terrain roughness or surface stability over time (Engeset and Weydahl, 1998; Floricioiu and Rott, 2001; Weydahl, 2001).

A number of permafrost-related processes involve surface changes over time, thus requiring multi-temporal methodologies for their assessment (Kääb and Haeberli, 2001; Kääb, 2005). This includes the group of spectral change detection methods, for instance to map changes over time in condition and extent of snow cover, vegetation, thaw lakes or rock glaciers. Remote sensing of vertical and horizontal surface deformations is covered under Section 13.2.3.

13.2.2   Generation of digital elevation models

Digital elevation models (DEMs) represent a key data set in most investigations of mountain permafrost. An efficient method of generating DEMs for virtually every terrestrial location on Earth is satellite along-track stereo from sensors such as ALOS PRISM, ASTER, SPOT-5, Ikonos, and QuickBird. Satellite stereo DEMs are produced using digital photogrammetric methods, with a vertical accuracy approximating the pixel size of the applied sensor (e.g., 15 m for ASTER), and with typical horizontal grid spacings equivalent to 2–4 pixels (Kääb, 2002; Toutin, 2002). Large errors in DEMs derived from optical satellite stereo can occur due to steep mountain flanks facing away from the oblique stereo sensor and shadowed areas. Large DEM errors also occur for particularly rough topography with sharp peaks. Insufficient visual contrast and self-similar features may lead to erroneous matching of stereo parallaxes and corresponding DEM errors. In particular, for regional-scale applications, the global DEM compiled from individual ASTER DEMs (GDEM) (Hirano *et al.*, 2003) is a valuable data set.

Sensors operating in the microwave region of the spectrum are able to overcome the limitations of weather and sunlight dependency of optical sensors. Interferometric SAR (InSAR) can be used to generate DEMs by converting the difference in the phase of the radar wave between two images to height information (Renouard *et al.*, 1995; Crosetto, 2002). DEMs from spaceborne repeat-pass interferometry, such as provided by Envisat, ERS-1/2, and RADARSAT, have a vertical accuracy of meters and a spatial resolution of tens of meters. Improved accuracy is expected from the new high-resolution SAR sensors, such TerraSAR-X and its TanDEM-X mission, and RADARSAT-2. DEMs can also be generated from multi-angular SAR data using photogrammetric principles (Toutin, 1995). In particular, the single-pass interferometric DEM from the Shuttle Radar Topography Mission (SRTM) is frequently used (Rabus *et al.*, 2003). Due to radar shadow, foreshortening, layover and insufficient interferometric coherence, the SRTM DEM has significant voids in high mountain regions.

The spatial resolution of airborne SAR DEMs is meters or less, and their vertical accuracy for mountainous terrain is decimeters to meters. Current systems use X- to P-band SAR (3 cm to 80 cm wavelength) (Vachon *et al.*, 1996; Nolan and Prokein, 2003; Stebler *et al.*, 2005). Most airborne InSAR sensors apply single-pass interferometry. Multiple overflights with different azimuths may be required to overcome limitations from radar shadow or layover effects in some mountain regions.

Aerial photogrammetry applies techniques similar to spaceborne optical methods, although higher accuracies are reached, due to the generally higher spatial image resolution. Digital photogrammetry, based on digitized hard-copy images or digital imagery, enables automatic DEM and orthoimage generation (Hauber *et al.*, 2000; Kääb and Vollmer, 2000). Depending on the image scale and pixel size, DEMs have a vertical accuracy of centimeters to meters and a spatial resolution of meters to decimeters (Kääb and Vollmer, 2000). Aerial photogrammetry is a particularly important tool, in view of the existing large archives of analogue air photos which, for many areas, have been the earliest and longest remotely sensed time series and, thus, are invaluable for quantifying temporal change. Most aerial photographs in archives are based on negative film. However, digital frame or linear array cameras are increasingly being used (Hauber *et al.*, 2000; Otto *et al.*, 2007).

DEM accuracy that is similar or better than that provided by airborne photogrammetry can be obtained from airborne laser scanning or light detection and ranging, LIDAR (Baltsavias *et al.*, 2001; Stockdon *et al.*, 2002; Geist *et al.*, 2003; Janeras *et al.*, 2004; Glenn *et al.*, 2006; van Asselen and Seijmonsbergen, 2006). Depending on the wavelength used, LIDAR provides good results over snow, where photogrammetric methods have problems due to the lack of radiometric contrast. The point density of LIDAR DEMs is on the order of one point per square meter, and feature vertical accuracies on the order of centimeters. The added details of the terrain can permit more sophisticated and more accurate geomorphic interpretation and analysis of terrain dynamics. Recording multiple return pulse maxima, or even the full return waveform, enables, for example, analysis of surface roughness. Airborne laser scanners are increasingly able to record the signal intensity (Lutz *et al.*, 2003) or are equipped with electro-optical imaging devices that enable combined analyses of image and elevation data.

13.2.3 *Terrain elevation change and displacement*

Two types of terrain displacement can be measured using remote sensing methods: displacement of surface particles, and elevation change at a specific location. Particle displacements are typically three-dimensional but can be preferentially vertical, for example for thaw settlement and frost heave.

Terrain elevation change over time can be determined from vertical differences between repeat DEMs. The changes can be indicators of processes such as thaw settlement and subsidence, frost heave and mass movements (Kääb *et al.*, 2005; Delacourt *et al.*, 2007). The accuracy of the calculated vertical changes is a function of the accuracy of the repeat DEMs that are used (Etzelmüller, 2000; Kääb, 2005). Pre- and post-processing procedures such as multi-temporal adjustment of photogrammetric blocks, or DEM co-registration, help to improve this accuracy (Pilgrim, 1996; Kääb and Vollmer, 2000; Li *et al.*, 2001; Kääb, 2005; Nuth and Kääb, 2011).

A well-established method for detecting terrain elevation changes is subtraction of repeat aero-photogrammetric DEMs. A large number of applications exist, including ground ice aggradation or degradation (Kääb *et al.*, 1997), thermokarst development, and slope instability (Kääb 2005; Kääb and Vollmer 2000). Repeat laser scanning is increasingly used in such studies (Geist *et al.*, 2003; Chen *et al.*, 2006).

Elevation changes from repeat satellite stereo can only be measured for a limited number of processes, such as pronounced mass advections from creep or landslides, due to the relatively low accuracy of such DEMs (Berthier *et al.*, 2004; Kääb, 2005). Lateral terrain movements are the main characteristics of slope movements such as rock glacier creep. If digital image correlation techniques are used, measurements are possible at the scale of the pixel size of the sensor and below (Figure 13.2).

Figure 13.2 Gruben rock glacier, Swiss Alps. Overlay of the horizontal surface velocities (vectors) and the color-coded changes in elevation, both for 1970 to 1995, derived from aerial photographs. To the northeast, a patchy distribution of horizontal velocities and high rates of thaw settlement indicate dead ice occurrences that are not in thermal equilibrium. To the southwest, a coherent flow field, and almost constant thickness, point to creeping permafrost in thermal equilibrium. Measurements were done in relation to hazard assessment associated with the larger of the thermokarst lakes.

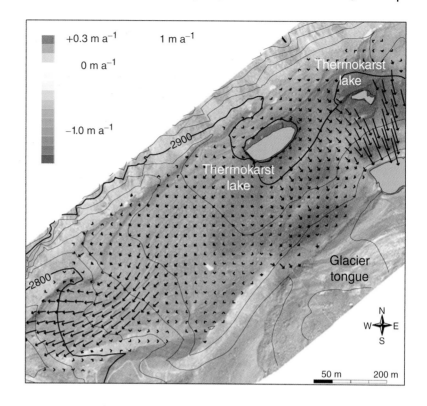

Image matching techniques can be applied to repeat terrestrial photos, aerial photos, optical satellite images, airborne and spaceborne SAR images, or high-resolution DEMs. Depending on the data and the technique employed, the horizontal, vertical or both components of surface displacement are measured (Kaufmann and Ladstädter, 2002; Berthier *et al.*, 2005; Kääb, 2005). Slow rock mass displacements or permafrost creep, with displacement rates of centimeters to meters per year, can be detected and measured using aerial photographs and high-resolution satellite images (Ikonos, QuickBird, WorldView, ALOS PRISM, SPOT-5) (Kääb and Vollmer, 2000; Kääb, 2002; Delacourt *et al.*, 2004; Daanen *et al.*, 2012). Matching of repeat high-resolution images or LIDAR-derived DEMs may also enable detection of lateral displacements.

Differential InSAR (D-InSAR), where the phase differences between at least three SAR images are exploited, enables measurements of terrain displacements of a few millimeters (Figure 13.3). Application of the method depends on interferometric coherence, topography and SAR imaging geometry; information is missing in layover and shadowed areas (Nagler *et al.*, 2002; Eldhuset *et al.*, 2003; Strozzi *et al.*, 2004). Coherence is lost, for example, where the terrain is eroded or the surface wetness changes significantly (Weydahl, 2001). Loss of interferometric phase coherence does not necessary mean that the method has failed. It may, rather, point to processes such as destruction of vegetation on a landslide, or ground ice

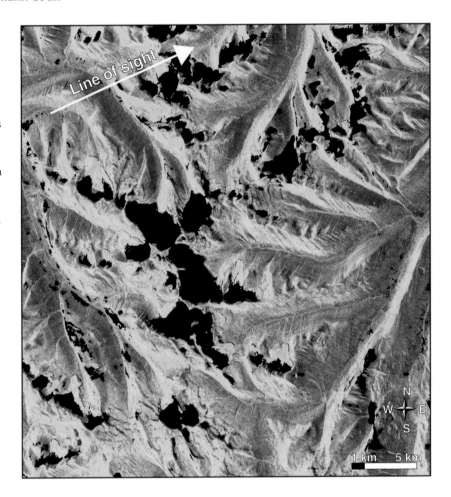

Figure 13.3 Line-of-sight displacements of rock glaciers in northern Iceland, close to the village of Akureyri, from radar interferometry based on ALOS PALSAR L-band data from 16 Aug and 1 Oct 2007. Background image is the SAR intensity image of the first epoch. Color-coded displacements over the 46 days range from 0 (cyan) to up to about 25–30 cm (3–4 cm/year; red). Black areas are masks of low phase coherence, for the most part due to coherence loss over glaciers, presumably due to melt. The areas with significant displacements fit well with rock glacier landforms as visible in the radar data and other image sources.

melt (Weydahl, 2001; Sjogren *et al.*, 2003). D-InSAR and optical image matching methods for permafrost assessments are highly complementary; as a very general rule, D-InSAR may work where image matching fails, and vice versa.

D-InSAR provides only the line-of-sight displacement directly, that is the projection of the actual terrain displacement vector on the line between the terrain point and the sensor. In theory, the horizontal and vertical displacement components can be separated by combining line-of-sight displacements measured from ascending and descending orbits (Joughin *et al.*, 1999). Application of this approach depends on the azimuth angle between ascending and descending orbits and the topography, which may limit terrain visibility from both orbits, particularly in mountain regions. Otherwise, the vertical and horizontal components must be estimated or modeled from the type of terrain movement under investigation (Wang and Li, 1999; Strozzi *et al.*, 2001).

Typical D-InSAR applications in permafrost hazard assessments are detection of rock mass movements, landslide movement and permafrost creep (Rott and Siegel, 1999; Nagler *et al.*, 2002; Strozzi *et al.*, 2004; Singhroy *et al.*, 2007).

D-InSAR can be applied to dominant and permanent microwave backscatterers, such as buildings or distinct rock formations, using a large number of SAR scenes, reducing atmospheric error effects and enabling detection of small movements due to landslides or settlement processes (Dehls *et al.*, 2002; Colesanti *et al.*, 2003; Hilley *et al.*, 2004).

For local-scale applications, repeat terrestrial laser scanning is increasingly gaining importance for detecting geometry changes at high detail, for instance at rock glacier fronts or cliffs of frozen ground (Bauer *et al.*, 2003). When the resulting repeat DEMs contain sufficient topographic detail, lateral movements can also be detected through matching techniques similar to the ones used for matching of repeat images (Bauer *et al.*, 2003).

The steep topography of rock walls severely limits the applicability of remote sensing methods and DEM generation techniques for monitoring frozen rock walls. Often, parts, or entire rock faces, are hidden in images due to adjacent topography, or are highly distorted due to the steep surface slope. The most promising remote sensing techniques for monitoring rock faces include aerial and terrestrial photogrammetry, terrestrial and airborne laser-scanning for deriving repeat high-resolution DEMs and geometry changes (Fischer *et al.*, 2011), and ground-based radar interferometry for deformation measurement. Spaceborne sensors are less useful for examining steep rock faces, with the exception of images taken with large off-nadir angles pointing towards the rock face under study (e.g., using the backward or forward channels of a stereo sensor).

13.3 Lowland permafrost – identification and mapping of surface features

There is some degree of overlap between the remote sensing methods employed for mountain permafrost and lowland permafrost. DEMs, for example, are a useful tool to characterize permafrost features, surface displacement, and mass movement in lowland areas as well but, usually, small topographic gradients pose high requirements on the vertical resolution and accuracy, so that InSAR and LIDAR are often preferable to stereo imaging for deriving DEMs. Spectral and object-oriented classification methods, similar to the ones presented in the previous section, are widely employed, but the target features are rather different in lowland areas.

In the following subsections, various remote sensing approaches are highlighted that link landcover data to permafrost presence/absence, map landforms indicating permafrost aggradation or presence, and observe changes in landforms or processes indicative of permafrost degradation.

13.3.1 Land cover and vegetation

Mapping of land cover can be employed as an indirect indicator for permafrost conditions in some environments. Vegetation classification and mapping, combined with topographic variables, were used in the past with varying degrees of success, to derive information about permafrost presence or absence (Peddle and Franklin, 1992; Leverington and Duguay, 1997) and late-summer active layer thickness (Peddle and Franklin, 1993; Leverington and Duguay, 1996). Nguyen *et al.* (2009) used SPOT-5 data to map riparian vegetation in the Mackenzie River delta to derive information on near-surface permafrost distribution. The authors reported high classification accuracies, greater than 80%.

More recent studies combine various sensors and topographic information in complex models to increase classification accuracies. For example, Panda *et al.* (2010) applied a maximum likelihood classification to derive vegetation classes from SPOT-5. They combined this information with airborne LIDAR-based digital elevation data, including slope and aspect, to develop a binary logistic regression model of permafrost probability along the Alaska Highway Corridor in interior Alaska. Very high classification accuracies (80% with ground data and >75% with an independent aerial photo-interpreted permafrost map) were achieved for both continuous and discontinuous permafrost types in the study area.

Chasmer *et al.* (2011) used aerial photography, Ikonos data, and airborne LIDAR to study the impact of vegetation canopy and shadowing on the radiation budget of the ground surface, and thus the ground thermal regime, in a peat plateau in the discontinuous permafrost zone near in the Northwest Territories, Canada. The authors found a positive feedback between permafrost thaw on peat plateau edges, collapse of trees, reduced canopy coverage and shadowing, and increased incoming shortwave radiation at the ground surface that warms the ground further.

13.3.2   Permafrost landforms

Permafrost landforms are useful indicators for the presence of permafrost. Remote sensing-based mapping is largely confined to high and medium resolution sensors, including panchromatic, multispectral, and hyperspectral systems on both airborne and satellite platforms. Many of these landforms have spatial dimensions on the order of decimeters (e.g., small-scale patterned ground) to a few hundreds of meters (e.g., thermokarst lakes and basins), often requiring a combination of different sensors and techniques to be used for characterization of the full variety of features in a study area. Heterogeneity of permafrost landforms requires different approaches, and examples are provided here for pingos and ice wedge polygons.

Pingos, large conical hills, formed by perennial subsurface ice accumulation and subsequent frost heave of the ground surface, are an indicator of ground water mobility in permafrost regions, and have traditionally been mapped using aerial imagery (Holmes, 1967; Carter and Galloway, 1979). In an effort to verify and enhance existing maps of pingo distribution in the Western Arctic Coastal Plain of northern Alaska, Jones *et al.* (2012a) used airborne color-infrared aerial imagery, in combination with a high-resolution elevation model derived from airborne interferometric synthetic aperture radar (InSAR) data, to semi-automatically detect and map pingos. Of the 1,247 pingos identified, more than 400 had not been recognized in previous datasets based on only aerial imagery.

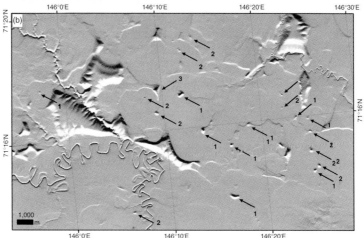

Figure 13.4 (a) Pingo in the Lena Delta, Siberia (M. Ulrich); (b) Mapping of pingos in a lowland landscape in NE Siberia using Terra ASTER late winter imagery (Grosse and Jones, 2011).
1. Pingos identified in both topographic map and ASTER data;
2. Pingos only in ASTER.
3. Pingos only in topographic map.

Similarly, Grosse and Jones (2011) extracted pingos from topographic maps for a 3.5×10^6 km^2 area in Russia and validated the map data for several sub-regions, using late winter Terra ASTER data for pingo mapping. The timing of these images was chosen to fall into the March to April period, when snow cover was relatively uniform across the landscape, resulting in hillshade-like images that enabled the easy visual identification of pingos, appearing as conical hill features with very bright, sunlit, south-facing slopes and long shadows on the north-facing slopes (Figure 13.4).

Ice wedge polygonal ground has been studied with aerial imagery to estimate ground ice contents in near-surface permafrost, and to derive information on sediment and environmental characteristics, using polygon morphology in a wide range of settings over the last decades (Pollard and French, 1980; Bode *et al.*, 2008). Recently, the availability of high resolution historic and new satellite imagery allows ice wedge polygons over even larger regions to be studied.

For example, Grosse *et al.* (2005) used 2.5 m ground resolution satellite photography of the Corona KH-4B satellite to study distribution of periglacial landforms, including ice wedge polygons, on the Bykovsky Peninsula, northeast Siberia. However, airborne surveys are still an important tool for studying ice wedge polygons, as was demonstrated by Ulrich *et al.* (2011) in their application of the airborne version of the High Resolution Stereo Camera (HRSC-AX) over ice wedge polygonal ground on Svalbard, Norway. The stereo data of the HRSC-AX and four spectral channels were merged to create very detailed data for morphometric analysis of ice wedge polygons.

13.3.3 *Landforms and processes indicating permafrost degradation*

In regions where permafrost contains excess ground ice, thawing of permafrost leads to surface deformation, such as surface subsidence (thermokarst), down-slope mass wasting, or lateral thermal erosion. Remote sensing-based observation of thermokarst landforms, and other features and processes indicating permafrost degradation, allows quantification of changes in permafrost distribution and vulnerability. Rates of thaw and mass wasting, and observations on the change in number and size of such features, can be used to assess impacts of permafrost thaw on hydrology, carbon cycle, and other affected components of the environment.

Thermokarst lakes, basins, and gullies are important indicators of permafrost degradation, cover very large regions, and substantially impact hydrology, geomorphology, habitats, and biogeochemical cycling in permafrost lowlands (Grosse *et al.*, 2013). These features were mapped early on with aerial imagery and first available ERTS-1 (LANDSAT-1) data (Sellmann *et al.*, 1975; Tarnocai and Kristof, 1976). Many studies since then have used LANDSAT data to map the distribution and morphology of these lakes and basins in different regions (Frohn *et al.*, 2005; Grosse *et al.*, 2006; Morgenstern *et al.*, 2011; Jones *et al.*, 2012b; Regmi *et al.*, 2012). Wang *et al.* (2011) used an InSAR-derived high resolution elevation model (see Section 13.2.2) to reconstruct the extent of drained thermokarst lake basins on the Alaska Arctic Coastal Plain.

Many studies using LANDSAT data focused on change in thermokarst lake area and tried to link this to either permafrost degradation, changes in precipitation/evapotranspiration, or even direct human influences (Smith *et al.*, 2005; Hinkel *et al.*, 2007; Plug *et al.*, 2008; Kravtsova and Bystrova, 2009; Arp *et al.*, 2011; Kravtsova and Tarasenko, 2011). However, LANDSAT-type data has limitations for the application of lake change detection due to its moderate spatial resolution and a temporal coverage starting only in 1975. Hence, aerial image and very high-resolution satellite (Ikonos-2, GeoEye-1, Quickbird, WorldView-1 and -2) time series are also frequently used to assess local and regional changes in thermokarst lakes in more detail, including lake disappearance from drainage

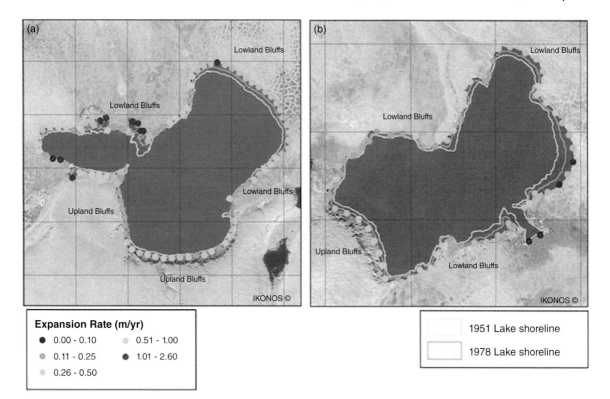

Figure 13.5 Expansion rates and historic shorelines for two thermokarst lakes on the Seward Peninsula, Alaska, based on aerial images from 1951 and 1978, and IKONOS high-resolution satellite imagery from 2006. Steep upland bluffs consist of thick syngenetic ice-rich windblown silt, while lowland bluffs consist of moderate ice-rich refrozen lake and peat deposits. The background image is from IKONOS. (Jones *et al.*, 2011. Reproduced with permission of John Wiley & Sons Ltd).

or drying (Yoshikawa and Hinzman, 2003; Riordan *et al.*, 2006; McGraw, 2008; Marsh *et al.*, 2009; Roach *et al.*, 2011) and lake expansion by thermal erosion (Sannel and Brown, 2010; Jones *et al.*, 2011; Sannel and Kuhry, 2011) (Figure 13.5).

Thermokarst features were also analyzed by Vallee and Payette (2007), who compared historical aerial images from 1957 with a 2001 field survey to study the increase in thermokarst ponds due to palsa collapse in northern Québec, Canada. Osterkamp *et al.* (2009) mapped development of various thermokarst landforms in the Healy area of Interior Alaska, using repeat aerial photography, from 1954–2005.

Retrogressive thaw slumps and active layer detachment slides are prominent features of permafrost degradation that can be mapped with remote sensing data. Both types occur in environments with high excess ground ice contents in near-surface permafrost and a sufficient slope gradient. Lacelle *et al.* (2010)

observed retrogressive thaw slump formation in the Aklavik Plateau, NW Canada, based on aerial images from 1950–2004. To quantify the volume of sediment and ice mobilized by retrogressive thaw slumps on Herschel Island, northwest Canada, Lantuit and Pollard (2005) derived a time series of digital elevation models from 1950 and 1970 aerial photographs and a 2004 Ikonos image using stereo-photogrammetry. In addition, Lantuit and Pollard (2008) mapped the distribution and activity of thaw slumps with aerial and Ikonos data on Herschel Island, and found a substantial increase in thaw slump number (+127%) and area (+160%) between 1952–2000, most likely linked to climate warming.

Lantz and Kokelj (2008) used aerial photographs from 1950, 1973, and 2004 to study the development of retrogressive thaw slumps in the Mackenzie River delta region, Canada, and found that significant increases in annual and summer air temperatures in the second period after 1973 resulted in increased thaw slump activity. In a follow-up study, using the 2004 aerial imagery, Kokelj *et al.* (2009) investigated the morphological character of 530 of these thaw slumps, particularly their orientation and whether they consist of multiple generations. They found that 97% are, indeed, polycyclic, with 20% still active at the time of image acquisition, and that slump orientation was significantly linked to aspect of the lake shore, with more slumps being found on north, west, and east-facing shores.

Aerial image time series have been used to map the distribution, morphology, and activity of active layer detachment slides on Fosheim Peninsula, Ellesmere Island, Nunavut, Canada (Lewkowicz and Harris, 2005a, 2005b). Similarly, SPOT panchromatic data was used to identify active layer detachment slides, among other permafrost features, in the same region (Lewkowicz and Duguay, 1999). Bowden *et al.* (2008) used 1980 aerial imagery to compare the distribution of historic active layer detachment slides with the location of recent GPS-surveyed features in the Brooks Range Foothills, North Alaska.

In addition to optical methods, mass wasting from thaw slumps and landslides can also be detected through high-resolution radar interferometry (see Section 13.2.3). Alasset *et al.* (2008) used RADARSAT InSAR to detect permafrost activity and landslide movement in the Canadian Mackenzie River region over 18 months. With a similar focus on monitoring of landslides and retrogressive thaw slumps, Singhroy *et al.* (2007) used RADARSAT-based InSAR in northwest Canada.

The melting of massive ice wedge bodies and subsequent surface collapse has been documented using high resolution aerial photography. Jorgenson *et al.* (2006) observed the strong increase of thaw pit area over degrading ice wedges from 0.5–4.4% of the studied landscape on the Alaska Arctic Coastal Plain between 1945 and 2001. Fortier *et al.* (2007) used vertical and oblique aerial imagery to observe and map the development of a rapidly evolving thermal erosion gully network formed by ice wedge melt and erosion on Bylot Island, Canadian Arctic Archipelago, from 1999 to 2002.

Lateral thaw of permafrost in peat plateaus has been traditionally observed with aerial photography (Laberge and Payette, 1995; Beilman *et al.*, 2001; Jorgenson *et al.*, 2001; Payette *et al.*, 2004). Quinton *et al.* (2011) investigated permafrost change in peat plateaus in the Hay River Lowlands of the Northwest Territories,

Canada, using aerial imagery from 1947–2008. They found a 38% decrease in permafrost extent in their 1 km^2 study area over the study period. Using LIDAR data and aerial and Ikonos imagery, Chasmer *et al.* (2010) conducted a detailed assessment of errors associated with high-resolution aerial data analysis for permafrost detection in a peatland complex near Fort Simpson. They found that accurate delineation of permafrost peatland plateaus from degraded historical aerial imagery can be exceedingly difficult and thus associated with substantial errors, and that even Ikonos multispectral data, with 4 m resolution, may not be sufficient to detect changes in permafrost peat plateau extent over the period covered with remote sensing data, if lateral thaw rates are not large enough.

Coastal erosion along Arctic permafrost coasts is observed using time series of high-resolution aerial and satellite images. Erosion rates are often very high and may exceed 10 m/year along coasts consisting of ice-rich permafrost. Jones *et al.* (2009b) used a time series of aerial images covering 1955–2007 to study rapid increase in erosion rates along the Alaska Beaufort Sea coast. The authors (Jones *et al.*, 2009a) complemented this time series by adding panchromatic high-resolution Corona, Hexagon, and Ikonos imagery for the analysis of erosion rates specifically at Cape Halkett.

Jorgenson and Brown (2005) used LANDSAT data to classify coastal segments according to geology, permafrost, and carbon content, and reviewed literature of remotely-sensed coastal erosion rates for the Alaska Beaufort Sea coast to derive carbon and sediment flux rates. Mars and Houseknecht (2007), working in the same region, found a substantial increase in permafrost-related coastal erosion rates using aerial photo-based topographic maps and LANDSAT data for the 1955–2005 period. Similarly, Solomon (2005) used aerial imagery to study permafrost coastal erosion rates in the Mackenzie Delta region. In NE Siberia, Lantuit *et al.* (2011) studied coastal erosion dynamics at the ice-rich permafrost-dominated Bykovsky Peninsula. Their observation period, from 1951–2006, comprised several time steps with observations from high-resolution panchromatic aerial, Corona, Hexagon, and SPOT-5 imagery, plus a medium resolution RESURS KFK 1000 image. Their data suggests no long-term trend in erosion rates, but strong interannual and interdecadal variability.

13.4 Lowland permafrost – remote sensing of physical variables related to the thermal permafrost state

In the vast permafrost lowland areas, low and very low resolution remote sensors allow to measure a number of physical variables that can directly or indirectly help to characterize the permafrost state on the regional to continental scale. Although subsurface temperature and ice and water contents cannot be accessed remotely, air- and spaceborne measurements of proxy variables, such as land surface temperature or freeze-thaw state of the surface, represent an important step towards quantitative permafrost monitoring.

13.4.1 Land surface temperature through thermal remote sensing

The land surface temperature (LST) is accessible through a number of sensors, ranging from low or very low (e.g., MODIS, AATSR, AVHRR) to medium (e.g., LANDSAT, ASTER) spatial resolution. The thermal regime of permafrost is determined by the temporal average of surface temperatures, since fast surface temperature fluctuations, such as the daily cycle, do not contribute to the temperatures of deep soil layers, due to buffering effects of the vegetation and upper soil layers. Therefore, thermal sensors must be able to deliver reliable surface temperature averages, which requires a dense time series of LST acquisitions. High and medium resolution sensors feature only a limited temporal coverage, so that they are not well suited for this purpose.

In contrast, low resolution sensors, such as MODIS, can provide several measurements per day under clear sky conditions, which is an excellent basis for calculating weekly to yearly averages. Despite considerable data gaps due to cloudiness in the MODIS time series, Langer *et al.* (2010) and Westermann *et al.* (2011) found an agreement generally better than 2°C between weekly averages, computed from *in situ* measurements and MODIS LST, for the snow free-period at sites in Siberia and Svalbard. For the winter season on Svalbard, however, a significant cold-bias of average MODIS LST of more than 3°C was found as a result of missing measurements during cloudy periods and cloud-contaminated LST measurements (Westermann *et al.*, 2011).

The average annual LST is not a direct measure for soil temperatures, as the winter snow cover insulates the ground from cold temperatures, while a similar effect is not present during summer. Hachem *et al.* (2012) pointed out the considerable differences between MODIS LST for snow surfaces and corresponding temperatures at the snow-soil interface, for a number of permafrost sites. To overcome this problem, simple degree-day-based permafrost models have been developed, which have been successfully employed with remotely sensed LST to reproduce larger-scale permafrost boundaries. Hachem *et al.* (2008) employed the ratio between the freezing and the thawing index (Carlson, 1952) to delineate the boundaries between continuous, discontinuous and sporadic permafrost in Québec and Labrador, Canada (Figure 13.6).

Low resolution sensors can not resolve potential local-scale LST differences caused by the considerable variability of the surface cover typical for permafrost areas (Hachem *et al.*, 2009). As an example, Figure 13.7 demonstrates the considerable spread of surface temperatures over distances of less than 10 m in highly structured polygonal tundra in Siberia. Using *in situ* measurements from a thermal camera system, Langer *et al.* (2010) showed that significant spread of LST present in single scenes (as in Figure 13.7) is strongly reduced in long-term averages, which are more relevant for permafrost temperatures. Nevertheless, differences in summer-average LST of several degrees over distances of less than 100 m have been reported (Westermann *et al.*, 2011), so that a better understanding of sub-grid variability of LST could be an important step towards improving the performance of LST-based permafrost monitoring.

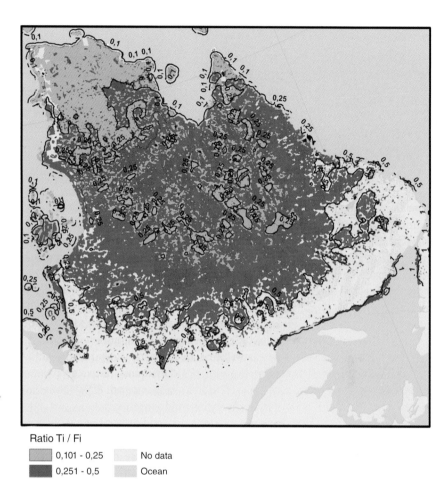

Figure 13.6 Map of the ratio of thawing index to freezing index (Ti/Fi), derived from MODIS (Terra and Aqua) LST data over northern Quebec, Canada. The 0.5 isoline (i.e., when Fi is twice as large as Ti) corresponds to the known southern limit of the sporadic permafrost zone, while the 0.25 isoline (i.e., when Fi is four times as large as Ti) corresponds to the southern limit of the discontinuous permafrost zone. Modified from Hachem et al., 2008.

Ratio Ti / Fi

▨ 0,101 - 0,25	□ No data	
■ 0,251 - 0,5	▨ Ocean	
□ 0,501 - 1	—— Treeline	
■ 1,001 - 1,5		

Figure 13.7 Surface temperature measured by an *in situ* thermal camera system (location marked by the red dot) in polygonal tundra in the Lena River Delta, Siberia (adapted from Langer *et al.*, 2010). The figure shows deviation from average scene temperature for two clear-sky situations in summer, 2008.

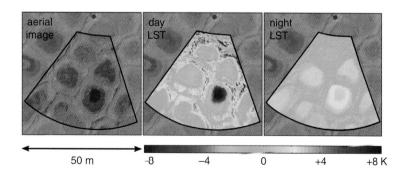

50 m   −8   −4   0   +4   +8 K

13.4.2 *Freeze-thaw state of the surface soil through microwave remote sensing*

The relative permittivity of soil is an important parameter for both emission and reflection of microwaves at the ground. While it is close to unity for air, and below 10 for both ice and typical soil minerals, it is larger than 80 for water, so that freezing of wet soil leads to a drastic decrease of its relative permittivity. Freezing and thawing of soil, as well as melting of snow, are hence visible in both passive and active microwave remote sensing (Judge *et al.*, 1997; Zhang *et al.*, 2003a; Bartsch *et al.*, 2007; Park *et al.*, 2011; Zhao *et al.*, 2011).

Several studies have demonstrated that microwave remote sensing can capture the patterns of soil freezing and thawing on the regional to continental scale (Kimball *et al.*, 2001, 2004). In regions subject to seasonal frost, microwave remote sensing has been applied to map the areal extent of frozen and snow-covered areas (Zhang *et al.*, 2003a). Furthermore, multi-annual variations and trends of the dates of freezing and thawing, as well as the length of the growing season, can be derived (Wismann, 2000; McDonald *et al.*, 2004; Smith *et al.*, 2004). However, the microwave signal from the ground is influenced by a number of additional factors, such as soil texture, soil temperature, surface roughness, and snow and vegetation parameters, so that determining the freeze-thaw state of the soil is not a trivial task. In addition, commonly employed sensors operate at wavelengths of a few centimeters (e.g., 2.14 cm for K_u-band of NSCAT, 5.7 cm for C-band of ERS), resulting in a signal typically from the uppermost centimeter of soil, so that freeze and thaw processes of deeper soil layers can not be captured.

The new SMOS (Soil Moisture and Ocean Salinity) mission and the planned Soil Moisture Active Passive (SMAP) mission (which was due for launch in 2014), operating at a lower frequency in the L-band (Kerr *et al.*, 2001; Entekhabi *et al.*, 2010; Kim *et al.*, 2010) may facilitate measurements that are influenced by the landcover to a lesser degree than previous microwave remote sensing products (Kontu *et al.*, 2010). In addition, the use of L-band (wavelength 21 cm) allows larger penetration depths, which may provide better insight in the dynamics of the active layer freezing and thawing.

13.4.3 *Permafrost mapping with airborne electromagnetic surveys*

In contrast to spaceborne sensors, airborne electromagnetic surveys are conducted at radio frequencies in the kilohertz range (compared to gigahertz for microwave remote sensing), which facilitates mapping a profile of electrical resistivity of the ground to depths of more than 100 m. With resistivity closely linked to the freeze-thaw state of the ground, the vertical extent of frozen ground can be mapped, thus allowing an unprecedented two- or even three-dimensional snapshot of the permafrost configuration. However, electrical resistivity also has a strong dependence on the soil type, which requires a careful interpretation of resistivity profiles inferred from electromagnetic surveys. In the discontinuous permafrost zone of Alaska, Minsley *et al.* (2012) mapped the occurrence and depth of frozen ground over an 1800 line-kilometer flight line, thus demonstrating the potential of this novel technique in permafrost investigations.

13.4.4   *Regional surface deformation through radar interferometry*

As explained in the previous sections, interferometric SAR is capable of measuring vertical surface displacements on the order of millimeters to centimeters. Drivers of elevation changes in permafrost regions are seasonal freeze-thaw cycles, and are short-term and long-term natural disturbances caused by climate change or anthropogenic disturbances. The active layer thaws every summer to depths ranging from a few centimeters to several meters, and completely refreezes every winter. Phase transitions of water during the freeze-thaw cycle result in soil volume changes that are expressed as soil surface heave during freezing and subsidence during thawing. Seasonal elevation differences due to this process may reach up to several 10 cm (Romanovsky *et al.*, 2008).

Wang and Li (1999) and Rykhus and Lu (2008) applied radar interferometry to detect heave in the refreezing active layer and thaw settlement in Northern Alaska. InSAR has also been used by Liu *et al.* (2010) to quantify heave and subsidence signals on the Alaska Arctic Coastal Plain. Beyond a seasonal active layer signal, these authors also found a multi-annual ground subsidence overlaying the seasonal signal, which is interpreted as the signature of thawing ice-rich near-surface permafrost. If a soil supersaturated with ice thaws, and the melt water is converted to runoff in the case of well-drained conditions, the ground subsides, relative to stable areas, e.g., a floodplain without ground ice (Liu *et al.*, 2010).

In a follow-up study, Liu *et al.* (2012) used long-term ERS-1 and -2 data for an InSAR analysis of active layer thickness in the same region. In their study, they combined the InSAR observations of surface deformation with an algorithm that models active layer conditions, using ground data from CALM sites as input to quantify changes in the active layer in a large region. While multi-annual ground subsidence may be direct evidence of degrading permafrost, the main shortcoming of interferometric SAR in permafrost areas is the lack of coherence between scenes, for example due to snowmelt or changes in surface soil moisture or vegetation. The planned Tandem-L mission scheduled for 2016 (Krieger *et al.*, 2009), which will combine global mapping with repeat times of weeks and a spatial resolution of less than 100 m, could significantly improve the monitoring of such dynamic ground processes with radar interferometry in the future.

13.4.5   *A gravimetric signal of permafrost thaw?*

In 2002, the Gravity Recovery and Climate Experiment (GRACE) launched two satellites capable of mapping the Earth's gravity field with unprecedented accuracy.

While the rather coarse spatial resolution of a few hundred kilometers makes GRACE only suitable for large-scale applications, its accuracy is sufficient to investigate changes in the water budget in the catchments of the large Arctic rivers, Ob, Yenisei, Lena, Mackenzie River and Yukon. In particular, the catchment of the Lena river, which is predominantly located within the continuous permafrost zone, shows a notable increase in mass during the operation period of GRACE (Muskett and Romanovsky, 2009).

On the other hand, watersheds dominated by discontinuous permafrost, such as the Mackenzie and Yukon Rivers, experienced a net mass loss during the same period (Muskett and Romanovsky, 2009, 2011). Several studies relate these mass changes to a change in terrestrial water storage which, in turn, is influenced by permafrost-related processes. As suggested by field evidence (Walvoord and Striegl, 2007; St Jacques and Sauchyn, 2009) and modeling studies (Bense *et al.*, 2009), thawing of permafrost leads to water infiltration in supra- and sub-permafrost taliks and, thus, to the formation and recharging of aquifers, which could explain the mass gain in the continuous permafrost zone. Conversely, in the discontinuous permafrost zone, thawing of hydraulic barriers could lead to the drainage of lakes, thereby decreasing the terrestrial water storage in these areas (Muskett and Romanovsky, 2011).

If this interpretation of GRACE data is correct, it would allow a direct quantification of large-scale changes of the water balance triggered by permafrost thaw. However, the still short time series of GRACE, and considerable uncertainties in the quantification of other processes leading to mass changes, make directly proving this hypothesis difficult (Syed *et al.*, 2007; Landerer *et al.*, 2010; Vey *et al.*, 2012).

13.5 Outlook – remote sensing data and permafrost models

In the last two decades, remote sensing missions have delivered a wealth of data sets that can be exploited in permafrost research. However, an operational permafrost monitoring scheme, based on remote sensors that could deliver permafrost temperatures on large scales, has not yet been demonstrated. While this is partly due to the lack of a sensor that can truly derive below-ground physical properties in permafrost regions, progress may be achieved by combining the information content of different sensors with thermal permafrost models of various degrees of complexity (Zhang *et al.*, 2003c; Riseborough *et al.*, 2008; Westermann *et al.*, 2013).

Remote sensing data is predestined to deliver valuable driving data sets for spatially distributed numerical permafrost models (Duguay *et al.*, 2005), which require input variables from the terrain surface over large regions that can be observed and monitored with satellite sensors. In particular, remotely sensed LST and snow water equivalent (Luojus *et al.*, 2011; see, for example, Chapter 5 in this book) could be well suited as forcing data for such permafrost models, while, for example, soil moisture information from microwave remote sensing or optical mapping of the land cover could help to estimate the thermal properties of the ground. Such data sets have recently been compiled in the ESA DUE permafrost project for large permafrost areas (Bartsch *et al.*, 2010).

As a proof-of-concept, Marchenko *et al.* (2009) presented a first remote sensing-based permafrost modeling scheme, using MODIS LST, which was capable of reproducing the known extent of permafrost in Northern Eurasia. Langer *et al.* (2013) performed a sensitivity study of a permafrost modeling scheme based on remotely sensed LST, snow water equivalent and snow extent as input data. At a site in Siberia, this approach could deliver permafrost temperatures as well as the depth of the active layer, which are important parameters for the release of greenhouse gases from permafrost landscapes.

Future studies should further benchmark the performance of such remote sensing-based permafrost monitoring schemes, and investigate their sensitivity and sources of error (e.g., the impact of biased long-term average LST (see Section 13.4.1) on the modeled soil temperatures). Another critical issue is the limited time series of most remotely sensed data sets, which only facilitate to reproduce permafrost conditions in equilibrium with current climate forcing, while especially deeper soil temperatures are determined by past climate conditions, for which no remote sensing data exist. Nevertheless, assimilating remotely sensed data sets in a thermal permafrost model is a promising approach towards quantitative permafrost monitoring. Establishing and operationalizing such a scheme could become crucial to investigating permafrost in a changing climate and its impacts on local to global scales.

References

ACIA (2005). *Arctic Climate Impact Assessment*. New York, Cambridge University Press.

Alasset, P.-J., Poncos, V., Singhroy, V. & Couture, R. (2008). *InSAR monitoring of permafrost activity in the lower Mackenzie valley, Canada*. International Geoscience and Remote Sensing Symposium, IGARSS '08 Proceedings, IEEE 2008 International, Boston, USA.

Arp, C.D., Jones, B.M., Urban, F.E. & Grosse, G. (2011). Hydrogeomorphic processes of thermokarst lakes with grounded-ice and floating-ice regimes on the Arctic coastal plain, Alaska. *Hydrological Processes* **25**, 2422–2438, doi: 10.1002/hyp.8019.

Baltsavias, E.P., Favey, E., Bauder, A., Boesch, H. & Pateraki, M. (2001). Digital surface modelling by airborne laser scanning and digital photogrammetry for glacier monitoring. *Photogrammetric Record* **17**, 243–273.

Bartsch, A., Kidd, R.A., Wagner, W. & Bartalis, Z. (2007). Temporal and spatial variability of the beginning and end of daily spring freeze/thaw cycles derived from scatterometer data. *Remote Sensing of Environment* **106**, 360–374.

Bartsch, A., Wiesmann, A., Strozzi, T., Schmullius, C., Hese, S., *et al.* (2010). *Implementation of a satellite data based permafrost information system – the DUE permafrost project*. Proceedings of the ESA Living Planet Symposium, Bergen, Norway, 2010.

Bauer, A., Paar, G. & Kaufmann, V. (2003). *Terrestrial laser scanning for rock glacier monitoring*. Proceedings of the Eighth International Conference on Permafrost, Zurich, 2003.

Beilman, D.W., Vitt, D.H. & Halsey, L.A. (2001). Localized permafrost Peatlands in Western Canada: Definition, distributions, and degradation. *Arctic Antarctic and Alpine Research* **33**, 70–77.

Bense, V.F., Ferguson, G. & Kooi, H. (2009). Evolution of shallow groundwater flow systems in areas of degrading permafrost. *Geophysical Research Letters* **36**, L22401, doi: 22410.21029/22009gl039225.

Berthier, E., Arnaud, Y., Baratoux, D., Vincent, C. & Remy, F. (2004). Recent rapid thinning of the "Mer de Glace" glacier derived from satellite optical images. *Geophysical Research Letters* **31**, L17401, doi: 10.1029/2004GL020706.

Berthier, E., Vadon, H., Baratoux, D., Arnaud, Y., Vincent, C., *et al.* (2005). Surface motion of mountain glaciers derived from satellite optical imagery. *Remote Sensing of Environment* **95**, 14–28.

Bode, J.A., Moorman, B.J., Stevens, C.W. & Solomon, S.M. (2008). *Estimation of Ice Wedge Volume in the Big Lake Area, Mackenzie Delta, NWT, Canada*. Proceedings of the Ninth International Conference on Permafrost, Fairbanks, Alaska, USA, 2008.

Bowden, W.B., Gooseff, M.N., Balser, A., Green, A., Peterson, B.J., *et al.* (2008). Sediment and nutrient delivery from thermokarst features in the foothills of the North Slope, Alaska: Potential impacts on headwater stream ecosystems. *Journal of Geophysical Research-Biogeosciences* **113**, G02026, doi: 10.1029/2007gl032433.

Brown, J., Ferrians Jr., O., Heginbottom, J. & Melnikov, E. (1997). *Circum-Arctic map of permafrost and ground-ice conditions*. US Geological Survey Circum-Pacific Map.

Carlson, H. (1952). Calculation of depth of thaw in frozen ground. *Frost Action in Soils: A Symposium*, 192–223.

Carter, L.D. & Galloway, J.P. (1979). Arctic Coastal Plain pingos in National Petroleum Reserve in Alaska, In: Johnson, K.M. & Williams, J.K. (eds). *US Geological Survey Circular, 804-B*, 33–35. The United States Geological Survey in Alaska.

Chasmer, L., Hopkinson, C. & Quinton, W. (2010). Quantifying errors in discontinuous permafrost plateau change from optical data, Northwest Territories, Canada: 1947–2008. *Canadian Journal of Remote Sensing* **36**, S211–S223.

Chasmer, L., Quinton, W., Hopkinson, C., Petrone, R. & Whittington, P. (2011). Vegetation Canopy and Radiation Controls on Permafrost Plateau Evolution within the Discontinuous Permafrost Zone, Northwest Territories, Canada. *Permafrost and Periglacial Processes* **22**, 199–213.

Chen, R.F., Chang, K.J., Angelier, J., Chan, Y.C., Deffontaines, B., et al. (2006). Topographical changes revealed by high-resolution airborne LiDAR data: The 1999 Tsaoling landslide induced by the Chi-Chi earthquake. *Engineering Geology* **88**, 160–172.

Colesanti, C., Ferretti, A., Prati, C. & Rocca, F. (2003). Monitoring landslides and tectonic motions with the Permanent Scatterers Technique. *Engineering Geology* **68**, 3–14.

Comiso, J.C. & Parkinson, C.L. (2004). Satellite-observed changes in the Arctic. *Physics Today* **57**, 38–44.

Crosetto, M. (2002). Calibration and validation of SAR interferometry for DEM generation. *ISPRS Journal of Photogrammetry and Remote Sensing* **57**, 213–227.

Daanen, R.P., Grosse, G., Darrow, M., Hamilton, T.D. & Jones, B.M. (2012). Rapid movement of frozen debris lobes: implications for permafrost degradation and slope instability in the Brooks Range, Alaska. *Natural Hazards and Earth Systems Sciences* **12**, 1521–1537, doi :10.5194/nhess-12-1521-2012.

Dehls, J.F., Basilico, M. & Colesanti, C. (2002). *Ground deformation monitoring in the Ranafjord area of Norway by means of the permanent scatterers technique.* Proceedings of the IEEE Geoscience and Remote Sensing Symposium, 2002.

Delacourt, C., Allemand, P., Casson, B. & Vadon, H. (2004). Velocity field of "La Clapière" landslide measured by the correlation of aerial and QuickBird satellite images. *Geophysical Research Letters* **31**, L15619, doi: 15610.11029/ 12004GL020193.

Delacourt, C., Allemand, P., Berthier, E., Raucoules, D., Casson, B., et al. (2007). Remote-sensing techniques for analysing landslide kinematics: a review. *Bulletin de la Societe Geologique de France* **178**, 89–100.

Delisle, G. (2007). Near-surface permafrost degradation: How severe during the 21st century? *Geophysical Research Letters* **34**, 9503, doi: 9510.1029/ 2007GL029323.

Duguay, C.R., Zhang, T., Leverington, D.W. & Romanovsky, V.E. (2005). Satellite remote sensing of permafrost and seasonally frozen ground, In: Duguay, C.R. & Pietroniro, A. (eds). *Remote sensing in northern hydrology: measuring environmental change*, 91–118. Washington, DC, American Geophysical Union.

Eldhuset, K., Andersen, P.H., Hauge, S., Isaksson, E. & Weydahl, D.J. (2003). ERS tandem INSAR processing for DEM generation, glacier motion estimation and coherence analysis on Svalbard. *International Journal of Remote Sensing* **24**, 1415–1437.

Engeset, R.V. & Weydahl, D.J. (1998). Analysis of glaciers and geomorphology on Svalbard using multitemporal ERS-1 SAR images. *IEEE Transactions on Geosciences and Remote Sensing* **36**, 1879–1887.

Entekhabi, D., Njoku, E.G., O'Neill, P.E., Kellogg, K.H., Crow, W.T., et al. (2010). *The Soil Moisture Active Passive (SMAP) Mission.* Proceedings of the IEEE '98, 704–716.

Etzelmüller, B. (2000). On the quantification of surface changes using grid-based digital elevation models (DEMs). *Transactions in GIS* **4**, 129–143.

Etzelmüller, B., Ødegård, R.S., Berthling, I. & Sollid, J.L. (2001). Terrain parameters and remote sensing data in the analysis of permafrost distribution and periglacial processes: principles and examples from Southern Norway. *Permafrost and Periglacial Processes* **12**, 79–92.

Fischer, L., Kääb, A., Huggel, C. & Noetzli, J. (2006). Geology, glacier retreat and permafrost degradation as controlling factors of slope instabilities in a high-mountain rock wall: the Monte Rosa east face. *Natural Hazards and Earth System Sciences* **6**, 761–772.

Fischer, L., Eisenbeiss, H., Kääb, A., Huggel, C. & Haeberli, W. (2011). Monitoring topographic changes in a periglacial high-mountain face using high-resolution DTMs, Monte Rosa East Face, Italian Alps. *Permafrost and Periglacial Processes* **22**, 140–152.

Floricioiu, D. & Rott, H. (2001). Seasonal and short-term variability of multifrequency, polarimetric radar backscatter of alpine terrain from SIR-C/X-SAR and AIRSAR data. *IEEE Transactions on Geoscience and Remote Sensing* **39**, 2634–2648.

Fortier, D., Allard, M. & Shur, Y. (2007). Observation of rapid drainage system development by thermal erosion of ice wedges on Bylot island, Canadian Arctic Archipelago. *Permafrost and Periglacial Processes* **18**, 229–243.

Frohn, R.C., Hinkel, K.M. & Eisner, W.R. (2005). Satellite remote sensing classification of thaw lakes and drained thaw lake basins on the North Slope of Alaska. *Remote Sensing of Environment* **97**, 116–126.

Geist, T., Lutz, E. & Stötter, J. (2003). *Airborne laser scanning technology and its potential for applications in glaciology*. Proceedings of the ISPRS Workshop on 3-D Reconstruction from Airborne Laserscanner and INSAR Data, Dresden, Germany, 2003.

Glenn, N.F., Streutker, D R., Chadwick, D.J., Thackray, G.D. & Dorsch, S.J. (2006). Analysis of LiDAR-derived topographic information for characterizing and differentiating landslide morphology and activity. *Geomorphology* **73**, 131–148.

Gorbunov, A.P. (1978). Permafrost Investigations in High-Mountain Regions. *Arctic and Alpine Research* **10**, 283–294.

Grosse, G. & Jones, B.M. (2011). Spatial distribution of pingos in northern Asia. *The Cryosphere* **5**, 13–33, doi: 10.5194/tc-5-13-2011.

Grosse, G., Schirrmeister, L., Kumitsky, V.V. & Hubberten, H.W. (2005). The use of CORONA images in remote sensing of periglacial geomorphology: An illustration from the NE Siberian Coast. *Permafrost and Periglacial Processes* **16**, 163–172, doi: 10.1002/ppp.509.

Grosse, G., Schirrmeister, L. & Malthus, T.J. (2006). Application of LANDSAT-7 satellite data and a DEM for the quantification of thermokarst-affected terrain types in the periglacial Lena-Anabar coastal lowland. *Polar Research* **25**, 51–67, doi: 10.1111/j.1751-8369.2006.tb00150.x.

Grosse, G., Harden, J., Turetsky, M., McGuire, A.D., Camill, P., et al. (2011). Vulnerability of high-latitude soil organic carbon in North America to disturbance. *Journal of Geophysical Research-Biogeosciences* **116**, G00K06, doi: 10.1029/2010jg001507.

Grosse, G., Jones, B. & Arp, C. (2013). Thermokarst Lakes, Drainage, and Drained Basins. In: Shroder, J. (ed. in chief), Giardino, R. & Harbor, J. (volume eds). Treatise on Geomorphology, Vol 8, *Glacial and Periglacial Geomorphology*, 325–353. San Diego: Academic Press.

Hachem, S., Allard, M. & Duguay, C. (2008). *A new permafrost map of Quebec-Labrador derived from near-surface temperature data of the moderate resolution imaging spectroradiometer (MODIS)*. Proceedings of the Ninth International Conference on Permafrost, Fairbanks, USA, 2008.

Hachem, S., Allard, M. & Duguay, C. (2009). Using the MODIS land surface temperature product for mapping permafrost: an application to northern Quebec and Labrador, Canada. *Permafrost and Periglacial Processes* **20**, 407–416.

Hachem, S., Duguay, C.R. & Allard, M. (2012). Comparison of MODIS-derived land surface temperatures with ground surface and air temperature measurements in continuous permafrost terrain. *The Cryosphere* **6**, 51–69.

Haeberli, W., Guodong, C., Gorbunov, A. & Harris, S. (1993). Mountain permafrost and climatic change. *Permafrost and Periglacial Processes* **4**, 165–174.

Hall, D.K. (1982). A Review of the Utility of Remote-Sensing in Alaskan Permafrost Studies. *IEEE Transactions on Geoscience and Remote Sensing* **20**, 390–394.

Harris, S.A. (1988). The alpine periglacial zone, In: Clark, M.J. (ed). *Advances in periglacial geomorphology*, 363–372. John Wiley & Sons.

Harris, S., French, H., Heginbottom, J., Johnston, G., Ladanyi, B., *et al.* (1988). *Glossary of permafrost and related ground ice terms*. Permafrost Subcommittee, Associate Committee on Geotechnical Research, National Research Council of Canada, Ottawa.

Hauber, E., Slupetzky, H., Jaumann, R., Wewel, F., Gwinner, K., *et al.* (2000). *Digital and automated high resolution stereo mapping of the Sonnblick glacier (Austria) with HRSC-A*. Proceedings of the EARSeL-SIG-Workshop Land Ice and Snow, Dresden, Germany, 2000.

Hilley, G.E., Burgmann, R., Ferretti, A., Novali, F. & Rocca, F. (2004). Dynamics of slow-moving landslides from permanent scatterer analysis. *Science* **304**, 1952–1955.

Hinkel, K.M., Jones, B.M., Eisner, W.R., Cuomo, C J., Beck, R.A., *et al.* (2007). Methods to assess natural and anthropogenic thaw lake drainage on the western Arctic coastal plain of northern Alaska. *Journal of Geophysical Research – Earth Surface* **112**, F02S16, doi: 10.1029/2006jf000584.

Hirano, A., Welch, R. & Lang, H. (2003). Mapping from ASTER stereo image data: DEM validation and accuracy assessment. *ISPRS Journal of Photogrammetry and Remote Sensing* **57**, 356–370.

Holmes, G.W. (1967). Location of pingos and pingo-like mounds observed from the ground, from aerial reconnaissance, and on aerial photographs in interior Alaska. *U.S. Geological Survey Open-File Report*, 13.

IPCC (2007). *Climate Change 2007 – The Physical Science Basis*. Working Group I Contribution to the Fourth Assessment Report of the IPCC, Cambridge University Press.

Janeras, M., Navarro, M., Arnó, G., Ruiz, A., Kornus, W., *et al.* (2004). *LIDAR applications to rock fall hazard assessment in Vall de Núria.* Proceedings of the 4th ICA Mountain Cartography Workshop, Vall de Núria, Catalonia, Spain, 2004.

Jones, B.M., Arp, C.D., Beck, R.A., Grosse, G., Webster, J.M., *et al.* (2009a). Erosional history of Cape Halkett and contemporary monitoring of bluff retreat, Beaufort Sea coast, Alaska. *Polar Geography* **32**, 129–142.

Jones, B.M., Arp, C.D., Jorgenson, M.T., Hinkel, K.M., Schmutz, J.A., *et al.* (2009b). Increase in the rate and uniformity of coastline erosion in Arctic Alaska. *Geophysical Research Letters* **36**, L03503, doi: 03510.01029/02008gl036205.

Jones, B.M., Grosse, G., Arp, C.D., Jones, M.C., Anthony, K.M.W., *et al.* (2011). Modern thermokarst lake dynamics in the continuous permafrost zone, northern Seward Peninsula, Alaska. *Journal of Geophysical Research – Biogeosciences* **116**, G00M03, doi: 10.1029/2005jg000150.

Jones, B.M., Grosse, G., Hinkel, K.M., Arp, C.D., Walker, S., *et al.* (2012a). Assessment of pingo distribution and morphometry using an IfSAR derived digital surface model, western Arctic Coastal Plain, Northern Alaska. *Geomorphology* **138**, 1–14, doi: 10.1016/j.geomorph.2011.08.007.

Jones, M., Grosse, G., Jones, B.M. & Walter Anthony, K.M. (2012b). Peat accumulation in a thermokarst-affected landscape in continuous ice-rich permafrost, Seward Peninsula, Alaska. *Journal of Geophysical Research – Biogeosciences* **117**, G00M07, doi: 10.1029/2011JG001766.

Jorgenson, M.T. & Brown, J. (2005). Classification of the Alaskan Beaufort Sea Coast and estimation of carbon and sediment inputs from coastal erosion. *Geo-Marine Letters* **25**, 69–80.

Jorgenson, M.T., Racine, C.H., Walters, J.C. & Osterkamp, T.E. (2001). Permafrost degradation and ecological changes associated with a warming climate in Central Alaska. *Climatic Change* **48**, 551–579.

Jorgenson, M.T., Shur, Y.L. & Pullman, E.R. (2006). Abrupt increase in permafrost degradation in Arctic Alaska. *Geophysical Research Letters* **33**, L02503, doi: 02510.01029/02005gl024960.

Joughin, I.R., Kwok, R. & Fahnestok, M.A. (1999). Interferometric estimation of three-dimensional ice-flow using ascending and descending passes. *IEEE Transactions on Geoscience and Remote Sensing* **36**, 25–37.

Judge, J., Galantowicz, J.F., England, A.W. & Dahl, P. (1997). Freeze/thaw classification for prairie soils using SSM/I radiobrightnesses. *IEEE Transactions on Geoscience and Remote Sensing* **35**, 827–832.

Kääb, A. (2002). Monitoring high-mountain terrain deformation from air- and spaceborne optical data: examples using digital aerial imagery and ASTER data. *ISPRS Journal of Photogrammetry and Remote Sensing* **57**, 39–52.

Kääb, A. (2005). *Remote Sensing of Mountain Glaciers and Permafrost Creep.* Zurich, University of Zurich.

Kääb, A. (2008). Remote sensing of permafrost-related problems and hazards. *Permafrost and Periglacial Processes* **19**, 107–136.

Kääb, A. & Haeberli, W. (2001). Evolution of a high-mountain thermokarst lake in the Swiss Alps. *Arctic, Antarctic, and Alpine Research* **33**, 385–390.

Kääb, A. & Vollmer, M. (2000). Surface geometry, thickness changes and flow fields on creeping mountain permafrost: automatic extraction by digital image analysis. *Permafrost and Periglacial Processes* **11**, 315–326.

Kääb, A., Haeberli, W. & Gudmundsson, G.H. (1997). Analysing the creep of mountain permafrost using high precision aerial photogrammetry: 25 years of monitoring Gruben rock glacier, Swiss Alps. *Permafrost and Periglacial Processes* **8**, 409–426.

Kääb, A., Huggel, C., Paul, F., Wessels, R., Raup, B., *et al.* (2003). Glacier monitoring from ASTER imagery: accuracy and applications. *EARSel eProceedings* **2**, 43–53.

Kääb, A., Huggel, C., Fischer, L., Guex, S., Paul, F., *et al.* (2005). Remote sensing of glacier- and permafrost-related hazards in high mountains: an overview. *Natural Hazards and Earth System Sciences* **5**, 527–554.

Kaufmann, V. & Ladstädter, R. (2002). Spatio-temporal analysis of the dynamic behaviour of the Hochebenkar rock glaciers (Oetztal Alps, Austria) by means of digital photogrammetric methods. *Grazer Schriften der Geographie und Raumforschung* **37**, 119–140.

Kerr, Y.H., Waldteufel, P., Wigneron, J.P., Martinuzzi, J.M., Font, J., *et al.* (2001). Soil moisture retrieval from space: The Soil Moisture and Ocean Salinity (SMOS) mission. *IEEE Transactions on Geoscience and Remote Sensing* **39**, 1729–1735.

Kim, S., van Zyl, J., McDonald, K. & Njoku, E. (2010). *Monitoring surface soil moisture and freeze-thaw state with the high-resolution radar of the Soil Moisture Active/Passive (SMAP) mission.* 2010 IEEE Radar Conference, 735–739.

Kimball, J.S., McDonald, K.C., Keyser, A.R., Frolking, S. & Running, S.W. (2001). Application of the NASA scatterometer (NSCAT) for determining the daily frozen and nonfrozen landscape of Alaska. *Remote Sensing of Environment* **75**, 113–126.

Kimball, J.S., McDonald, K.C., Frolking, S. & Running, S.W. (2004). Radar remote sensing of the spring thaw transition across a boreal landscape. *Remote Sensing of Environment* **89**, 163–175.

Kokelj, S.V., Lantz, T.C., Kanigan, J., Smith, S.L. & Coutts, R. (2009). Origin and Polycyclic Behaviour of Tundra Thaw Slumps, Mackenzie Delta Region, Northwest Territories, Canada. *Permafrost and Periglacial Processes* **20**, 173–184.

Kontu, A., Lemmetyinen, J., Pulliainen, J., Rautiainen, K., Kainulainen, J., *et al.* (2010). *L-Band Measurements of Boreal Soil.* 2010 IEEE International Geoscience and Remote Sensing Symposium, 706–709.

Kravtsova, V.I. & Bystrova, A.G. (2009). Changes in thermokarst lake size in different regions of Russia for the last 30 years. *KriosferaZemli* **13**, 16–26.

Kravtsova, V.I. & Tarasenko, T.V. (2011). The dynamics of thermokarst lakes under climate changes since 1950, Central Yakutia. *KriosferaZemli* **15**, 31–42.

Krieger, G., Hajnsek, I., Papathanassiou, K., Eineder, M., Younis, M., *et al.* (2009). *The Tandem-L Mission Proposal: Monitoring Earth's Dynamics with High Resolution SAR Interferometry.* 2009 IEEE Radar Conference, Vols 1 and 2, 94–99.

Laberge, M.J. & Payette, S. (1995). Long-Term Monitoring of Permafrost Change in a Palsa Peatland in Northern Quebec, Canada – 1983–1993. *Arctic and Alpine Research* **27**, 167–171.

Lacelle, D., Bjornson, J. & Lauriol, B. (2010). Climatic and Geomorphic Factors Affecting Contemporary (1950–2004) Activity of Retrogressive Thaw Slumps on the Aklavik Plateau, Richardson Mountains, NWT, Canada. *Permafrost and Periglacial Processes* **21**, 1–15.

Landerer, F.W., Dickey, J.O. & Guntner, A. (2010). Terrestrial water budget of the Eurasian pan-Arctic from GRACE satellite measurements during 2003–2009. *Journal of Geophysical Research-Atmospheres* **115**, D23115, doi: 23110.21029/22010jd014584.

Langer, M., Westermann, S. & Boike, J. (2010). Spatial and temporal variations of summer surface temperatures of wet polygonal tundra in Siberia – implications for MODIS LST based permafrost monitoring. *Remote Sensing of Environment* **114**, 2059–2069.

Langer, M., Westermann, S., Heikenfeld, M., Dorn, W., Boike, J. (2013). Satellite-based modeling of permafrost temperatures in a tundra lowland landscape. *Remote Sensing of Environment*, **135**, 12–24, 2013.

Lantuit, H. & Pollard, W.H. (2005). Temporal stereophotogrammetric analysis of retrogressive thaw slumps on Herschel Island, Yukon Territory. *Natural Hazards and Earth System Sciences* **5**, 413–423.

Lantuit, H. & Pollard, W.H. (2008). Fifty years of coastal erosion and retrogressive thaw slump activity on Herschel Island, southern Beaufort Sea, Yukon Territory, Canada. *Geomorphology* **95**, 84–102.

Lantuit, H., Atkinson, D., Overduin, P.P., Grigoriev, M., Rachold, V., *et al.* (2011). Coastal erosion dynamics on the permafrost-dominated Bykovsky Peninsula, north Siberia, 1951–2006. *Polar Research* **30**, 7341, doi: 10.3402/Polar.V3430i3400.7341.

Lantz, T.C. & Kokelj, S.V. (2008). Increasing rates of retrogressive thaw slump activity in the Mackenzie Delta region, NWT, Canada. *Geophysical Research Letters* **35**, L06502, doi: 10.1029/2007gl032433.

Larsen, P.H., Goldsmith, S., Smith, O., Wilson, M., Strzepek, K., *et al.* (2007). *Estimating future costs for Alaska public infrastructure at risk from climate change.* Project Report, Institute of Social and Economic Research, University of Alaska, Anchorage, 108.

Lawrence, D.M., Slater, A.G., Romanovsky, V.E. & Nicolsky, D.J. (2008). Sensitivity of a model projection of near-surface permafrost degradation to soil column depth and representation of soil organic matter. *Journal of Geophysical Research* **113**, F02011, doi: 02010.01029/02007JF000883.

Leverington, D.W. & Duguay, C.R. (1996). Evaluation of three supervised classifiers in mapping "depth to late-summer frozen ground", central Yukon Territory. *Canadian Journal of Remote Sensing* **22**, 163–174.

Leverington, D.W. & Duguay, C.R. (1997). A neural network method to determine the presence or absence of permafrost near Mayo, Yukon Territory, Canada. *Permafrost and Periglacial Processes* **8**, 207–217.

Lewkowicz, A.G. & Duguay, C.R. (1999). Detection of Permafrost Features Using SPOT Panchromatic Imagery, Fosheim Peninsula, Ellesmere Island, NWT. *Canadian Journal of Remote Sensing* **25**, 33–44.

Lewkowicz, A.G. & Harris, C. (2005a). Morphology and geotechnique of active-layer detachment failures in discontinuous and continuous permafrost, northern Canada. *Geomorphology* **69**, 275–297.

Lewkowicz, A.G. & Harris, C. (2005b). Frequency and magnitude of active-layer detachment failures in discontinuous and continuous permafrost, northern Canada. *Permafrost and Periglacial Processes* **16**, 115–130.

Li, Z., Xu, Z., Cen, M. & Ding, X. (2001). Robust surface matching for automated detection of local deformations using least-median-of-squares estimator. *Photogrammetric Engineering and Remote Sensing* **67**, 1283–1292.

Liu, L., Zhang, T. & Wahr, J. (2010). InSAR measurements of surface deformation over permafrost on the North Slope of Alaska. *Journal of Geophysical Research* **115**, F03023, doi: 03010.01029/02009jf001547.

Liu, L., Schaefer, K., Zhang, T.J. & Wahr, J. (2012). Estimating 1992–2000 average active layer thickness on the Alaskan North Slope from remotely sensed surface subsidence. *Journal of Geophysical Research – Earth Surface* **117**, F01005, doi: 01010.01029/02011jf002041.

Luojus, K., Pulliainen, J., Takala, M., Lemmetyinen, J., Derksen, C., Metsämäki, S. & Bojkov, B. (2011). *Investigating hemispherical trends in snow accumulation using GlobSnow snow water equivalent data*. Geoscience and Remote Sensing Symposium (IGARSS), 2011 IEEE International, 3772–3774, doi: 10.1109/IGARSS.2011.6050051.

Lutz, E., Geist, T. & Stötter, J. (2003). *Investigations of airborne laser scanning signal intensity on glacial surfaces – Utilizing comprehensive laser geometry modelling and orthophoto surface modelling (A case study: Svartisheibreen, Norway)*. Proceedings of the ISPRS Workshop on 3-D Reconstruction from Airborne Laserscanner and INSAR Data, Dresden, Germany, 2003.

Marchenko, S., Hachem, S., Romanovsky, V. & Duguay, C. (2009). Permafrost and Active Layer Modeling in the Northern Eurasia using MODIS Land Surface Temperature as an input data. *Geophysical Research Abstracts* **11**, EGU2009–11077.

Mars, J.C. & Houseknecht, D.W. (2007). Quantitative remote sensing study indicates doubling of coastal erosion rate in past 50 yr along a segment of the Arctic coast of Alaska. *Geology* **35**, 583–586.

Marsh, P., Russell, M., Pohl, S., Haywood, H. & Onclin, C. (2009). Changes in thaw lake drainage in the Western Canadian Arctic from 1950 to 2000. *Hydrological Processes* **23**, 145–158.

McDonald, K.C., Kimball, J.S., Njoku, E., Zimmermann, R. & Zhao, M.S. (2004). Variability in Springtime Thaw in the Terrestrial High Latitudes: Monitoring a Major Control on the Biospheric Assimilation of Atmospheric CO_2 with Spaceborne Microwave Remote Sensing. *Earth Interactions* **8**, 1–23.

McGraw, M. (2008). *The Degradation of Ice Wedges in the Colville River Delta and Their Role in Pond Drainage*. Proceedings of the Ninth International Conference on Permafrost, Fairbanks, USA, 2008.

Minsley, B.J., Abraham, J.D., Smith, B.D., Cannia, J.C., Voss, C.I., *et al.* (2012). Airborne electromagnetic imaging of discontinuous permafrost. *Geophysical Research Letters* **39**, L02503, doi: 02510.01029/02011gl050079.

Morgenstern, A., Grosse, G., Gunther, F., Fedorova, I. & Schirrmeister, L. (2011). Spatial analyses of thermokarst lakes and basins in Yedoma landscapes of the Lena Delta. *The Cryosphere* **5**, 849–867, doi: 10.5194/tc-5-849-2011.

Muskett, R.R. & Romanovsky, V.E. (2009). Groundwater storage changes in arctic permafrost watersheds from GRACE and *in situ* measurements. *Environmental Research Letters* **4**, 045009, doi: 10.1088/1748-9326/1084/1084/045009.

Muskett, R.R. & Romanovsky, V.E. (2011). Alaskan permafrost groundwater storage changes derived from GRACE and ground measurements. *Remote Sensing* **3**, 378–397.

Nagler, T., Mayer, C. & Rott, H. (2002). *Feasibility of DInSAR for mapping complex motion fields of Alpine ice- and rock-glaciers.* Proceedings of the Third International Symposium on Retrieval of Bio- and Geophysical Parameters from SAR Data for Land Application, Sheffield, UK, 11–14 September 2001.

Nelson, F., Shiklomanov, N., Hinkel, K. & Brown, J. (2008). *Decadal results from the Circumpolar Active Layer Monitoring (CALM) Program.* Proceedings of the Ninth International Conference on Permafrost, Fairbanks, USA, 2008.

Nguyen, T.N., Burn, C.R., King, D.J. & Smith, S.L. (2009). Estimating the Extent of Near-surface Permafrost using Remote Sensing, Mackenzie Delta, Northwest Territories. *Permafrost and Periglacial Processes* **20**, 141–153.

Nolan, M. & Prokein, P. (2003). *Evaluation of a new DEM of the Putuligayuk watershed for Arctic hydrologic applications.* Proceedings of the Eighth International Conference on Permafrost, Zurich, Switzerland, 2003.

Nuth, C. & Kääb, A. (2011). Co-registration and bias corrections of satellite elevation data sets for quantifying glacier thickness change. *The Cryosphere* **5**, 271–290.

Osterkamp, T.E., Jorgenson, M.T., Schuur, E.A.G., Shur, Y.L., Kanevskiy, M.Z., *et al.* (2009). Physical and Ecological Changes Associated with Warming Permafrost and Thermokarst in Interior Alaska. *Permafrost and Periglacial Processes* **20**, 235–256.

Otto, J.C., Kleinod, K., Konig, O., Krautblatter, M., Nyenhuis, M., *et al.* (2007). HRSC-A data: a new high-resolution data set with multipurpose applications in physical geography. *Progress in Physical Geography* **31**, 179–197.

Overland, J.E., Wang, M. & Salo, S. (2008). The recent Arctic warm period. *Tellus A* **60**, 589–597.

Panda, S.K., Prakash, A., Solie, D.N., Romanovsky, V.E. & Jorgenson, M.T. (2010). Remote Sensing and Field-based Mapping of Permafrost Distribution along the Alaska Highway Corridor, Interior Alaska. *Permafrost and Periglacial Processes* **21**, 271–281.

Park, S.E., Bartsch, A., Sabel, D., Wagner, W., Naeimi, V., *et al.* (2011). Monitoring freeze/thaw cycles using ENVISAT ASAR Global Mode. *Remote Sensing of Environment* **115**, 3457–3467.

Parker, W.B. (2001). *Effect of permafrost changes on economic development, environmental security and natural resource potential in Alaska*. Permafrost response on economic development, environmental security and natural resources, NATO Science Series Volume 76, 2001, pp 293–296.

Payette, S., Delwaide, A., Caccianiga, M. & Beauchemin, M. (2004). Accelerated thawing of subarctic peatland permafrost over the last 50 years. *Geophysical Research Letters* **31**, L18208, doi: 18210.11029/12004gl020358.

Peddle, D.R. & Franklin, S.E. (1992). Multisource evidential classification of surface cover and frozen ground. *International Journal of Remote Sensing* **13**, 3375–3380.

Peddle, D.R. & Franklin, S.E. (1993). Classification of Permafrost Active Layer Depth from Remotely Sensed and Topographic Evidence. *Remote Sensing of Environment* **44**, 67–80.

Pilgrim, L.J. (1996). Surface matching and difference detection without the aid of control points. *Survey Review* **33**, 291–303.

Plug, L.J., Walls, C. & Scott, B.M. (2008). Tundra lake changes from 1978 to 2001 on the Tuktoyaktuk Peninsula, western Canadian Arctic. *Geophysical Research Letters* **35**, L03502, doi: 10.1029/2007gl032303.

Pollard, W.H. & French, H.M. (1980). A first approximation of the volume of ground ice, Richards Island, Pleistocene Mackenzie delta, Northwest Territories, Canada. *Canadian Geotechnical Journal* **17**, 509–516.

Prowse, T.D., Furgal, C., Chouinard, R., Melling, H., Milburn, D., *et al.* (2009). Implications of climate change for economic development in Northern Canada: energy, resource, and transportation sectors. *AMBIO: A Journal of the Human Environment* **38**, 272–281.

Quinton, W.L., Hayashi, M. & Chasmer, L.E. (2011). Permafrost-thaw-induced land-cover change in the Canadian subarctic: implications for water resources. *Hydrological Processes* **25**, 152–158.

Rabus, B., Eineder, M., Roth, A. & Bamler, R. (2003). The shuttle radar topography mission – a new class of digital elevation models acquired by spaceborne radar. *ISPRS Journal of Photogrammetry and Remote Sensing* **57**, 241–262.

Regmi, P., Grosse, G., Jones, M.C., Jones, B.M. & Anthony, K.W. (2012). Characterizing Post-Drainage Succession in Thermokarst Lake Basins on the Seward Peninsula, Alaska with TerraSAR-X Backscatter and LANDSAT-based NDVI Data. *Remote Sensing* **4**, 3741–3765, doi: 10.3390/rs4123741.

Renouard, L., Perlant, F. & Nonin, P. (1995). Comparison of DEM generation from SPOT stereo and ERS interferometric SAR data. *EARSeL Advances in Remote Sensing – Topography from Space* **4**, 103–109.

Riordan, B., Verbyla, D. & McGuire, A.D. (2006). Shrinking ponds in subarctic Alaska based on 1950–2002 remotely sensed images. *Journal of Geophysical Research–Biogeosciences* **111**, G04002, doi: 10.1029/2005jg000150.

Riseborough, D., Shiklomanov, N., Etzelmueller, B., Gruber, S. & Marchenko, S. (2008). Recent advances in permafrost modeling. *Permafrost and Periglacial Processes* **19**, 137–156.

Roach, J., Griffith, B., Verbyla, D. & Jones, J. (2011). Mechanisms influencing changes in lake area in Alaskan boreal forest. *Global Change Biology* **17**, 2567–2583.

Romanovsky, V.E., Marchenko, S.S., Daanen, R., Sergeev, D.O. & Walker, D.A. (2008). *Soil climate and frost heave along the Permafrost/Ecological North American Arctic Transect.* Proceedings of the Ninth International Conference on Permafrost, Fairbanks, USA, 2008.

Romanovsky, V.E., Smith, S.L. & Christiansen, H.H. (2010). Permafrost Thermal State in the Polar Northern Hemisphere during the International Polar Year 2007–2009: a Synthesis. *Permafrost and Periglacial Processes* **21**, 106–116.

Romanovsky, V.E., Smith, S.L., Christiansen, H.H., Shiklomanov, N.I., Streletskiy, D.A., Drozdov, D.S., Oberman, N.G., Kholodov, A.L., Marchenko, S.S. (2012): *NOAA Artic Report Card – Permafrost.* November 9, 2012. http://www.arctic .noaa.gov/reportcard/permafrost.html.

Rott, H. & Siegel, A. (1999). *Analysis of mass movement in alpine terrain by means of SAR interferometry.* Proceedings of the IEEE Geoscience and Remote Sensing Symposium, IGARSS'99, Hamburg, 1999.

Rykhus, R.P. & Lu, Z. (2008). InSAR detects possible thaw settlement in the Alaskan Arctic Coastal Plain. *Canadian Journal of Remote Sensing* **34**, 100–112.

Salzmann, N., Kääb, A., Huggel, C., Allgöwer, B. & Haeberli, W. (2004). Assessment of the hazard potential of ice avalanches using remote sensing and GIS-modelling. *Norwegian Journal of Geography* **58**, 74–84.

Sannel, A.B.K. & Brown, I.A. (2010). High-resolution remote sensing identification of thermokarst lake dynamics in a subarctic peat plateau complex. *Canadian Journal of Remote Sensing* **36**, S26–S40.

Sannel, A.B.K. & Kuhry, P. (2011). Warming-induced destabilization of peat plateau/thermokarst lake complexes. *Journal of Geophysical Research – Biogeosciences* **116**, G03035, doi: 10.1029/2010jg001635.

Schaefer, K., Zhang, T.J., Bruhwiler, L. & Barrett, A.P. (2011). Amount and timing of permafrost carbon release in response to climate warming. *Tellus Series B – Chemical and Physical Meteorology* **63**, 165–180.

Schuur, E., Bockheim, J., Canadell, J., Euskirchen, E., Field, C., *et al.* (2008). Vulnerability of permafrost carbon to climate change: Implications for the global carbon cycle. *Bioscience* **58**, 701–714.

Schuur, E.A.G., Vogel, J.G., Crummer, K.G., Lee, H., Sickman, J.O., *et al.*, (2009). The effect of permafrost thaw on old carbon release and net carbon exchange from tundra. *Nature* **459**, 556–559.

Sellmann, P.V., Brown, J., Lewellen, R.I., McKim, H.L. & Merry, C.J. (1975). *The classification and geomorphic implications of thaw lakes on the arctic coastal plain, Alaska.* DTIC Document, U.S. Army Cold Regions Research and Engineering Laboratory, Hanover, New Hampshire, 21.

Serreze, M.C., Walsh, J.E., Chapin, F.S., Osterkamp, T., Dyurgerov, M., *et al.* (2000). Observational evidence of recent change in the northern high-latitude environment. *Climatic Change* **46**, 159–207.

Shiklomanov, N., Nelson, F., Streletskiy, D., Hinkel, K. & Brown, J. (2008). *The Circumpolar Active Layer Monitoring (CALM) program: Data collection, management, and dissemination strategies.* Proceedings of the Ninth International Conference on Permafrost, Fairbanks, USA, 2008.

Singhroy, V., Couture, R., Alasset, P.J. & Poncos, V. (2007). *InSAR monitoring of landslides on permafrost terrain in Canada.* Igarss: 2007 IEEE International Geoscience and Remote Sensing Symposium, Vols. 1–12, 2451–2454.

Sjogren, D.B., Moorman, B.J. & Vachon, P.W. (2003). *The importance of temporal scale when mapping landscape change in permafrost environments using Interferometric Synthetic Aperture Radar.* Proceedings of the Eighth International Conference on Permafrost, Zurich, Switzerland, 2003.

Smith, L.C., Sheng, Y., MacDonald, G.M. & Hinzman, L.D. (2005). Disappearing Arctic lakes. *Science* **308**, 1429–1429.

Smith, N.V., Saatchi, S.S. & Randerson, J.T. (2004). Trends in high northern latitude soil freeze and thaw cycles from 1988 to 2002. *Journal of Geophysical Research – Atmospheres* **109**, D12101, doi: 10.1029/2003jd004472.

Solomon, S.M. (2005). Spatial and temporal variability of shoreline change in the Beaufort-Mackenzie region, northwest territories, Canada. *Geo-Marine Letters* **25**, 127–137.

St Jacques, J.M. & Sauchyn, D.J. (2009). Increasing winter baseflow and mean annual streamflow from possible permafrost thawing in the Northwest Territories, Canada. *Geophysical Research Letters* **36**, L12402, doi: 12410.11029/12007gl030216.

Stebler, O., Schwerzmann, A., Luthi, M., Meier, E. & Nuesch, D. (2005). Pol-InSAR observations from an Alpine glacier in the cold infiltration zone at L- and P-band. *IEEE Geoscience and Remote Sensing Letters* **2**, 357–361.

Stockdon, H.F., Sallenger, A.H., List, J.H. & Holman, R.A. (2002). Estimation of shoreline position and change using airborne topographic LiDAR data. *Journal of Coastal Research* **18**, 502–513.

Strozzi, T., Wegmüller, U., Tosi, L., Bitelli, G. & Spreckels, V. (2001). Land subsidence monitoring with differential SAR interferometry. *Photogrammetric Engineering and Remote Sensing* **67**, 1261–1270.

Strozzi, T., Kääb, A. & Frauenfelder, R. (2004). Detecting and quantifying mountain permafrost creep from *in situ*, airborne and spaceborne remote sensing methods. *International Journal of Remote Sensing* **25**, 2919–2931.

Syed, T.H., Famiglietti, J.S., Zlotnicki, V. & Rodell, M. (2007). Contemporary estimates of Pan-Arctic freshwater discharge from GRACE and reanalysis. *Geophysical Research Letters* **34**, L19404, doi: 19410.11029/12007gl031254.

Tarnocai, C. & Kristof, S. (1976). Computer-aided classification of land and water bodies using LANDSAT data, Mackenzie Delta area, NWT, Canada. *Arctic and Alpine Research* **8**, 151–159.

Tarnocai, C., Canadell, J.G., Schuur, E. A. G., Kuhry, P., Mazhitova, G., *et al.* (2009). Soil organic carbon pools in the northern circumpolar permafrost region. *Global Biogeochemical Cycles* **23**, 2023, doi: 2010.1029/2008gb003327.

Toutin, T. (1995). Generating DEM from stereo images with a photogrammetric approach: examples with VIR and SAR data. *EARSeL Journal Advances in Remote Sensing* **4**, 110–117.

Toutin, T. (2002). DEM from stereo LANDSAT 7 ETM data over high relief areas. *International Journal of Remote Sensing* **23**, 2133–2139.

Ulrich, M., Hauber, E., Herzschuh, U., Hartel, S. & Schirrmeister, L. (2011). Polygon pattern geomorphometry on Svalbard (Norway) and western Utopia Planitia (Mars) using high-resolution stereo remote-sensing data. *Geomorphology* **134**, 197–216.

Vachon, P.W., Geudtner, D., Gray, A.L., Mattar, K., Brugman, M., *et al.* (1996). *Airborne and Spaceborne SAR Interferometry: Application to the Athabasca Glacier Area.* Proceedings of the IGARSS '96, Lincoln, Nebraska.

Vallee, S. & Payette, S. (2007). Collapse of permafrost mounds along a subarctic river over the last 100 years (northern Quebec). *Geomorphology* **90**, 162–170.

van Asselen, S. & Seijmonsbergen, A.C. (2006). Expert-driven semi-automated geomorphological mapping for a mountainous area using a laser DTM. *Geomorphology* **78**, 309–320.

Vey, S., Steffen, H., Müller, J. & Boike, J. (2012). Inter-annual water mass variations from GRACE in central Siberia. *Journal of Geodesy*, doi: 10.1007/s00190-012-0597-9.

Walter, K., Zimov, S., Chanton, J.P., Verbyla, D. & Chapin III, F. (2006). Methane bubbling from Siberian thaw lakes as a positive feedback to climate warming. *Nature* **443**, 71–75.

Walvoord, M.A. & Striegl, R.G. (2007). Increased groundwater to stream discharge from permafrost thawing in the Yukon River basin: Potential impacts on lateral export of carbon and nitrogen. *Geophysical Research Letters* **34**, L12402, doi: 12410.11029/12007gl030216.

Wang, J., Sheng, Y., Hinkel, K.M. & Lyons, E.A. (2011). Drained thaw lake basin recovery on the western Arctic Coastal Plain of Alaska using high-resolution digital elevation models and remote sensing imagery. *Remote Sensing of Environment* **19**, 325–336, doi: 10.1016/j.rse.2011.1010.1027.

Wang, Z. & Li, S. (1999). *Detection of winter frost heaving of the active layer of arctic permafrost using SAR differential interferometry.* Proceedings of the IEEE Geoscience and Remote Sensing Symposium, Hamburg, Germany, 1999.

Wessels, R., Kargel, J.S. & Kieffer, H.H. (2002). ASTER measurement of supraglacial lakes in the Mount Everest region of the Himalaya. *Annals of Glaciology* **34**, 399–408.

Westermann, S., Langer, M. & Boike, J. (2011). Spatial and temporal variations of summer surface temperatures of high-arctic tundra on Svalbard – Implications for MODIS LST based permafrost monitoring. *Remote Sensing of Environment* **115**, 908–922.

Westermann, S., Langer, M. & Boike, J. (2012). Systematic bias of average winter-time land surface temperatures inferred from MODIS at a site on Svalbard, Norway. *Remote Sensing of Environment* **118**, 162–167.

Westermann, S., Schuler, T.V., Gisnås, K. & Etzelmüller, B. (2013). Transient thermal modeling of permafrost conditions in Southern Norway. *The Cryosphere*, 7, 719–739.

Weydahl, D.J. (2001). Analysis of ERS tandem SAR coherence from glaciers, valleys, and Fjord Ice on Svalbard. *IEEE Transactions on Geoscience and Remote Sensing* **39**, 2029–2039.

Wismann, V. (2000). Monitoring of seasonal thawing in Siberia with ERS scatterometer data. *IEEE Transactions on Geoscience and Remote Sensing* **38**, 1804–1809.

WMO (2010). GCOS: Global Climate Observing System, http://gcos.wmo.int.

Yoshikawa, K. & Hinzman, L.D. (2003). Shrinking thermokarst ponds and groundwater dynamics in discontinuous permafrost near Council, Alaska. *Permafrost and Periglacial Processes* **14**, 151–160.

Zhang, T., Armstrong, R.L. & Smith, J. (2003a). Investigation of the near-surface soil freeze-thaw cycle in the contiguous United States: Algorithm development and validation. *Journal of Geophysical Research – Atmospheres* **108**, 8860, doi: 10.1029/2003jd003530.

Zhang, T., Barry, R.G., Knowles, K., Ling, F. & Armstrong, R.L. (2003b). Distribution of seasonally and perennially frozen ground in the Northern Hemisphere. *Permafrost, Vols 1 and 2*, 1289–1294.

Zhang, T., Barry, R.G. & Armstrong, R.L. (2004). Application of satellite remote sensing techniques to frozen ground studies. *Polar Geography* **28**, 163–196.

Zhang, Y., Chen, W.J. & Cihlar, J. (2003c). A process-based model for quantifying the impact of climate change on permafrost thermal regimes. *Journal of Geophysical Research–Atmospheres* **108**, 4695, doi: 10.1029/2002jd003354.

Zhao, T.J., Zhang, L.X., Jiang, L.M., Zhao, S.J., Chai, L.N., *et al.* (2011). A new soil freeze/thaw discriminant algorithm using AMSR-E passive microwave imagery. *Hydrological Processes* **25**, 1704–1716.

Zimov, S.A., Schuur, E.A.G. & Chapin III, F.S. (2006). Permafrost and the global carbon budget. *Science* **312**, 1612–1613.

Acronyms

AATSR	Advanced Along-Track Scanning Radiometer
ALOS PRISM	Panchromatic Remote-sensing Instrument for Stereo Mapping
ASTER	Advanced Spaceborne Thermal Emission and Reflection Radiometer
AVHRR	Advanced Very High Resolution Radiometer
CALM	Circumpolar Active Layer Monitoring
DEMs	digital elevation models
D-InSAR	Differential InSAR
Envisat	Environmental Satellite

ERS-1/2	European remote sensing satellite-1/2
ERTS-1	Earth Resources Technology Satellite - LANDSAT-1
ESA DUE	Permafrost project European Space Agency, Data User Element permafrost project
GCMs	General circulation models
GCOS	Global Climate Observing System
GDEM	Global Digital Elevation Map
GRACE	Gravity Recovery and Climate Experiment http://www.csr.utexas.edu/grace/
HRSC-AX	High Resolution Stereo Camera
InSAR	Interferometric synthetic aperture radar
IPCC	Intergovernmental Panel on Climate Change
LIDAR	Airborne laser scanning (light detection and ranging)
LST	Land surface temperature
MODIS	Moderate Resolution Imaging Spectroradiometer
SAR	Synthetic aperture radar
SMAP	Soil Moisture Active Passive mission
SMOS	Soil Moisture and Ocean Salinity mission
SPOT-5	Système Pour l'Observation de la Terre (System for Earth Observation)
SRTM	Shuttle Radar Topography Mission
TSP	Thermal State of Permafrost program

14 Field measurements for remote sensing of the cryosphere

Hans-Peter Marshall[1], Robert L. Hawley[2] and Marco Tedesco[3]

[1] Boise State University, USA
[2] Dartmouth College, Hanover, USA
[3] The City College of New York, City University of New York, New York, USA

Summary

Remote sensing observations of the cryosphere, like any other target of interest, require ground-based measurements for both calibration and validation, as inversion algorithms are usually underdetermined and uncertainties in the retrieval are needed for application. Field-based observations are performed in selected representative locations, and typically involve both direct *in situ* measurements of the physical properties of interest, as well as ground-based remote sensing techniques.

New state-of-the-art modern techniques for measuring physical properties rapidly and at high spatial resolution have recently given us a new view of spatiotemporal variability. These are important, as large variability at scales below the typical footprint of spaceborne sensors often exists. Simulating remote sensing measurements using ground-based sensors provides the ability to perform both *in situ* and remote sensing measurements at the same scale, providing insight into the dominant physical processes that must be accounted for in inversion models and retrieval schemes.

While direct *in situ* measurements provide the most accurate information about the properties of interest, they are time-consuming and expensive and are, therefore, only practical at relatively few locations, and often with low temporal resolution. Spatial sampling strategies, designed specifically for the remote sensing observation of interest, can reduce uncertainties in comparisons between ground-based and airborne/spaceborne estimates. Intensive remote sensing calibration and validation campaigns, often associated with an upcoming or recent satellite launch, provide unique opportunities for detailed characterization at a wide range of scales, and these are typically large international collaborative efforts.

This chapter reviews standard *in situ* manual field measurements for snow and ice properties, as well as newer high-resolution techniques and instruments used to simulate airborne and spaceborne remote sensing observations. Sampling strategies and example applications from recent international calibration and validation experiments are given.

Remote Sensing of the Cryosphere, First Edition. Edited by M. Tedesco.
© 2015 John Wiley & Sons, Ltd. Published 2015 by John Wiley & Sons, Ltd.
Companion Website: www.wiley.com/go/tedesco/cryosphere

Field measurements are a crucial component of remote sensing of the cryosphere, as they provide both the necessary direct observations of the variables of interest, as well as measurements that simulate the particular remote sensing technique at scales that can be characterized accurately. Ground-based observations provide the information needed to:

1 improve and develop new retrieval algorithms;
2 calibrate algorithms; and
3 validate results to provide accurate uncertainty assessments.

14.1 Introduction

Remote sensing observations of the cryosphere, like any other target of interest, require ground-based measurements for calibration and validation (Figure 14.1). For example, many remote sensing inversion algorithms require estimates of surface and subsurface properties for calibration, and field-based observations are consequently performed in locations assumed to be as representative as possible of the target studied from space. Moreover, to evaluate the accuracy and limitations of remote sensing results, direct measurements of the physical properties of interest are required and performed in the field. Such field measurements are typically made manually by observers, or by automatic instrumentation.

This chapter begins with an overview of the physical properties of the cryosphere that are of interest and have a significant effect on remote sensing observations. We then cover techniques for measuring these properties, beginning with established standard techniques. We also describe new state-of-the-art modern techniques for measuring physical properties rapidly and at high spatial resolution, which have recently given us a new view of the variability of the sought parameters at different spatial scales, and often reveal large variability at scales below the typical footprint of spaceborne sensors.

A major complication with relating ground-based point measurements of physical properties to remote sensing observations is the mismatch in scale.

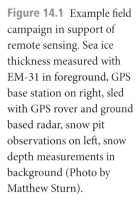

Figure 14.1 Example field campaign in support of remote sensing. Sea ice thickness measured with EM-31 in foreground, GPS base station on right, sled with GPS rover and ground based radar, snow pit observations on left, snow depth measurements in background (Photo by Matthew Sturn).

Indeed, an individual *in situ* measurement can be seen as representative of physical properties at a scale of the order of a few meters or even centimeters, while remote sensing observations are affected by areas and targets within the field of view of the instruments, ranging from several meters to tens, or even hundreds of kilometers. Simulating remote sensing measurements using ground-based sensors provides the ability to perform both measurements at the same scale, also providing insight into the dominant physical processes that must be accounted for in inversion models and retrieval schemes.

In order to relate ground-based measurements of physical properties to actual remote sensing observations at larger scales, sampling strategies must be developed to adequately cover the scale of the remote sensor footprint. These sampling strategies are discussed, and the chapter concludes with examples of field-based calibration and validation efforts made in recent years.

14.2   Physical properties of interest

It is obviously impossible to cover all the physical properties of interest for remote sensing of the cryosphere in a single chapter. In this section, we highlight some of the most important ones, hoping that this represents a valuable introduction for the reader interested in expanding his/her knowledge through the references reported below.

Remote sensing observations can be used to estimate a wide range of the physical properties of snow and ice, using both passive and active electromagnetic sensors and measurements of the Earth's gravitational field (http://climate2 .jpl.nasa.gov/ice/missions/; See also Chapters 10 and 15, for example). High-frequency electromagnetic waves can accurately probe the upper layers of firn (e.g., snow that has been left over from past seasons and has been recrystallized) and snow at high spatial resolution (Marshall and Koh, 2008), while lower frequencies can be used to probe the entire thickness of glaciers and ice sheets (Conway *et al.*, 2009). At a very large spatial scale, variations in the gravitational field can be used to estimate changes in the mass of snow and ice.

Any remote sensing estimate requires independent estimates of the physical property of interest for development and calibration. The most accurate estimates of these physical properties come from direct, ground-based *in situ* observations. These observations are usually destructive, time-consuming, and costly, and therefore can only be made at very limited temporal and spatial resolutions.

Geophysical instruments, which measure electromagnetic or acoustic waves, can be used to estimate properties from the ground at high spatial resolution and over much larger spatial scales than direct *in situ* measurements, and they improve the accuracy of estimates scaled from point observations to remote sensing footprints. These techniques are usually non-destructive. Sensors with the same characteristics as spaceborne instruments are also used on the ground and, due to their small footprint, the estimates deriving from such measurements can be more easily assessed and validated compared to *in situ* observations.

14.2.1   Surface properties

Accurate measurements of the elevation of snow and ice are used to monitor changes in mass of ice sheets (Abdalati and Krabill, 1999; Howat *et al.*, 2007), glaciers (Larsen *et al.*, 2007; Paul and Haeberli, 2008), and seasonal snowpacks (Deems *et al.*, 2006; Trujillo *et al.*, 2007). With estimates of elevation at many locations and at a range of scales, other topographic parameters can be calculated, including slope, aspect, and roughness. These are all important for estimating the energy balance for modeling, for example, melt and wind redistribution of snow. Slope estimates are required for ice flow modeling and calculating shear stress for avalanche studies, and surface roughness has a strong effect on passive and active microwave remote sensing, due to scattering of electromagnetic waves.

The ratio of outgoing to incoming solar radiation, termed *albedo*, is one of the most important parameters needed for energy balance modeling. Uncertainties in albedo often dominate model estimates of melt, so direct measurements are extremely valuable. Albedo measurements are usually produced as an integrated result over the shortwave portion of the electromagnetic spectrum. However, more accurate spectral measurements, which produce albedo estimates at many different wavelengths, are more useful, as they can provide validation and calibration data for the different channels at which spaceborne or airborne sensors collect information.

Surface temperature estimates can determine if snow or ice is melting and are the primary control on emitted radiation. The grain size of snow on the surface changes on a daily timescale and has a strong effect on albedo. The composition of the surface which, for the cryosphere, could be snow or ice, possibly mixed with water, dust or dirt, is also of interest. These observations can be used to indicate whether melt is occurring, and impurities are important as they can reduce the albedo drastically (e.g., darker material over the brighter snow and ice) and change the amount of solar radiation absorbed by the snowpack. Permafrost extent and depth would also be very valuable, but are still difficult to estimate with remote sensing (See Chapter 13).

Ice velocity is an important surface measurement for glaciers and ice sheets. These observations are used for ice flow modeling, for mass balance studies, and for monitoring the stability of ice (see Chapter 9). The height of sea ice above the water surface, called *sea ice freeboard*, is also an important surface observation, as it can be used to estimate sea ice thickness (see Chapter 11).

14.2.2   Sub-surface properties

Ground-based measurements of snow and ice properties below the surface are more difficult, as they require probing the sub-surface either destructively (e.g., digging snow pits) or using some type of short-range remote sensing. Destructive measurements are time-consuming and expensive, while remote observations require inversion algorithms that are often complex and can contain large uncertainties.

The most important subsurface property of a seasonal snowpack, from a hydrology perspective, is the snow water equivalent (SWE), or the amount of water resulting from melting all of the snow, which is defined as:

$$\text{SWE} = \int_{ground}^{surface} \frac{\rho(z)}{\rho_w} dz \qquad (14.1)$$

where:

$\rho(z)$ is the snow density
ρ_w is the density of water
z is the direction normal to the surface
SWE is the water equivalent, in the same units as z.

The integration is performed from the ground (or previous summer surface) to the snow surface. The annual SWE is of interest for glaciologists and hydrologists, for example, as the SWE at the end of winter is input for the net mass balance at any point on a glacier or ice sheet, and represents the amount of water from snow available at the beginning of the melting season. Snow depth is a much easier measurement to make than SWE and has a larger variability than density over relevant spatial and temporal scales. Therefore, in general, often few density measurements are combined with a much higher number of depth observations, in order to produce many estimates of SWE.

The temperature profile within the snow and firn, another important sub-surface parameter, is the primary driver for snow metamorphism (McClung and Schaerer, 1993) in the near surface, where overburden pressures are low. Deep in the firn, in locations where temperatures remain below zero, the temperature profile can provide information about past climate. In areas where the melting point is reached, temperatures in the snow and firn can indicate melt regions, and these are necessary for estimating the energy input required to melt all of the mass. The deformation of ice also depends strongly on ice temperature, and ice temperatures can be used to infer past climate, due to the low thermal conductivity of snow and ice (Cuffey and Paterson, 2010).

The snowpack is made of many different layers, as each individual storm deposits a unique snow type. The saying *"No two snowflakes are alike"* comes from the fact that snow crystal formation is an incredibly complex process, with growth patterns that are extremely sensitive to temperature and humidity conditions. Snowflakes that reach the surface have passed through a wide range of conditions on their journey from the clouds. Once they reach the ground, they continue to change on a time scale of hours, depending on the environmental conditions. Changes occur faster at higher temperatures, and temperature gradients control the direction of snow grain growth.

While no two snowflakes from even an individual storm are alike, snowflakes from the same storm are much more similar to each other than to snow grains anywhere else in the snowpack. Before negligible melt occurs in the spring, the snowpack holds a stratigraphic record of the weather conditions during the previous fall and winter. In places where the snow never melts, such as the major ice sheets in the Arctic and Antarctic, this record of local weather is preserved. Ice cores from the major ice sheets provide information about climate going back more than 400 000 years (Cuffey and Paterson, 2010).

Layering in snow and firn, termed stratigraphy, manifests through distinct changes in snow hardness, density, and microstructure. The major changes in these properties occur at boundaries between snowfall events, but they also vary within layers as well, although to a much lesser extent. These variations in snow morphology at the millimeter scale control the basic physical processes of snow heat flow, snow mechanics, and snow melt. Changes in grain size and shape, and the bonds between them, affect remote sensing observations, control avalanche formation, and cause complex water flow patterns within the snow. Due to capillary forces interacting with variations in snow microstructure, meltwater pathways form as isolated vertical columns, often with slope parallel flow at stratigraphic boundaries (Figure 14.2).

Liquid water content is a primary snow property of interest, and it can vary at the centimeter scale. Liquid water has an enormous effect on microwave remote sensing, as the electrical properties of snow change by more than an order of magnitude when snow becomes wet (see Chapter 2). Microwave sensors can detect the presence of melt easily, but estimating mass (SWE) in wet snow is not possible, due to the complexity of snowmelt processes (see Chapters 5 and 6).

Remote sensing observations of snow are also incredibly sensitive to variations in snow microstructure. Existing and proposed methods for measuring SWE from space are focused on measuring changes in passive or active radar measurements caused by electromagnetic signals being scattered by snow grains – termed *volume scattering*. As the mass of snow increases, more volume scattering occurs, which generally decreases passive microwave brightness temperatures and increases active microwave backscatter. There is a relationship between scattering and SWE but, clearly, there is also a very strong relationship between scattering

Figure 14.2 Frozen melt channels in polar firn. Capillary forces cause preferential flow paths within the percolation zone. Re-frozen melt channels shown as solid vertical columns; dry snow between has been removed. Meltwater is pooled at a stratigraphic boundary in foreground (Photo by Michael Demuth).

and snow microstructure. Information about snow grain size, type, and specific surface area, are therefore of interest throughout the snow and firn.

A glacier is divided up into different facies, depending on whether and for how long the location on the glacier has experienced melt. In the *dry snow facies*, the snow never melts; in the *percolation facies*, snow melts during the year but refreezes after melting; and in the *wet snow facies*, the snow experiences melt which moves mass beyond the current year in the stratigraphic record. At the lower end of the wet snow facies is the *slush facies*, where the snow is completely saturated with water, and the *superimposed ice facies*, where there is only ice in the summer. Active radar observations show a distinct change at the upper boundary of the percolation zone, due to the complex geometry of the liquid water pathways. Field observations at the end of summer are often used to determine glacier facies.

The thickness of ice, both on land and on sea, is of great interest to scientists studying the cryosphere. Estimates of ice thickness are needed for ice flow modeling, and sea ice thickness estimates are required for simulating annual sea ice distribution. Changes in the surface elevation of ice and snow on land can be measured with laser and radar altimeters. With an estimate of freeboard and accurate digital elevation models (DEMs), sea ice thickness can be estimated. More recently, intensive airborne remote sensing missions have measured land ice thickness using low frequency radar not available on current spaceborne platforms (http://www.nasa.gov/mission_pages/icebridge/index.html). Remote sensing missions which use repeat LiDAR can map snow depth by differencing surface elevations measured with and without snow cover (Deems *et al.*, 2006; Trujillo *et al.*, 2009). Other airborne missions use microwave radar to measure snow depth on sea ice (Farrell *et al.*, 2012).

First year sea ice is typically relatively smooth, while multi-year ice is often cracked and buckled, therefore the age of the sea ice is of interest. Crevasses and moulins on glaciers can control the glacier hydrology, give information about flow patterns, and cause scattering of electromagnetic signals (Conway *et al.*, 2002).

14.3 Standard techniques for direct measurements of physical properties

14.3.1 Topography

Surface topography is one of the most common remote-sensing measurements. Many traditional survey techniques have been employed to obtain surface topography in support of remote-sensing measurements. These include optical survey techniques, aneroid barometry, TRANSIT satellite (also known as NAVSAT, Navy Navigation Satellite System) and, more recently, the use of the Global Positioning System (GPS), one of several Global Navigation Satellite Systems (GNSSs).

When relative topography is desired, or the surface to be measured is geographically close to a known benchmark elevation, optical survey techniques can be used.

Figure 14.3 Sight leveling on Peyto Glacier, Alberta, Canada. The sight level is mounted on the tripod. The operator views a stadia rod through the telescope on the level, reading the markings on the rod to determine the distance from the reference plane to the surface. A handheld tape measure determines the distance between stadia rod placements. Another stadia rod is visible in the background, used by another survey team (Photo courtesy Dartmouth College Off-Campus programs).

Standard optical techniques include sight leveling and theodolite and Electronic Distance Measurement (EDM) surveying. In sight leveling, a surveyor's level is typically used. This device is essentially a telescope with crosshairs that rotates on a fixed plane (Figure 14.3).

The plane of the level is oriented normal to the geoid surface by using one or more bubble levels attached to the instrument. Once leveled, the telescope can rotate, and anything in the telescope crosshairs is at the same level. When combined with the use of a stadia rod or level staff (a staff marked with measurements along its length) and a standard surveyor's tape measure, profiles of elevation can be measured. To measure a profile, the level is placed near a high point in the profile and leveled. The stadia rod is placed on the first point to be surveyed. A reading of the elevation markings on the stadia rod through the telescope provides the elevation of the plane above the snow/ice surface. The rod is then moved to the next point. A second reading through the telescope provides the elevation of the reference plane above the surface at this point – the difference between the two is the difference in elevation between the two points. The distance between the points is measured with a standard surveyor's tape measure. After a series of measurements, the level can be moved to a new location, keeping the stadia rod fixed for one measurement, and a new reference plane is established. In this way, a long profile of surface topography can be constructed.

Sight leveling generally produces surface topography measurements that are inherently two-dimensional; elevation measurements are taken along a straight line. If elevation measurements of arbitrary points (not falling along a straight line) are desired, or if exact geographical co-ordinates are required for each elevation point, a more complete positioning system is needed. In optical surveying, a *theodolite* is used to make precise measurements of the angles (both vertical and horizontal) between target points of interest. In common use, the distance between the theodolite and each target is determined with *Electronic Distance*

Measurement (EDM), using a corner-cube reflector and electromagnetic radiation at several frequencies. When combined with angle measurement, this combination is often referred to as a *Total Station* (e.g., Kavenagh and Bird, 1996). When the angles, both vertical and horizontal, and the distances to each target point are both measured, a vector can be constructed to each surveyed point and, thus, each point can be located in 3D space relative to the others.

Optical techniques provide relative measures of topography, and a reference point is needed to determine absolute elevations. If a suitable rock benchmark is near enough to the survey area, an optical survey can be connected to the benchmark for absolute positioning. In some surveys, the closest benchmark is too far away to be practical, as with central locations on ice sheets. In such situations, an independent measure of surface elevation is required. Before satellite navigation became popular, elevations in remote locations were frequently measured using atmospheric pressure. Aside from the complications ensuing due to local pressure variations and weather patterns, an *aneroid barometer* can serve as a useful indicator of elevation. These instruments are sometimes termed *pressure altimeters*.

While early attempts at surveying via satellite positioning were cumbersome (e.g., TRANSIT/NAVSAT), the more modern GPS (as mentioned, one of the GNSS system) is now commonly used to make both absolute and relative measurements of topography. A GNSS receiver determines its position in space by measuring the distance between itself and the satellites in the GNSS "constellation". The distance measurement is made by analyzing "pseudo-random" code transmitted by the satellite to determine the time-of-flight between the satellite and receiver (http://en.wikipedia.org/wiki/Global_Positioning_System). With distances to at least four satellites, a position fix can be made. While this is a minimum for an accurate fix, in typical usage, a receiver tracks many more satellites of the 24-satellite GPS constellation at any given time.

Normal GNSS receivers can produce a fix in real time, but the position information is generally only accurate to within several meters or even tens of meters, even with the best receivers. There are several reasons for this level of accuracy. The positions of the satellites themselves, or *ephemeris*, are measured precisely by ground stations, but these precise measurements are not available in real time. Instead, navigation receivers rely on *broadcast ephemeris* transmitted by the satellite itself to determine positions. In addition to errors in the ephemeris data, the atomic clocks on each satellite experience slight drift which, while it can be measured, cannot be determined in real time. In addition to these measurable errors, an additional error is introduced by the propagation of the GPS signal through the Earth's ionosphere. This causes path delays to the signal, and these delays are affected by the changing patterns in the ionosphere.

To achieve survey-grade (cm-level) accuracy, rivaling the measurements made by optical techniques, a *survey grade* GNSS receiver is used. These receivers continuously record the GPS/GNSS signals from the satellites in view, to be downloaded for further post-processing. In post-processing, software can use *precise ephemeris*, measured as the satellite passes over ground stations of known

Figure 14.4 GPS survey setup used by Seigfried *et al.* (2011). The survey-grade GPS antenna is mounted high on the sled towed behind the snowmobile, which provides a clear sky view and prevents the bouncing that would be associated with a mounting point on the back of the snowmobile itself. Additionally, the length of the sled integrates surface roughness at the scale of the sled (Photo by Bob Hawley).

location, rather than the less accurate broadcast ephemeris. This enhances the accuracy considerably. In addition, since the receiver continuously records satellite data, it is possible to occupy a position for a very long time, (12–24 hours), long enough for ionospheric disturbances to change, and thus for the delays associated with them to change. This long occupation is known as a *static* measurement, and can result in centimeter accuracy.

In situations where multiple points are to be surveyed, it is common to survey a single point using the static method, and then measure the *baselines* between that point and every other point, using *differential* processing. In differential processing, two or more receivers are used to simultaneously record satellite data at different points. Because, at the scale of the survey, the signals traveling from the satellite to each receiver are passing through the same ionosphere, errors in the *difference* between the two positions due to the ionosphere can be eliminated. Additionally, because the length and angle of the baseline is required in a differential measurement, rather than absolute position, errors from satellite ephemeris are also eliminated. One result of this is that accurate baselines (1–3 cm) can be measured with a much shorter occupation time, often 10–20 minutes. This measurement style is known as *fast static*. In recent years, high spatial resolution surveys of topography have become possible using *kinematic* data collection and processing (King, 2004), described in section 14.4.1. Figure 14.4 shows a survey-grade GPS unit mounted to the top of a sled for fast static and kinematic measurements.

14.3.2 Snow depth

Snow depth is monitored, in the simple case, at a fixed point by frequently making a reference measurement throughout the year. In a seasonal snowpack, this can be typically done with an ultrasonic sensor which monitors the location of the snow surface (DeWalle and Rango, 2011). On a glacier, measurements of the height of an ablation stake are made (Cuffey and Paterson, 2010). Destructive measurements are made spatially with a probe at lower temporal resolution, and are possible up to a depth of approximately 10 meters. These measurements can be made in the ablation zone on a glacier, but it can be very difficult in the accumulation zone to distinguish the previous year's surface by detecting a change in hardness. In this situation, snow pits or shallow cores (up to 15 meters) are required to estimate snow depth.

14.3.3   Snow water equivalent and density

As mentioned, snow density is required, in addition to depth, for estimating the mass or SWE. This involves taking snow samples of known volume and weighing them, by carefully filling density cutters with appropriate procedures, for a direct destructive measurement of density. Since density is much more time-consuming to measure, and is assumed to vary spatially and temporally less than depth, an order of magnitude more depth measurements than density measurements are usually available. Density cutters vary in size from $10\,cm^3$ to $1000\,cm^3$. Small sizes are used to measure density in small thin layers, usually for avalanche applications, while large cutter sizes do a better job of accurately sampling the average density of the snowpack. These cutters are used to take samples at equal depth intervals in a snow pit, with depth varying from approximately 30 cm to 3 meters.

Snow density can be also measured with a coring device. In seasonal snow, the core is pushed into the dirt at the base of the snowpack, which acts as a plug to prevent snow from falling out of the tube, and the dirt plug is removed before weighing. The core is weighted, with and without the snow, and the SWE can be estimated from the differences in the weights. This works well in cold conditions where soil exists, but is much more challenging in warm weather and at snow-rock interfaces. At several hundred locations around the western USA, SWE is measured continuously by weighing a location of several square meters with a glycol-filled "pillow" and a pressure sensor (http://www.wcc.nrcs.usda.gov/snow/).

In the polar regions, shallow cores, typically 10–30 meters long, are extracted using a larger diameter core barrel (see Figure 14.5). These samples are cut into sections and weighed for density. Density observations are also made on deeper ice cores, providing density measurements throughout the firn.

14.3.4   Temperature

Temperature observations in snow are straightforward. Typically, a dial stem thermometer is inserted into the sidewall of a snow pit, and a temperature measurement is made every 5 cm. Near ice layers and weak layers, temperature might be measured in more detail to estimate the temperature gradient influencing metamorphism. Strings of thermistors are often deployed at weather stations

Figure 14.5 Removing the core from the core barrel at a shallow firn coring operation at Kongsvagen, Svalbard. The drilling head is to the right, and the core is extracted through the opposite end of the barrel into the waiting tray. The core protruding from the drilling head is a result of slippage of the "core dogs" when extracting the core after a drill run. The density data from this core is shown in Figure 14.12 (Photo by Bob Hawley).

Figure 14.6 Thermocouples for measuring surface and subsurface temperature. Brown wires are individual thermocouples leading to the black box containing the datalogger and multiplexer. The setup is powered by the single deep-cycle battery to the right of the box, which is charged by a solar panel (not pictured) (Photo by Bob Hawley).

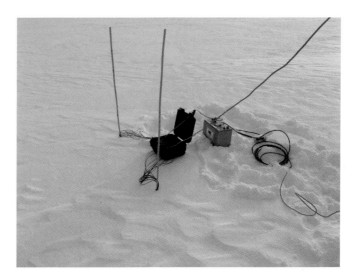

to make subsurface temperature measurements in snow and firn continuously. Temperature measurements within boreholes are made to monitor temperatures in the snow, firn, and ice to much greater depths (Figure 14.6).

14.3.5 Stratigraphy

Stratigraphy is much harder to observe and the observations are more subjective. Typically, individual storm layers are identified using a hand-hardness test, on a scale from 1–5, using a fist, fingers, and a pencil with a small amount of pressure. Within each layer, snow grain size and shape are estimated using a hand lens. Stratigraphy in snow pits is measured from the surface to the ground in seasonal snowpacks, and within the upper two meters typically in polar regions (Figure 14.7a). The International Classification for Seasonal Snow on the Ground (Fierz *et al.*, 2009) and other observational guidelines (Greene *et al.*, 2010) are valuable references for anyone performing ground-based snow observations. Sometimes, samples are cast by filling the void space with a chemical liquid that freezes immediately and preserves microstructure (Figure 14.7b). These samples are analyzed in detail in a lab using thin sectioning, surface sectioning techniques and, more recently, X-ray tomography. This process is very time-consuming and less common, but it provides detailed information on the snow microstructure.

14.3.6   Sea ice depth and ice thickness

Sea ice depth is commonly measured by coring the ice for a manual measurement. Drilling to the bed on glaciers and ice sheets also provides a destructive method for measuring ice thickness. Both are very time-consuming, so this is performed at relatively few locations. Low frequency radars are so commonly used for measuring land ice thickness that they are a standard technique. We will discuss this measurement strategy in the next section with the other radar techniques.

14.4   New techniques for high spatial resolution measurements

14.4.1   Topography

Advances in technology have made very high spatial resolution measurements of topography possible. Although some of these techniques technically qualify as remote sensing in themselves, we deal with them here because they are often used in remote-sensing ground-truthing campaigns. One such technique is terrestrial Light Detection And Ranging (LiDAR). Many terrestrial LiDAR systems are available; most mount on a standard survey tripod and can be automated to scan areas from tens of meters to several kilometers. As the scanning head of the LiDAR unit moves, it collects range, azimuth, and elevation information from a given reflector, thus recording a three-dimensional vector from the scan head to the ground point.

The difference between terrestrial LiDAR and theodolite measurements is that, instead of surveying specialized survey targets which have to be manually moved, the LiDAR receives reflections from almost all surfaces and objects around, allowing measurements of more than 100 000 points to be collected in a matter of hours. A set of these vectors defines the relationship of each ground point to the others, and a digital elevation model (DEM) can be constructed from these points.

Figure 14.7 Left panel (A): typical snow pit size. Note ruler and thermometers on pit wall to the left (Photo by Bob Hawley). Right panel (B): Casting a snow sample. The sample is placed in a liquid-tight container. Dimethyl phthalate is added slowly to a corner of the extracted sample to allow the chemical to fill from the bottom, preserving the microstructure for later laboratory analysis (Photo by HP Marshall).

This DEM has several uses in the context of larger remote sensing campaigns, and it has a much higher spatial resolution than remotely sensed elevation data from airborne and spaceborne platforms. Surface roughness is a critical parameter in many remote-sensing applications, and a terrestrial LiDAR-based DEM can be queried to determine roughness on any number of scales, in any number of directions.

Advances in processing strategies have also allowed GPS/GNSS surveys to become not only higher-resolution, but also simpler to run in the field. *Kinematic* processing of GPS data allows a pair of receivers to measure positions at very high frequency, eliminating the stops required by the fast static method described in the previous section. As with other differential GPS measurements, kinematic GPS determines a tridimensional vector between two receivers. One receiver is fixed for the duration of the survey, and is known as the *base* station. The other receiver moves over the terrain of interest, and is known as the *rover* station. Kinematic processing can be performed in post-processing (*post-processed kinematic or PPK*) or, using receivers equipped with a radio link, in real time (*real-time kinematic or RTK*).

For PPK surveys, the field procedure is very simple; start the two receivers, allow some time to elapse while both are collecting data (initialization time), and then move the rover over the terrain of interest. PPK data can then be processed into an X, Y, and Z position for each GPS measurement collected. Common survey strategies using this technique are to establish a base station receiver, allow it to collect enough data (6–24 hours) for *static* processing (see previous section), allowing the survey to be connected to a global reference frame, and then walk, ski, or drive the roving receiver in a grid pattern over the surface of interest. The exact nature of the grid is dependent on both the surface topography to be measured and the remote-sensing platform (see section on sampling strategies below).

14.4.2   Surface properties

Surface properties to be measured in the field can be divided into two categories:

- First, simultaneous measurement of electromagnetic properties (simulating an airborne or spaceborne instrument) and the physical properties inferred from the electromagnetic properties (e.g., measuring spectral albedo and grain size simultaneously).
- Second, manual measurements of properties that *affect* remote sensing measurements, but for which no simultaneous measurement is possible (for example, measuring surface roughness to see how it affects a radar altimeter).

There are many instruments for collecting measurements of the surface electromagnetic emission or reflectance. The *radiometer* is the most general of these. Radiometers carried aboard aircraft and satellites are the basis for many remote sensing techniques. Smaller versions of the same instrument have taken many forms. Broadband radiometers are commonly found on Automated Weather Stations (AWSs), where they measure both the *downwelling* (incoming) radiative flux (when pointing up) and the *upwelling* (reflected or emitted) flux (when pointing down). The ratio of these measurements is a measure of *albedo,* an important property in itself and for remote sensing studies.

Narrow-band radiometers for specific ground-truthing measurements in satellite frequency bands, such as the LandSat bands (see Figure 7.1), are also available. These allow measurements of specific surfaces at these frequencies, to assist in classification in the remotely-sensed image. Since many remote-sensing classifications employ band ratios to avoid the problems caused by path radiance, *ratioing radiometers* allow these ratios to be measured more directly (http://archive.org/details/nasa_techdoc_19840071065). In such a measurement, the user selects the bands to be ratioed and aims the instrument at a target, and the ratio is displayed.

To measure a wide spectrum of electromagnetic reflectance of a target, handheld *spectrometers* are used. Similar in measurement technique to the ratioing radiometer, the spectrometer is aimed at a target and a spectrum is acquired – a measurement of the intensity of electromagnetic radiation emanating from the target, in a large number of narrow bands. This allows a much more detailed classification of the material to be analyzed. Since the spectrum of a target is inherently dependent on the spectrum of the electromagnetic source, it is advisable to collect also the spectrum of the direct incoming electromagnetic radiation – essentially collecting a measurement "looking up". Differencing or ratioing these two measurements results in an actual *reflectance spectrum*, a property of the material itself, and predominantly independent of the electromagnetic source. Since this type of measurement is common and important, this differencing is frequently done in the acquisition software. Calibration measurements are also made in the field using targets with known albedo.

Measuring physical properties of the surface can be critical to any ground-truthing effort. Many physical property measurements undertaken in snow pits (see section 14.3.2) can be duplicated at the snow surface. An important measure for many ground-truthing studies is small-scale topography, or surface roughness. As discussed in the previous section, high-resolution local topography can be measured with a ground-based LiDAR instrument. Simpler techniques are also available. Albert and Hawley (2002) used a simple blade, 3 m in length, inserted into the snow. The top surface of the blade was made level, and a profile of the distance from the top of the blade to the snow surface, along the length of the blade, provides a measurement of the snow surface topography at the local scale. Lacroix *et al.* (2008) used a laser ranging device mounted on tracks to measure a similar profile of surface roughness.

Surface roughness can be determined digitally from standard visible-light photography of the snow surface using a board placed in the snow vertically (Fassnacht *et al.*, 2009). Time-lapse studies of changing snow surfaces can provide a time series of the evolution of surface roughness. New technologies, such as the Xbox360® Kinect® sensor (Mankoff *et al.*, 2011), may soon be adopted to obtain quantitative time-lapse information of the changing surface.

14.4.3 Sub-surface properties

High-resolution ground-based techniques for subsurface measurements either 1) require access to the subsurface via a snow pit or borehole, or 2) utilize a ground-based remote sensing method. The first requirement means that these techniques are effectively *in situ*, and the measurement is made very close to the snow it samples. The second requirement means that inversion algorithms and interpretation are necessary before subsurface properties are estimated. They allow rapid, non-destructive measurements, but require extensive data analysis and post-processing. The major advantage over airborne and spaceborne remote sensing is that the footprint is small and, therefore, ground calibration and validation is much more accurate and easier to interpret.

14.4.3.1 *In situ measurements of sub-surface properties*

At the smallest spatial scale and highest spatial resolution is the relatively new technique of X-ray tomography on snow samples (Kaempfer *et al.*, 2005; Lomonaco *et al.*, 2011; Schneebeli and Sokratov, 2004). This, indeed, provides a spatial resolution of micrometers, on samples up to the 10 cm scale. This high-resolution technique improves upon previous methods for studying snow at this scale, which includes surface and thin sectioning of cast snow samples (see Figure 14.7(B)). X-ray tomography can be performed on undisturbed samples, and environmental conditions can be controlled during time-lapse experiments. These studies are giving snow scientists a new view of microstructure and metamorphism.

At the snow pit scale, observations have advanced significantly with respect to spatial resolution, with the aid of near-infrared (NIR) photography for recording stratigraphy through differences in grain size-dependent reflectivity in the NIR band (Matzl and Schneebeli, 2006). Figure 14.8 shows a set-up for NIR photography on the Greenland ice sheet. Careful calibration can lead to estimates of Specific Surface Area (SSA), or the surface area to mass ratio, which controls both albedo and chemical reactions between the snow and the atmosphere. However. estimates are also sensitive to grain shape, which also needs to be taken into account (Gallet *et al.*, 2011). Broadband contact spectrometry has also been used by researchers to measure the spectral albedo of snow on a snow pit wall (Painter *et al.*, 2007). As in the description of spectrometry for surface characterization, this technique involves measuring the wall of a snow pit with a known light source, and can provide a vertical profile of SSA.

Borehole measurements allow some of the observations normally carried out in snow pits to be performed at greater depths. Measuring stratigraphy can be accomplished using a simple borehole video camera. Borehole Optical Stratigraphy (BOS) is a combination of a downward-looking video camera and post-processing software that allows interpretation of visual stratigraphy in a borehole. Hawley *et al.* (2003, 2008) used BOS to determine annual layering in Antarctica and Greenland, though its initial design goal was to track optical features through time. Because a borehole can remain accessible for long time periods, it is possible to make repeated measurements. As the firn column compacts, optical features move closer together. Hawley and Waddington (2011) demonstrated that optical features could be tracked through time to determine a profile of vertical strain and thus firn compaction. Figure 14.9 shows BOS in action on the Greenland ice sheet.

Neutron scattering in boreholes has been shown to provide accurate estimates of snow density (Morris and Cooper, 2003). A radioactive source and detector

Figure 14.8 Near-infrared photography set-up on the Greenland Ice Sheet. Note the layering visible at the bottom of the pit wall, the mount in the foreground to keep the camera still, the ruler for georeferencing, and the calibration targets and flat field surface on the pit wall. The pit is covered with fabric that lets diffused light only into the pit (Photo by Eric Lutz).

are lowered into the borehole, and neutrons scattered back to the detector are measured. The instrument, called a Wallingford Probe, was originally designed to measure soil moisture. Fast neutrons are scattered by hydrogen atoms, and the number of neutrons measured at the detector is related to the density of the medium. A similar technique is used in seasonal snowpacks, in which scattered gamma radiation is used to measure SWE either at a fixed location or from an aircraft. The difference with the gamma sensor is that an integrated value is estimated and, therefore, the scan can be performed remotely (see section on SWE or snow depth retrieval from more details).

Firn cores are typically scanned with optical sensors after they are extracted to locate annual layering, as the fall surface typically can be distinguished from

Figure 14.9 Borehole logging with a Borehole Optical Stratigraphy tool. The cable runs from the spool over the pulley mounted on the survey tripod, and down the hole to the instrument. The depth of the instrument is measured by an optical encoder mounted on the pulley (Photo by Gifford Wong).

the snow deposited during the winter. Dielectric measurements of conductivity on cores have been used to estimate density and changes in chemistry and impurities. More sophisticated techniques for measuring properties on a firn core have been developed, which include the Maine Automated Density Gauge Experiment (MADGE) and Mostly Automated Borehole Logging Experiment (MABLE). MADGE uses scattered gamma radiation, similar to the technique used in seasonal snow, to measure density at high resolution (3.3 mm) and precision (0.003 g/cm^3). MABLE is a down-hole technique that measures the hardness of the borehole wall and NIR reflectance, and it is used as a tool to determine the location of missing firn core sections after they have been extracted (Breton *et al.*, 2009).

Figure 14.10 illustrates length-scale issues, comparing density measurements made with standard coring techniques to density measured with a Wallingford Probe, density measured on the core using Dielectric Profiling, and optical stratigraphy on the core.

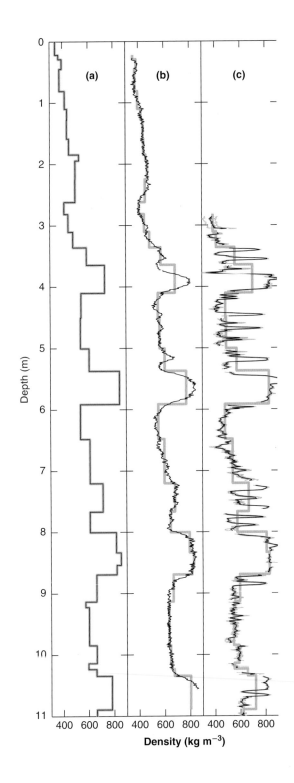

Figure 14.10 Figure 2 from Hawley *et al.* (2008); measurement of density on multiple length scales. Panel A: density measurements made on individual core sections in the field. Panel B: Grey line, same as panel A. Black line: density as measured in the borehole by a Wallingford neutron-scattering probe (Morris and Cooper, 2003). Note the slight smoothing of the stratigraphic signal. Panel C: Grey line, same as panel A. Black line: density as calculated from the dielectric profile (DEP) measured in the lab. Note how DEP resolves fine-scale ice layers that are only subtly seen in panel B. Panel D: optical imagery of core sections, for stratigraphic reference. Clear ice sections appear dark, and snow and firn appear light. (Hawley *et al.*, 2008. Reproduced with permission of International Glaciological Society).

In seasonal snow, a penetrometer technique has been developed, which measures snow hardness at 0.004 mm resolution (Johnson and Schneebeli, 1999; Schneebeli *et al.*, 1998). Statistical methods have been developed that use the hardness profile to estimate density and texture index (Pielmeier *et al.*, 2001), and the hardness profile has been used for ground truth of ground-based radar stratigraphy (Marshall *et al.*, 2007). The recorded hardness profile also contains detailed information about grain bond strength, and the signal can be inverted for microstructural and micromechanical properties (Johnson and Schneebeli, 1999; Loewe and Herwijnen, 2012; Marshall and Johnson, 2009).

Figure 14.11 shows an example of snow micro-penetrometer (SMP) profile, with the hardness of the entire profile and a zoomed-in region showing the sub-millimeter detail in this data. These microstructural and micromechanical properties have been used to classify the stability of a slope (Lutz *et al.*, 2009; Pielmeier and Marshall, 2009), and to statistically classify grain type (Havens *et al.*, 2013; Satyawali & Schneebeli, 2010; Satyawali *et al.*, 2009). While developed for a snow avalanche application, this instrument is valuable for remote sensing calibration and validation field campaigns, as it provides accurate, quantitative measurements that are sensitive to snow microstructure.

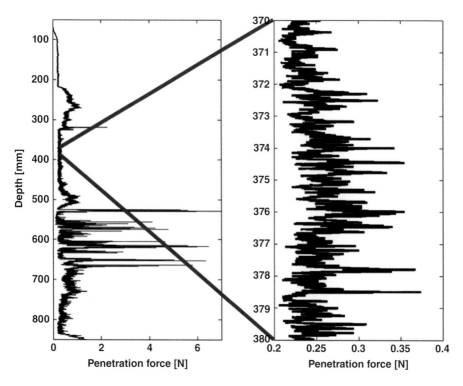

Figure 14.11 Example SMP profile in 85 cm snowpack showing layer hardness on the left, and 1 cm of data shown in detail on the right. Note the large ice crusts between 500 and 700 mm, and the individual ruptures in the zoomed-in area.

14.4.3.2 Ground-based remote sensing of sub-surface properties

The observation that snow and ice was fairly transparent to radio waves dates back to 1933. In 1957, after complaints from pilots about radar altimeters not working over snow and ice, the US Army conducted a study which showed that ice thickness of polar glaciers could be measured with a 440 MHz radar altimeter. In the 1960s and 70s, many ice-penetrating radars were built, and the technique is now standard for mapping ice thickness, both on Alpine glaciers and on polar ice sheets (Evans and Robin, 1972; Robin *et al.*, 1969; Waite and Schmidt, 1962).

Recent ground-based and airborne radar campaigns have mapped ice thickness and internal stratigraphy of glaciers and ice sheets (Gogineni *et al.*, 2007; Holt *et al.*, 2006; Vaughan *et al.*, 2008; http://www.nasa.gov/mission_pages/icebridge/index .html), and radar data have been used to infer past patterns of accumulation and histories of ice-sheet dynamics (Conway *et al.*, 2002; Nereson and Raymond, 2001; Waddington *et al.*, 2007). More recently, microwave radar has been used to measure snow stratigraphy and snow water equivalent in both seasonal and polar snowpacks (Marshall and Koh, 2008). Figure 14.12 shows a microwave radar measurement set-up, which is lightweight and allows two skiers to perform measurements with the radar over undisturbed snow.

When using ground-based radars for measuring snow and ice properties, there is a trade-off between penetration depth and resolution. The penetration depth is a function of frequency, with frequencies in the low MHz range capable of penetrating kilometers of ice. Frequencies in the GHz region only penetrate the upper tens of meters, and the upper meter or less in wet snow as, in wet snow, attenuation

Figure 14.12 Typical microwave radar measurement set-up. Radar is the white box suspended on a pole between two skiers, with control unit and batteries in a backpack. This allows measurements to be made easily in rugged alpine terrain (Photo by James McCreight).

is a strong function of frequency. Resolution is a function of bandwidth, or the range of frequencies over which a radar emits power. Bandwidths greater than an octave are difficult to achieve in practice, so therefore the bandwidth of most radar systems for cryosphere studies is at most of the same order of magnitude as the center frequency. Therefore, high-frequency systems can achieve much higher bandwidths, and provide therefore data at a typically higher vertical resolution. The center frequency also determines the sensitivity of the system to internal layering, heterogeneities in the subsurface, and liquid water, and ground-based radars operating at high frequencies typically have more narrow beams and, therefore, higher horizontal resolution.

In ground-based radar applications in the Cryosphere, 1–10 MHz radar systems are used to image the thickness of ice sheets and glaciers, 50–500 MHz systems are used in the firn and small glaciers, and microwave (1–18 GHz) radars are optimal for the upper 20 meters. Microwave systems also cover the range of current and planned spaceborne radar missions for snow (X- and Ku-bands, 8–18 GHz), so these measurements can provide data for direct comparison with airborne and spaceborne scatterometers.

Note that, from space, due to engineering limitations and Federal Communications Commission (FCC) regulations, active radar systems are limited to fairly low bandwidths. Internal reflections are difficult to separate from surface and ground returns, and inversion algorithms use integrated amplitude measurements at different frequencies and polarizations. Ground-based systems with larger bandwidths can help to provide information about the location and cause of the major contributions to the integrated backscatter amplitude. A typical microwave radar profile is shown in Figure 14.13, with power shown in dB.

14.5 Simulating airborne and spaceborne observations from the ground

One of the primary difficulties in relating ground-based field measurements to remote sensing observations is the difference in the *support*, or the footprint size of the measurements. Remote sensing products are often averages at the kilometer scale, while field measurements are performed at the sub-meter scale. Since snow properties change significantly over small spatial scales (e.g., 1–100 m), many field measurements are required for estimating average snow properties at remote sensing footprint scales, hence making the comparison between ground- and satellite-based estimates more complicated. Moreover, atmospheric parameters affect spaceborne measurements (with the effect depending on frequency and atmospheric parameters), which should be accounted for in the comparison.

In order better to understand the way electromagnetic signals interact with snow, ground-based instruments have been developed to simulate airborne and spaceborne observations, with the advantage of a very small footprint (e.g., Cline, 2000; Hardy *et al.*, 2008). The sensor must be located high enough above the snow surface

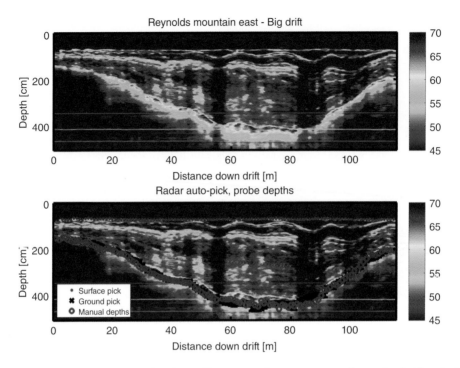

Figure 14.13 Typical microwave (2–10 GHz) radar profile. Top figure shows radar data, bottom figure shows locations of automatically picked surface and ground returns, and 118 manual measurements of snow depth. Note the internal stratigraphy that is resolved, and regions of low reflectivity likely caused by liquid water concentrations.

to be in the far-field, which is approximately $\frac{D^2}{\lambda}$, where D is the largest dimension of the antenna aperture and λ is the wavelength. Over this small footprint, detailed snow characterization can be performed and both forward electromagnetic models and inversion algorithms can be tested. Figure 14.14 shows a microwave radar suspended two meters above the snow surface in the far field for simulating airborne and spaceborne radar observations.

14.5.1 Active microwave

Experiments with tower mounted active radar scatterometers were performed in Europe and the US, and form the bulk of the existing data of radar backscatter in snow (Matzler *et al.*, 1997; Strozzi and Matzler, 1998; Ulaby *et al.*, 1977). Since those measurements, there have been few efforts to measure radar backscatter in a wide range of snow conditions, and the lack of data of this kind has severely limited progress with radar retrieval algorithms. Expanding the available data of radar backscatter in snow has been the focus of more recent snow remote sensing campaigns.

During the NASA Cold Land Processes Experiment (CLPX), in 2002 and 2003, several different tower-mounted radar systems were maintained at the Local Scale

Figure 14.14 Microwave radar measuring at 35 degrees incidence in far field. Radar is the white box on the far side of the sled. Control unit and batteries are located in boxes on the sled. Survey-grade GPS is mounted to the near side of the sled to avoid interference from radar (Photo by Hans-Peter Marshall).

Observation Site (LSOS) intensive study site. More recent campaigns, funded by the European Space Agency (ESA) for the proposed CoreH20 satellite mission, have furthered this effort in Finland and Canada. During CLPX-II, in 2007 and 2008, a sled-mounted radar system, with antennas located two meters above the snow surface, was used to collect far-field backscatter measurements throughout study sites in Colorado and Alaska.

One of the main challenges with such a measurement is stability of the radar system. Frequent calibration measurements must be made, along with ground-truth observations. The measurement must be performed in the far-field, requiring a tower platform, and these systems usually require a large amount of power (Hardy *et al.*, 2008). This complicates deployment, as most well-instrumented snow sites generally run on solar power and are remote.

14.5.2 Passive microwave

In situ passive microwave observations date back to the 1970s and, since then, there have been many field campaigns performed by different groups. Microwave radiometers on the ground have usually been collecting information at the frequencies of the sensors flying into space, namely $\approx 6-$, $\approx 10-$, $\approx 19-$, $\approx 37-$ and ≈ 85 GHz. Such measurements have provided crucial information for the development and testing of electromagnetic models (and theories which, in turn, support the development and refinement of retrieval schemes. Microwave radiometers are generally mounted on a tripod or a tower, and are pointed toward the target with the same incidence angle as that at which satellite data is collected ($\approx 50°$).

Measurements can also be performed in a scanning mode by changing the incidence angle from $0°$ to $90°$. Usually, the scanning incidence angles are kept between $15–20°$ and $60–70°$ for practical reasons. For low incidence angles, the feet of the tripod might be falling within the field of view of the instrument whereas, for high incidence angles, other features along the horizon (such as mountains or trees) might also fall within the field of view. In general, when selecting the site where passive microwave observations are performed, a relatively flat topography and a homogeneous snowpack is preferable. Also, in order to minimize the effects of specular reflection (e.g., contamination of emission coming from features in front of the radiometer which reflects on the snow surface and are recorded by the instrument), it is also preferable to select the measurement area at a certain distance from surrounding mountains and/or forest. Obviously, this is not always possible, and these effects should be accounted for in the post-processing phase of the data. Measurements of the sky brightness temperature should be taken regularly for the same purposes.

Wiesmann *et al.* (1996) contains data collected from a multi-frequency (11, 21, 35, 48 and 94 GHz) system based on portable radiometers operated on several sites in the Swiss Alps. The temporal and spatial behavior of the emissivity and brightness temperature is investigated for different snow and snow-free conditions, and the passive microwave measurements are complemented by ground observations and radar measurements. The data collected within the framework of CLPX also represents a unique data set. It was collected at the LSOS site at 18.7, 23.8, 36.5, and 89 GHz (both vertical and horizontal polarizations), using the Ground Based Microwave Radiometer (GBMR-7, http://nsidc.org/data/nsidc-0165.html). Three different measurement techniques were used:

1 scans of undisturbed total snow cover;
2 angular scans with varying incidence angles (between 30° and 70°); and
3 scans of bare soil and new snow.

Snow properties (density, snow temperature, stratigraphy, snow crystal size, soil moisture) and meteorological forcing data (wind speed, wind direction, air temperature, relative humidity, downward long-wave radiation, downward short-wave radiation and precipitation) were also collected. This data was used to study melting and refreezing cycles and evaluate electromagnetic models (e.g., Tedesco *et al.*, 2006).

During CLPX, multiband polarimetric brightness temperature images over the three Meso-cell Study Areas (MSAs) were also collected (http://nsidc.org/data /nsidc-0155.html). One of the goals was to collect data at a spatial resolution representative of the topography and vegetation cover, and to provide a simulated AMSR-E microwave data set. This data set was also used to perform scaling analysis and, for example, to study the impact of vegetation on brightness temperature (Tedesco *et al.*, 2005).

Passive microwave radiometers have also been used to study to collect information over the Greenland and Antarctica ice sheets. Over Antarctica, in particular, this is performed to provide a calibration data set for a relatively *stable* target on Earth (e.g., Dome C, Antarctica). Over Greenland, microwave radiometers have been used to collect passive microwave data in proximity to the Summit station in order to understand the spatial variability of brightness temperature within a satellite footprint over relatively stable targets.

14.6 Sampling strategies for remote sensing field campaigns: concepts and examples

When comparing ground-based measurements with remote sensing observations, one must take into account the difference in the scale triplet (Bloschl, 1999):

1 the *support*, or area over which the measurement integrates;
2 the *spacing*, or distance between measurements; and
3 the spatial *extent* of the measurements.

For example, a depth probe measurement has a support of 1 cm, a typical spacing of 50 meters, and a typical extent of several kilometers. A spaceborne passive microwave brightness temperature measurement has a support of 5–50 km, a spacing equal to the support, and global extent.

In order to account for these differences in scale triplet, efficient sampling strategies are required. This section describes strategies from several example calibration/validation field campaigns. A useful reference for better understanding problems and solutions associated to the collection on ground for remote sensing validation is McCoy (2004). In the following, some examples related to specific problems in the cryosphere are reported, focusing on major campaigns. Figure 14.15 shows a suite of measurements made during a validation and calibration campaign in Svalbard. The mismatch in support, spacing, and extent, require many people on the ground during airborne and spaceborne overpasses. Performing enough measurements to accurately characterize the mean snow and ice properties at the necessary scale is challenging. Example campaigns and sampling strategies are described in the sections below.

14.6.1 Ice sheet campaigns

The CryoSat Validation EXperiment (CryoVEX) campaigns were initiated in advance of the launch of the ESA satellites CryoSat and CryoSat-2. These satellites have a single mission goal – to measure fluctuations in the thickness of sea and land ice. The key instrument on these satellites is a radar altimeter, the Synthetic aperture Interferometric Radar Altimeter (SIRAL). Because SIRAL was a new radar altimeter design, never before flown in space, simulated measurements were desirable from the variety of ice and snow conditions likely to be observed by SIRAL. This was achieved through the construction of an airborne simulator for SIRAL. The Airborne Synthetic Aperture Interferometric Radar Altimetry System (ASIRAS) was a low-altitude sensor using the same radar technology as SIRAL, flown from a fixed-wing aircraft (frequently a De Havilland Twin Otter or Dornier 228–200).

CryoVex campaigns were carried out in 2003, 2004, 2006, 2007, 2008, 2011, and 2012. The basic scheme of the campaigns was to place ground teams at representative sites around the Arctic, including several sea ice locations with different types of sea ice, small ice caps and glaciers, the margins of the Greenland Ice

Figure 14.15 An intensive ground-truthing campaign on Kongsvegen, Svalbard. Visible in the photo are many field-measurement instruments. The sled in the center of the scene carries GPR and GPS antennas. To the left of the sled, shallow core drilling is in progress. To the right of the sled, a logging cable goes over a pulley to a Wallingford probe, measuring density in a previously drilled borehole. At the far left, a probe for measuring snow depth is ready for use.

Sheet, and its interior. At each location, tight coordination between the aircraft and ground teams allowed precise overflights of the ground measurements. Each ground team erected a corner reflector (e.g., a metallic object of specific known shape and dimensions), designed to create a powerful return, positioned above the snow surface at a measured distance. This was to determine the difference between the snow surface, as re-tracked by the radar altimeter, and the snow surface, as measured on the ground by the field teams. Absolute snow surface elevation was also measured *in situ*, using differential GPS.

In addition to the measurement of true surface height, teams carried out numerous *in situ* experiments, including detailed stratigraphy measurements in snow pits, ground-based radar measurements, detailed density profiles with shallow cores or a Neutron Probe technique (Morris and Cooper, 2003; Morris, 2008), and firn densification measurements with the "coffee can" technique (Hamilton and Whillans, 2000). On sea ice, teams measured sea ice thickness, snow depth over sea ice, absolute surface elevation, and profiles of snow and sea ice physical properties from snow pits and cores. To extend the spatial extent of sea ice thickness measurements, a helicopter with a low-altitude electromagnetic induction system flew surveys to determine sea ice thickness at larger spatial scales. The sea ice surface is extremely dynamic, changing on timescales as short as minutes. Thus, extremely precise coordination was required to ensure that ground teams and aircraft sampled the same sea ice location at the same time.

During the CryoVEX campaigns, one key question was the *spatial variability* of the properties to be measured on the ground. Put simply, how representative is a point measurement made on the ground, in context of the larger footprint (tens of kilometers) of a spaceborne radar altimeter? To assess spatial variability on a variety of scales, a *nested grid* layout was chosen for sampling sites. In the nested grid, measurements (snow pits, cores, etc.) were collected along two perpendicular lines. At the intersection of these two lines was the first measurement. Further measurements were made along each line at distances of 1 m, 10 m, 100 m, and 1000 m. In this way, differences at several length scales, and in differing directions, could be determined.

After the launch of ICESat-1, a satellite carrying the GeoScience Laser Altimetry System (GLAS), an elevation validation campaign, was undertaken at Summit station in the center of the Greenland Ice Sheet. An orbit track (track 412) passed within 5 km of the station, and it was determined that a time series of elevation along this track would be beneficial for validation. The design of this validation measurement considered several factors. First, it was desirable for science technicians to be able to make measurements on a monthly basis, all year round. This meant the length of the measurement along-track needed to be limited to a distance that could be reasonably covered in a day, including preparation time.

Because the *support* of an ICESat-1 measurement is relatively small ($\approx$70 m spot size on the ground), and the actual pointing direction of the satellite was not precise enough to target the same track within 70 m on each orbit, a survey in a straight line along the nominal ground track would not always be coincident with satellite measurements. Furthermore, because slope is so important in interpreting elevation measurements, a characterization of the slope is desirable. The sampling strategy chosen used kinematic GPS to measure surface topography. The GPS collected data on 1 second intervals, and the antenna was mounted on a sled pulled by a snow machine, along a 10 km track that crossed the nominal ground track in a square-wave pattern. In this way, crossovers between ground-based GPS and satellite laser measurements were assured for each survey. This eliminated the need to interpolate positions because, for every campaign sampled, multiple laser "shots" were transected by intersecting GPS data (Siegfried *et al.*, 2011).

14.6.2 *Seasonal snow campaigns*

The Cold Lands Processes eXperiment (CLPX; http://www.nohrsc.nws.gov/~cline /clpx.html) was the first large coordinated snow remote sensing calibration/validation effort. It was designed to address questions about processes understanding, spatial and temporal variability, and uncertainty in snow estimates in the terrestrial cryosphere. This large field campaign involved researchers from many different universities and government labs, and took place in the Colorado Rockies in 2002 and 2003. In particular, CLPX aimed to develop a strong synergism between process-oriented understanding, land surface models, and microwave remote sensing (Cline, 2000).

The primary goals of this sampling were to get accurate sample mean and variance estimates at the 1 km^2 scale in six different intensive study areas (ISAs), with a secondary goal of characterizing the spatial structure within these sites at a 100 m resolution.

Four Intensive Observing Periods (IOPs) were chosen in 2002 and 2003 (February and March) to provide information about both wet and dry snowpacks in north-central Colorado. Six different aircrafts operated during this field campaign, with both passive and active microwave sensors, gamma radiation sensors, and LiDAR. In addition, data acquisition from 12 different satellite sensors was acquired during these intensive field campaigns, using optical and passive and active microwave sensors. Ground-based observations included intensive snow sampling, soil and vegetation observations, ground-based passive and active microwave sensors, and Frequency Modulated Continuous Wave (FMCW radar. Large field teams performed coincident observations during the airborne and satellite overpasses. This dataset is archived at the National Snow and Ice Data Center, and it represents the largest single field-based snow measurement campaign to date (http://nsidc.org/data/clpx/). Figure 14.16 shows some example results from CLPX, in which ground-based radar, SMP, and standard snow pit observations are compared.

Building on the lessons learned during CLPX, a second series of smaller field campaigns took place in 2007 in Colorado and 2008 in Alaska (CLPX-II), which focused on one primary airborne radar system (NASA POLSCAT), one satellite-based sensor (TerraSAR-X), and attempted to extend the range of snowpack conditions measured during CLPX. A smaller third campaign (CLPX-III) took place in Grand Mesa, Colorado, to add a high-elevation, deeper snowpack to the database. For snow depth, wetness and surface roughness measurements, each site was stratified using a 100 m interval grid, with two orthogonal transects in each cell. The transect starting point and resolution (5, 10, 15, 20, or 25 meters) were chosen at random, with the orthogonal transects starting at the same location and with a direction based on the starting point to ensure that the transect remained within the cell. Two of the grid cells were chosen at random, and additional 25 measurements were made on a 20 m sub-grid, with the location of the measurement within each 20 m grid cell chosen at random. For snow pit observations, which are much more time-consuming, the 1 km^2 area was divided into 250 m grid cells, and a snow pit location within each cell was chosen at random.

14.6.3 Sea ice campaigns

Two major field campaigns for testing sea ice retrieval algorithms from the Advanced Microwave Scanning Radiometer (AMSR-E) were undertaken in 2003 and 2006 (AMSR-Ice03, AMSR-Ice06, http://nsidc.org/data/amsr_validation /cryosphere/amsrice03/index.html). These campaigns took place near Barrow, Alaska, on the Chukchi and Beaufort Seas and in Elson Lagoon. Extensive

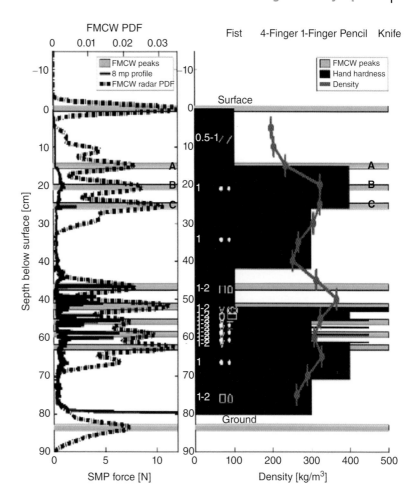

Figure 14.16 Left panel shows radar results as dashed line, SMP results as solid line, and location of major radar reflections highlighted in gray. Right panel shows snow hardness in black, density in gray, and grain size and type. Note that the major radar reflections correlate well with changes detected with other measurement techniques. (Marshall *et al.*, 2007. Reproduced with permission of Elsevier).

field-based measurements of sea ice conductivity, ocean salinity, snow depth, sea ice thickness, temperatures of the air, ice, and snow/ice interface, and surface roughness were performed. Snow pits, with stratigraphy and density, were also recorded. In 2006, the experiment was repeated in the same location, this time with an airborne microwave radar from the Center for Remote Sensing of Ice Sheets (CReSIS) for measuring snow. A similar suite of ground-based measurements was made as in 2003, with the addition of a ground-based FMCW radar mounted to a sled.

14.7   Conclusions

In this chapter, we have provided an overview of some of the physical properties of the cryosphere that are of interest for remote sensing observations. Because of the extensive number of quantities that should be discussed, and the limited space at our disposal in a general book such as this one, we acknowledge that we might have missed some quantities that could be of interest to some readers. We sincerely hope that the references provided in the chapter will help readers in this regard.

We have made an effort in describing both established standard techniques and new state-of-the-art modern techniques for measuring physical properties rapidly and at high spatial resolution. Although *in situ* observations can provide high quality but are obviously limited in both space and time, major complications in relating ground-based point measurements of physical properties to remote sensing observations stem from the mismatch in scale. An individual *in situ* measurement can be seen as representative of physical properties at a scale of the order of a few meters or even centimeters, while remote sensing observations are affected by processes at much larger spatial scales (e.g., tens of meters to tens of kilometers). In this regard, we have discussed sampling strategies and concluded the chapter with examples of field-based calibration and validation efforts made in recent years.

In situ observations represent a crucial tool for improving, validating and calibrating remote sensing algorithms and their outputs. In the case of the cryosphere, the collection of such data is particularly challenging, in view of the difficult access to test sites locations, to the harsh conditions and cold temperatures characterizing many field sites, and to the absence or lack of logistical support. The development and advancement of techniques for collecting *in situ* data is a key factor for the progress of the remote sensing of the cryosphere, together with the training of scientists to properly plan field campaigns and perform fieldwork activities. We hope this chapter can provide a contribution in this direction.

References

Abdalati, W. & Krabill, W. (1999). Calculation of ice velocities in the Jakobshavn Isbrae area using airborne laser altimetry. *Remote Sensing of Environment*, **67**(2), 194–204.

Albert, M. & Hawley, R. (2002). Seasonal changes in snow surface roughness characteristics at Summit, Greenland: implications for snow and firn ventilation. *Annals of Glaciology* **35**(35), 510–514.

Bloschl, G. (1999). Scaling issues in snow hydrology. *Hydrological Processes*, **13**(14–15), 2149–2175.

Breton, D.J., Hamilton, G.S. & Hess, C.T. (2009). Design, optimization and calibration of an automated density gauge for firn and ice cores. *Journal of Glaciology* **55**(194), 1092–1100.

Cline, D. (2000). The NASA Cold Land Processes Mission. *Eos, Transactions, American Geophysical Union* **81**(48).

Conway, H., Catania, G., Raymond, C., Gades, A., Scambos, T. & Engelhardt, H. (2002). Switch of flow direction in an antarctic ice stream. *Nature* **419**(6906), 465–467.

Conway, H., Smith, B., Vaswani, P., Matsuoka, K., Rignot, E. & Claus, P. (2009). A low-frequency ice-penetrating radar system adapted for use from an airplane: test results from Bering and Malaspina glaciers, Alaska, USA. *Annals of Glaciology* **50**(51), 93–97.

Cuffey, K. & Paterson, S. (2010). *The physics of glaciers*, 4th edition. Academic Press.

Deems, J., Fassnacht, S. & Elder, K. (2006). Fractal distribution of snow depth from LiDAR data. *Journal of Hydrometeorology* **7**(2), 285–297.

DeWalle, D. & Rango, A. (2011). *Principles of snow hydrology*, 2nd Edition. Cambridge University Press.

Evans, S. & Robin, G.D.Q. (1972). Ice thickness measurement by radio echo sounding, 1971–1972. *Antarctic Journal of the United States* **7**(4), 108.

Farrell, S.L., Kurtz, N., Connor, L.N., *et al.* (2012). A first assessment of IceBridge snow and ice thickness data over Arctic sea ice. *IEEE Transactions On Geoscience and Remote Sensing* **50**(6), 2098–2111.

Fassnacht, S.R., Stednick, J.D., Deems, J.S. & Corrao, M.V. (2009). Metrics for assessing snow surface roughness from digital imagery. *Water Resources Research* **45**, W00D31.

Fierz, C., Armstrong, R.L., Durand, Y., *et al.* (2009). The international classification for seasonal snow on the ground. *IHP-VII Technical Documents in Hydrology* **83**.

Gallet, J.-C., Domine, F., Arnaud, L., Picard, G. & Savarino, J. (2011). Vertical profile of the specific surface area and density of the snow at Dome C and on a transect to Dumont D'Urville, Antarctica – albedo calculations and comparison to remote sensing products. *Cryosphere* **5**(3), 631–649.

Gogineni, S., Braaten, D., Allen, C., *et al.* (2007). Polar radar for ice sheet measurements (PRISM). *Remote Sensing of Environment* **111**(2–3), 204–211.

Greene, E., Atkins, D., Birkeland, K., *et al.* (2010). *Snow, weather and avalanches: observational guidelines for avalanche programs in the U.S*, 2nd Edition. American Avalanche Association.

Hamilton, G. & Whillans, I. (2000). Point measurements of mass balance of the Greenland Ice Sheet using precision vertical global positioning system (GPS) surveys. *Journal of Geophysical Research – Solid Earth* **105**(B7), 16295–16301.

Hardy, J., Davis, R., Koh, Y., *et al.* (2008). NASA Cold Land Processes Experiment (CLPX 2002/03): Local Scale Observation Site. *Journal of Hydrometeorology* **9**(6), 1434–1442.

Hawley, R., Waddington, E., Alley, R. & Taylor, K. (2003). Annual layers in polar firn detected by borehole optical stratigraphy. *Geophysical Research Letters* **30**(15), 1788.

Hawley, R.L., Brandt, O., Morris, E.M., Kohler, J., Shepherd, A.P. & Wingham, D.J. (2008). Techniques for measuring high-resolution firn density profiles: case study from Kongsvegen, Svalbard. *Journal of Glaciology* **54**(186), 463–468.

Hawley, R.L. & Waddington, E.D. (2011). *In situ* measurements of firn compaction profiles using borehole optical stratigraphy. *Journal of Glaciology* **57**(202), 289–294.

Havens, S., Marshall, H.-P., Pielmeier, C. & Elder, K. (2013). Automatic grain type classification of snow micro penetrometer signals with random forests. *Transactions on Geoscience and Remote Sensing* **51**(6), 3328–3335.

Holt, J., Blankenship, D., Morse, D., *et al.* (2006). New boundary conditions for the West Antarctic Ice Sheet: subglacial topography of the Thwaites and Smith glacier catchments. *Geophysical Research Letters.* **33**(9), L09502.

Howat, I.M., Joughin, I. & Scambos, T.A. (2007). Rapid changes in ice discharge from Greenland outlet glaciers. *Science* **315**(5818), 1559–1561.

Johnson, J.B. & Schneebeli, M. (1999). Characterizing the microstructural and micromechanical properties of snow, *Cold Regions Science and Technology*, **30**, 91–100.

Kavanagh, B.F. & Glenn Bird, S.J. (1996). *Surveying principles and applications*, 4th edition. Prentice Hall, pp. 257–264. ISBN 0-13-438300-1.

Kaempfer, T., Schneebeli, M. & Sokratov, S. (2005). A microstructural approach to model heat transfer in snow. *Geophysical Research Letters* **32**(21), L21503.

King, M. (2004). Rigorous GPS data-processing strategies for glaciological applications. *Journal of Glaciology* **50**(171), 601–607.

Lacroix, P., Legresy, B., Langley, K., *et al.* (2008). *In situ* measurements of snow surface roughness using a laser profiler. *Journal of Glaciology* **54**(187), 753–762.

Larsen, C.F., Motyka, R.J., Arendt, A.A., Echelmeyer, K.A. & Geissler, P.E. (2007). Glacier changes in Southeast Alaska and Northwest British Columbia and contribution to sea level rise. *Journal of Geophysical Research – Earth Surface* **112**(F1), F01007.

Loewe, H. & van Herwijnen, A. (2012). A poisson shot noise model for micro-penetration of snow. *Cold Regions Science and Technology* **70**, 62–70.

Lomonaco, R., Albert, M. & Baker, I. (2011). Microstructural evolution of fine-grained layers through the firn column at Summit, Greenland. *Journal of Glaciology* **57**(204), 755–762.

Lutz, E., Birkeland, K. & Marshall, H.-P. (2009). Quantifying changes in weak layer microstructure associated with artificial load changes. *Cold Regions Science and Technology* **59**(2–3), 202–209.

Mankoff, K.D., Russo, T.A., Norris, B.K., *et al.* (2011). Kinects as sensors in earth science: glaciological, geomorphological, and hydrological applications. *Eos, Transactions, American Geophysical Union* (Abstract #C41D-0442).

Marshall, H.P., Schneebeli, M. & Koh, G. (2007). Snow stratigraphy measurements with high-frequency FMCW radar: comparison with snow micropenetrometer. *Cold Regions Science and Technology* **47**, 108–117.

Marshall, H.-P. & Koh, G. (2008). FMCW radars for snow research. *Cold Regions Science and Technology* **52**(2), 118–131.

Marshall, H.-P. & Johnson, J.B. (2009). Accurate inversion of high-resolution snow penetrometer signals for microstructural and micromechanical properties. *Journal of Geophysical Research – Earth Surface* **114**, F04016.

Matzl, M. & Schneebeli, M. (2006). Measuring specific surface area of snow by near-infrared photography. *Journal of Glaciology* **52**(179), 558–564.

Matzler, C., Strozzi, T., Weise, T., Floricioiu, D. & Rott, H. (1997). Microwave snowpack studies made in the Austrian Alps during the SIR-C/X-SAR experiment. *International Journal of Remote Sensing* **18**(12), 2505–2530.

McClung, D.M. & Schaerer, P.A. (1993). *The Avalanche Handbook*. The Mountaineers, Seattle, WA.

McCoy, R. (2004). *Field Methods in Remote Sensing*. The Guilford Press, ISBN10 1593850794.

Morris, E.M. (2008). A theoretical analysis of the neutron scattering method of measuring snow and ice density. *Journal of Geophysical Research – Earth Surface* **113**(F4), F04099.

Morris, E. & Cooper, J. (2003). Instruments and methods – density measurements in ice boreholes using neutron scattering. *Journal of Glaciology* **49**(167), 599–604.

Nereson, N. & Raymond, C. (2001). The elevation history of ice streams and the spatial accumulation pattern along the Siple Coast of West Antarctica inferred from ground-based radar data from three inter-ice-stream ridges. *Journal of Glaciology* **47**(157), 303–313.

Painter, T.H., Molotch, N.P., Cassidy, M., Flanner, M. & Steffen, K. (2007). Instruments and methods – contact spectroscopy for determination of stratigraphy of snow optical grain size. *Journal of Glaciology* **53**(180), 121–127.

Paul, F. & Haeberli, W. (2008). Spatial variability of glacier elevation changes in the Swiss Alps obtained from two digital elevation models. *Geophysical Research Letters* **35**(21), L21502.

Pielmeier, C., Schneebeli, M. & Stucki, T. (2001). Snow texture: a comparison of empirical versus simulated texture index for alpine snow. *Annals of Glaciology* **32**, 7–13.

Pielmeier, C. & Marshall, H.-P. (2009). Rutschblock-scale snowpack stability derived from multiple quality-controlled snowmicropen measurements. *Cold Regions Science and Technology* **59**(2–3), 178–184.

Robin, G.D.Q., Evans, S. & Bailey, J.T. (1969). Interpretation of radio echo sounding in polar ice sheets. *Philosophical Transactions of the Royal Society of London Series A – Mathematical and Physical Sciences* **265**(1166), 437–505.

Satyawali, P.K. & Schneebeli, M. (2010). Spatial scales of snow texture as indicator for snow class. *Annals of Glaciology* **51**, 55–63.

Satyawali, P.K., Schneebeli, M., Pielmeier, C., Stucki, T. & Singh, A.K. (2009). Preliminary characterization of alpine snow using snowmicropen. *Cold Regions Science and Technology* **55**, 311–320.

Schneebeli, M., Coleou, C., Touvier, F. & Lesaffre, B. (1998). Measurement of density and wetness in snow using time-domain reflectometry. *Annals of Glaciology* **26**, 69–72.

Schneebeli, M. & Sokratov, S. (2004). Tomography of temperature gradient metamorphism of snow and associated changes in heat conductivity. *Hydrological Processes* **18**(18), 3655–3665.

Siegfried, M.R., Hawley, R.L. & Burkhart, J.F. (2011). High-resolution ground-based GPS measurements show intercampaign bias in ICESat elevation data near Summit, Greenland. *IEEE Transactions On Geoscience and Remote Sensing* **49**(9), 3393–3400.

Strozzi, T. & Matzler, C. (1998). Backscattering measurements of alpine snowcovers at 5.3 and 35 GHz. *IEEE Transactions On Geoscience and Remote Sensing* **36**(3), 838–848.

Tedesco, M., Kim, E., Gasiewski, A., Klein, M. & Stankov, B. (2005). Analysis of multiscale radiometric data collected during the Cold Land Processes Experiment-1 (CLPX-1). *Geophysical Research Letters* **32**(18), L18501.

Tedesco, M., Kim, E.J., England, A.W., De Roo, R.D. & Hardy, J.P. (2006a). Brightness temperatures of snow melting/refreezing cycles: observations and modeling using a multilayer dense medium theory-based model. *IEEE Transactions On Geoscience and Remote Sensing* **44**(12), 3563–3573.

Tedesco, M., Kim, E., Cline, D., *et al.* (2006b). Comparison of local scale measured and modelled brightness temperatures and snow parameters from the CLPX 2003 by means of a dense medium radiative transfer theory model. *Hydrological Processes* **20**(4), 657–672.

Trujillo, E., Ramirez, J.A. & Elder, K.J. (2007). Topographic, meteorologic, and canopy controls on the scaling characteristics of the spatial distribution of snow depth fields. *Water Resources Research* **43**(7), W07409.

Trujillo, E., Ramirez, J.A. & Elder, K.J. (2009). Scaling properties and spatial organization of snow depth fields in sub-alpine forest and alpine tundra. *Hydrological Processes* **23**(11), 1575–1590.

Ulaby, F.T., Stiles, W.H., Dellwig, L.F. & Hanson, B.C. (1977). Experiments on radar backscatter of snow. *IEEE Transactions On Geoscience and Remote Sensing* **15**(4), 185–189.

Vaughan, D.G., Corr, H.F.J., Smith, A.M., Pritchard, H.D. & Shepherd, A. (2008). Flow-switching and water piracy between Rutford Ice Stream and Carlson Inlet, West Antartica. *Journal of Glaciology* **54**(184), 41–48.

Waddington, E.D., Neumann, T.A., Koutnik, M.R., Marshall, H.-P. & Morse, D.L. (2007). Inference of accumulation-rate patterns from deep layers in glaciers and ice sheets. *Journal of Glaciology* **53**(183), 694–712.

Waite, A.H. & Schmidt, S.J. (1962). Gross errors in height indication from pulsed radar altimeters operating over thick ice or snow. *Proceedings of the Institute of Radio Engineers* **50**(6), 1515.

Wiesmann, A., Strozzi, T. & Weise, T. (1996). *Passive microwave signature catalogue of snowcovers at 11, 21, 35, 48 and 94 GHz.* Institute of Applied Physics (IAP), University of Bern, IAP-Report 96-8.

Acronyms

AMSR-E	Advanced Microwave Scanning Radiometer
ASIRAS	Airborne Synthetic Aperture Interferometric Radar Altimetry System
AWSs	Automated Weather Stations
BOS	Borehole Optical Stratigraphy

CLPX	NASA Cold Lands Processes Experiment
CryoVEX	CryoSat Validation EXperiement
DEMs	Digital elevation models
EDM	Electronic Distance Measurement
ESA	European Space Agency
FCC	Federal Communications Commission
GNSSs	Global Navigation Satellite Systems
GPS	Global Positioning System
IOPs	Four Intensive Observing Periods
ISAs	Intensive study areas
LiDAR	Light Detection And Ranging
LSOS	Local Scale Observation Site
MABLE	Mostly Automated Borehole Logging Experiment
MADGE	Maine Automated Density Gauge Experiment
MSAs	Meso-cell Study Areas
NAVSAT	Navy Navigation Satellite System
NIR	Near InfraRed
PPK	Post-processed kinematic
RTK	Real-time kinematic
SIRAL	Synthetic aperture Interferometric Radar Altimeter
SMP	Snow micro-penetrometer
SSA	Specific Surface Area
SWE	Snow water equivalent

Websites cited

http://climate2.jpl.nasa.gov/ice/missions/
http://www.nasa.gov/mission_pages/icebridge/index.html
http://en.wikipedia.org/wiki/Global_Positioning_System
http://www.wcc.nrcs.usda.gov/snow/
http://www.nasa.gov/mission_pages/icebridge/index.html
http://nsidc.org/data/nsidc-0155.html
http://www.nohrsc.nws.gov/~cline/clpx.html
http://nsidc.org/data/clpx/
http://nsidc.org/data/amsr_validation/cryosphere/amsrice03/index.html
http://nsidc.org/data/amsr_validation/cryosphere/amsrice03/index.html
http://nsidc.org/data/amsr_validation/cryosphere/amsrice03/index.html

15 Remote sensing missions and the cryosphere

Marco Tedesco[1], Tommaso Parrinello[2], Charles Webb[3] and Thorsten Markus[4]

[1] The City College of New York, City University of New York, New York, USA
[2] European Space Agency, ESRIN, Frascati, Italy
[3] NASA Headquarters, Washington DC, USA
[4] NASA Goddard Space Flight Center, Greenbelt, USA

Summary

This last chapter provides a brief overview of spaceborne remote sensing missions and airborne operations that are either cryosphere-oriented or that have potential application to the cryosphere. A general description of such ongoing or planned missions is provided for the reader's convenience, reporting information on the sensors, techniques and applications of the different missions. In particular, the following missions are discussed in the chapter: SMOS, SMAP, CoreH20, ICESat and ICESat-2, Operation IceBridge and CryoSat-2.

15.1 Introduction

In this chapter, we report and discuss current and planned remote sensing missions that specifically focus on, or have supported, studies concerning the cryosphere. In general, though a mission or a sensor may be designed for a specific target or objective, the creativity of scientists, together with the support of the different funding agencies, has allowed data to be used for applications that were different from those originally conceived. This is also true for the cryosphere. An example is represented by QuikSCAT, originally planned to support wind and ocean studies but then also used, for example, to study melting and accumulation over Greenland and Antarctica. Remote sensing missions focused on Earth or ice studies can be found at http://climate2.jpl.nasa.gov/ice/missions/, at http://en.wikipedia.org/wiki/List_of_Earth_observation_satellites and at http://science.nasa.gov/missions/.

The past decade has been a very resourceful period for cryosphere-oriented remote sensing missions. In the following, we briefly describe some of the past, current and planned remote sensing missions and airborne operations that focus on studying several components of the cryosphere.

15.2   SMOS and SMAP

The Soil Moisture and Ocean Salinity satellite (SMOS) is a part of ESA's living planet program intended to provide, among other things, improved weather forecasting and monitoring of snow and ice accumulation (http://www.esa.int /Our_Activities/Observing_the_Earth/The_Living_Planet_Programme/Earth_ Explorers/SMOS/ESA_s_water_mission_SMOS, http://www.cesbio.ups-tlse.fr/us /indexsmos.html). The project was proposed in November 1998, and the launch occurred on 2 November 2009 from Plesetsk Cosmodrome on a *Rockot* rocket. The goal of the SMOS mission is to monitor surface soil moisture with an accuracy of 4% (at 35 – 50 km spatial resolution). A potential cryospheric application concerns the study of thawing-freezing soil cycles. The SMOS satellite carries a new type of instrument called Microwave Imaging Radiometer with Aperture Synthesis (MIRAS), collecting the radiation emitted in the microwave L-band (1.4 GHz), with an eight-meter aperture reminiscent of helicopter rotor blades (Figure 15.1).

The Soil Moisture Active and Passive (SMAP) is a planned NASA mission that will provide measurements of the land surface soil moisture and freeze-thaw state, with near-global revisit coverage in 2 – 3 days (O'Neill *et al.*, 2011). SMAP launch is planned for 2014 (http://smap.jpl.nasa.gov/mission/status/). The SMAP measurement system consists of a radiometer instrument and a synthetic aperture radar instrument operating with multiple polarizations in the L-band range (1.20 – 1.41 GHz). The active and passive sensors provide coincident measurements of the surface emission and backscatter, hence, for example, allowing study of the freeze-thaw soil state at global scale, taking advantage of the spatial resolution of the radar and the sensing accuracy of the radiometer. The radar and radiometer share a single feed and deployable mesh reflector antenna system, with a diameter of 6 m making conical scans of the surface.

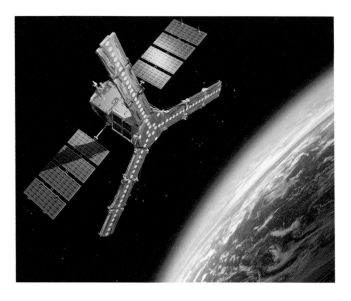

Figure 15.1 Artist's view of SMOS (ESA).

15.3 CoReH20

The Cold Regions Hydrology High-resolution Observatory (CoReH2O) satellite mission was selected by ESA as one of the three Earth Explorer Core mission candidates that are currently going through extended feasibility studies (Phase-A). However, in May 2013, the Biomass mission concept was chosen to become the next in the series of satellites developed to further our understanding of Earth, and CoReH2O was not selected. Nevertheless, activities connected to the planning of CoReH2O have been providing important insights into active and passive remote sensing of snow and ice. Major objectives of CoReH20 were the improvement of modeling and prediction of water balance and streamflow for snow-covered and glaciated basins, the modeling of water and energy cycles at high latitudes, and forecasting water supply from snow cover and glaciers.

The mission concept uses two synthetic aperture radars (9.6 and 17.2 GHz) to deliver all-weather, year-round information on regional and continental-scale snow-water equivalent (http://epic.awi.de/18999/1/Rot2008a.pdf, https://earth.esa .int/web/guest/content?p_r_p_564233524_assetIdentifier=coreh20). Basic measurements of CoReH2O include extent and melting state of the seasonal snow cover, snow accumulation and diagenetic facies of glaciers, permafrost features, and sea ice types. The dual frequency, dual polarization approach should enable the decomposition of the scattering signal for retrieving physical properties of snow and ice.

15.4 ICESat and ICESat-2

ICESat (Ice, Cloud, and land Elevation Satellite) was a satellite mission of the NASA Earth Observing System (EOS) for measuring ice sheet mass balance, cloud and aerosol heights, as well as land topography and vegetation characteristics. The satellite was launched in January 2003 on a Delta II rocket from Vandenberg Air Force Base in California, into a near-circular, near-polar orbit, at an altitude of approximately 600 km. One of the main goals of ICESat was to generate elevation data that could be used to determine ice sheet mass balance. The Geoscience Laser Altimeter System (GLAS), a LiDAR, was the only instrument onboard ICESat, emitting laser pulses at 1064 and 532 nm wavelengths with a spatial resolution of 70 m (diameter of laser spots separated by nearly 170 m along the spacecraft's ground track). Unfortunately, during initial on-orbit operations, the first GLAS laser failed prematurely, in March 2003. This forced a change to the operational plan for GLAS, and the instrument was subsequently operated for one-month periods, every three to six months, to extend the time series of measurements, particularly for the ice sheets. The last laser failed in October 2009 and, following attempts to restart it, the satellite was decommissioned and de-orbited in February 2010.

ICESat provided invaluable information on both Greenland and Antarctica ice sheets, allowing quantitative analysis of surface elevation changes related to, for example, ablation and accumulation. Also, the combination of ICESat data with data from GRACE has been crucial towards understanding the temporal trends of mass balance of the two ice sheets and other glaciated areas.

Scheduled for launch in 2016, the ICESat-2 mission will employ the Advanced Topographic Laser Altimeter System (ATLAS) to build on the ice-sheet elevation record established by its groundbreaking predecessor during the original ICESat mission (2003–2009). Although the orbit for ICESat-2 also has a 91-day repeat cycle, it will operate at a lower altitude, slightly increasing the number of reference ground tracks, and at a lower inclination, allowing coverage to 88° latitude in both hemispheres. Furthermore, the ATLAS design will provide significantly denser spatial coverage by using one of two redundant lasers to generate three pairs of beams (Figure 15.2). The pairs are separated by approximately 3 km at the surface, and the beams within each pair are maintained to within 90 meters, to allow for determination of local surface slopes. The laser will operate at a 532 nm (green) wavelength, and send pulses 10 000 times per second. Finally, ATLAS will detect and report the round-trip travel times for individual photons, providing enhanced resolution of surface features, including canopy heights in vegetated areas.

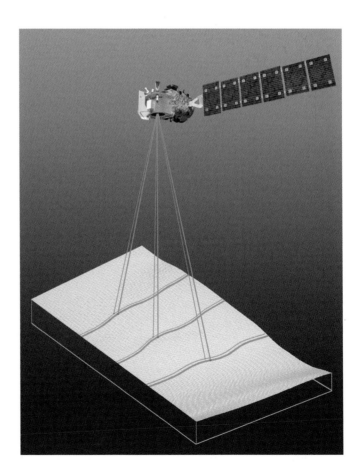

Figure 15.2 Artist's view of ATLAS on ICESat-2 generating three pairs of beams (NASA).

The increased spatial density and continuous operation offered by the ICESat-2 mission promise to reveal unprecedented details about the elevation changes occurring in the Greenland and Antarctic ice sheets, allowing for more accurate estimates of mass balance and insight into the mechanisms that control it. Furthermore, by the end of the ICESat-2 mission, comparisons with ICESat observations, cross-calibrated through IceBridge data (see the following section), will provide a clear picture of the changes in ice-sheet surface elevation across a period of more than 15 years. Similarly, for sea ice, the resolution and accuracy of these measurements will yield improved estimates of freeboard and, thus, ice thickness, and its temporal variation. Mapping sea ice thickness and its changes are critical to understanding the energy and mass exchanges between the ocean and the atmosphere, and the influence of freshwater fluxes associated with melting ice. More can be found at http://icesat.gsfc.nasa.gov/.

15.5 Operation IceBridge

The progressive decline in laser energy observed near the end of the ICESat mission prompted NASA to plan a series of airborne missions that would extend the time series of elevation data to fill in the gap between the eventual failure of GLAS and the launch of the ICESat-2 mission in 2016. In March 2009, Operation Ice-Bridge began with 20 science flights over the Greenland Ice Sheet and Arctic sea ice, using NASA's P-3B airborne laboratory, equipped with two scanning laser altimeters – the Airborne Topographic Mapper (ATM) and the Land, Vegetation and Ice Sensor (LVIS). In addition, it carried an ultra-wideband microwave radar, known as the Snow Radar, from the University of Kansas Center for Remote Sensing of Ice Sheets (CReSIS), and the Pathfinder Advanced Radar Ice Sounder (PARIS) from Johns Hopkins University, which measured snow and ice thickness, respectively.

The IceBridge mission continued in October 2009, with 21 science flights over western Antarctica, the Antarctic Peninsula, and Antarctic sea ice, using NASA's DC-8 aircraft (Figure 15.3). The instrument suite again included the ATM and LVIS altimeters and the Snow Radar, which were supplemented with a Ku-band radar altimeter and the Multichannel Coherent Radar Depth Sounder/Imager (MCoRDS/I), both from CReSIS. To infer the bathymetry beneath ice shelves and the ice-sheet bed topography beneath outlet glaciers, which cannot be mapped by radar, a gravimeter was also flown to measure subtle variations in the local gravity field. Lastly, a Digital Mapping System (DMS) provided high-resolution stereographic imagery to facilitate the interpretation of the data collected by other sensors. This larger suite of instruments has been used consistently in most of the subsequent campaigns, in both the Arctic and Antarctic, which have routinely taken place each year, in March – April and October – November, respectively.

Operation IceBridge will continue collecting data over the next years, providing an elevation data set that will link the measurements of ICESat and ICESat-2, as well as those collected by Europe's Envisat and CryoSat-2 satellites, allowing for

Figure 15.3 NASA's DC-8 aircraft (NASA).

the development of a long-term ice elevation record. Together with the additional geophysical parameters measured by its other instruments, IceBridge data will help to improve our understanding of the mechanisms governing mass balance and the dynamics of the Greenland and Antarctic ice sheets, and those that drive sea ice cover, particularly in the Arctic. They will also be used to validate and to improve predictive models of land-based ice contributions to sea level, and of sea ice cover, during this century.

15.6 CryoSat-2

CryoSat-2 was launched in April 2010 by a Russian Dnepr rocket from Baikonour Cosmodrome in Kazakhstan. The launch came four and half years after the original CryoSat satellite was destroyed in a launch failure of a Russian *Rockot* over the North Pole. The current satellite is therefore CryoSat-2, but the mission is also known simply as CryoSat.

CryoSat-2 orbits the planet at an altitude of 720 km, with a retrograde orbit inclination of 92° and a repeat cycle of 369 days (30 days sub-cycle). This offers a high-density coverage over the polar regions. The orbit is not sun-synchronous and allows the orbit plane to drift naturally around the sun direction at about 0.75° per day. Thanks to this orbit geometry, the satellite is able to reach latitudes up to 88°, covering more than 4.6 million km^2 of unexplored areas over the poles when compared with previous altimeters.

The CryoSat-2 mission has two important goals. Firstly, the mission seeks to build a detailed picture of the trends and natural variability in Arctic sea ice. Secondly, the mission seeks to completely observe the trend in the thinning rate of the Antarctica and Greenland ice sheets. The primary payload for CryoSat-2 is a radar altimeter called SIRAL (Synthetic Aperture Radar and Interferometric Radar Altimeter). The SIRAL radar altimeter is derived from the conventional pulse-width-limited altimeter called Poseidon-2 and the US-French Jason missions. SIRAL is a single-frequency Ku-band radar, featuring some new design characteristics that enable it to provide data that can be more elaborately processed on the ground.

SIRAL can be operated in three main modes. As well as the conventional pulse-width-limited mode (LRM), which offers continuity with earlier altimeter missions (like ERS and ENVISAT), the instrument can also operate in so-called synthetic aperture mode (SAR, Figure 15.4). This increases the along-track resolution, enabling it, for example, to more readily distinguish the narrow leads of open water between sea-ice floes. Over the rough terrain at the edges of the major ice sheets, this increased along-track resolution is further augmented by across-track interferometry (SARIN), using the second antenna and receiving channel mounted on the same optical bench. The derived angle of arrival of the radar echoes allows more precise identification of the point from which the echo came.

Figure 15.4 SIRAL SAR mode over ice floes (ESA).

Thanks to the high-density sampling, the averaging of many such measurements brings the system performance to the level needed to satisfy the mission objectives. However, the precise measurement of range alone is insufficient. The exact position of the satellite at the time of each measurement is needed to convert this simple measurement of range into something scientifically meaningful, which is the height of the surface above some known reference. To achieve this, CryoSat-2 flies a DORIS Receiver, a special radio receiver that picks up signals from a network of more than 50 transmitting stations evenly spread around the Earth. By measuring the Doppler shifts of these signals, the range-rate to each one can be determined. The DORIS receiver is augmented by a passive laser retro-reflector, which allows precise range measurements to be taken by ground-based laser-ranging stations. The final item in this collection of high-precision payload equipment is a set of three identical star trackers, needed to complement the SIRAL interferometer measurements and to identify the baseline orientation with high accuracy.

All of the data generated by CryoSat-2's scientific payload is recorded onboard, as the satellite is only in contact with the receiving station (Kiruna) for brief periods. Typically, there are ten passes of 5–10 minutes duration each day, occurring on consecutive orbits where the data is downlinked at 100 Mbps. These contacts are followed by a gap of three or four *blind orbits*, where there is no contact with the ground and the data is accumulated in the onboard solid-state recorder. There are other data flows in the system, since CryoSat-2, like any altimeter mission, needs auxiliary data from a variety of sources. For example, the precise orbits are computed by an expert group at CNES in Toulouse, France. They need to get the data from the DORIS instrument as soon as it is available, and it takes around 30 days to compute and check the orbits to the highest accuracy. This data is sent back to Kiruna and incorporated into the CryoSat data products.

Table 15.1 CryoSat products available to the users community.

SIRAL products	Main characteristics	Volume	Distribution
Level 1B full bit rate	Time-ordered, coherent synthetic radar beams for raw SAR and SARIn modes. In LRM, echoes are incoherently multi-looked on board the satellite prior to download.	50 Gb/day	Limited
Level 1B multi-looked wave	Multi-looked echoes. SARIn data contains multi-looked phase. Full engineering and geophysical corrections are applied.	2.5 Gb/day	To all users
Level 2 GDR	These are consolidated products on orbit basis containing LRM, SAR and SARIn. These are time-ordered elevation values (of ice, ocean or land), and, in the case of sea ice, ice thickness.	60 Mb/day	To all users
Level 2 FDM	Fast delivery ocean for meteorologist use.	35 Mb/day	To all users

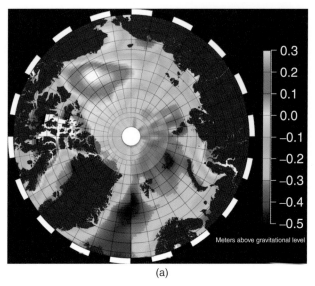

(a)

Sea ice thickness in the Arctic ocean
(January/February 2011)

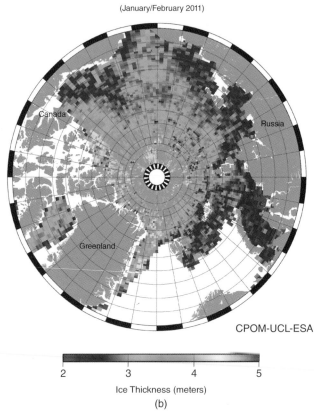

CPOM-UCL-ESA

(b)

Figure 15.5 (a) Ocean dynamic topography and currents. (b) Sea ice thickness map in Arctic basin. (ESA).

To satisfy the existing meteo-ocean community, who require a subset of altimeter data in near real time (which is, to say, within three hours of the measurement), there is a further dedicated product. This so-called FDM product is processed in the same way as the other products, but uses the real-time orbit solution computed by DORIS and normally used for on-board satellite control. Its main content is ocean elevation, wind speed and wave height, and it is only made from the SIRAL LRM mode data over the oceans. Clearly, it is not always available in near-real time because of the 3–4 blind orbits. A new set of products, dedicated to the oceanographic community, was due to become available at the beginning of 2014. Table 15.1 summarizes the main products available to the user community.

Two examples of results obtained from the relatively short life of Cryosat-2 are reported in Figure 15.5. Measurements from the SIRAL altimeter were used to make a map of ocean circulation (Figure 15.5a) across the Arctic basin. In sensing the surface of the water, CryoSat-2 also becomes a powerful tool to study ocean behavior, especially up to latitudes never reached by an altimeter mission. In fact, the interest of the oceanography community in the CryoSat-2 mission has always been very high since the mission was conceived.

In June 2010, CryoSat mission delivered the first sea-ice map relative to January-February 2011 (Figure 15.5b). This was the first fully processed map of the Arctic sea-ice using CryoSat-2 data that proved all the potentiality of the SIRAL instrument. The obtained thickness information of the sea-ice was validated with a number of independent *in situ* and airborne measurements showing very good matches. In January 2013, CryoSat delivered the first two-year measurements of the trend of the sea ice thickness over the Arctic basin, validated with ground and airborne measurements (Laxon *et al.*, 2013).

More information on the CryoSat mission can be found at http://earth.esa.int /cryosat. Documents and resources on CryoSat can be found at https://wiki .services.eoportal.org/tiki-index.php?page=CryoSat%20Wiki

References

CryoSat Science Report (2003). ESA SP-1272. European Space Agency.

Francis, R. (2010). *ESA's ice mission, CryoSat: more important than ever*. ESA Bulletin-141. European Space Agency.

Laxon, S.W., Giles, K.A., Ridout, A.L., Wingham, D.J. *et al.* (2013). CryoSat-2 estimates of Arctic sea ice thickness and volume. *Geophysical Research Letters* **40**(4), 732–737, doi: 10.1002/grl.50193.

O'Neill, P., Entekhabi, D., Njoku, E. & Kellogg, K. (2011). *The NASA Soil Moisture Active Passive (SMAP) Mission: Overview*. NASA, Goddard Space Flight Center, Jet Propulsion Laboratory.

Ratier, G., *et al.* (2005). *The CryoSat System – The satellite and its radar altimeter*. ESA Bulletin-122, European Space Agency.

Wingham, D. (2005). CryoSat: a mission to the ice fields of Earth, ESA Bulletin-122. European Space Agency.

Acronyms

ATLAS	Advanced Topographic Laser Altimeter System
ATM	Airborne Topographic Mapper
CNES	Centre National d'Etudes Spatiales/National Centre for Space Studies
CoReH2O	COld REgions Hydrology High-resolution Observatory
CReSIS	University of Kansas Center for Remote Sensing of Ice Sheets
DMS	Digital Mapping Sensor
EOS	NASA Earth Observing System
GLAS	Geoscience Laser Altimeter System
ICESat	Ice, Cloud, and land Elevation Satellite
LiDAR	Light Detection And Ranging
LVIS	Land, Vegetation and Ice Sensor
MCoRDS/I	Multichannel Coherent Radar Depth Sounder/Imager
MIRAS	Microwave Imaging Radiometer with Aperture Synthesis
PARIS	Pathfinder Advanced Radar Ice Sounder
SIRAL	Synthetic Aperture Radar and Interferometric Radar Altimeter
SMAP	Soil Moisture Active and Passive
SMOS	Soil Moisture and Ocean Salinity satellite

Websites cited

http://climate2.jpl.nasa.gov/ice/missions/

http://en.wikipedia.org/wiki/List_of_Earth_observation_satellites

http://science.nasa.gov/missions/

http://www.esa.int/Our_Activities/Observing_the_Earth/The_Living_Planet_Programme/Earth_Explorers/SMOS/ESA_s_water_mission_SMOS

http://www.cesbio.ups-tlse.fr/us/indexsmos.html

http://smap.jpl.nasa.gov/mission/status/

http://epic.awi.de/18999/1/Rot2008a.pdf, https://earth.esa.int/web/guest/content?p_r_p_564233524_assetIdentifier=coreh20

http://icesat.gsfc.nasa.gov/

http://esamultimedia.esa.int/docs/Cryosat/Mission_and_Data_Descrip.pdf

http://earth.esa.int/cryosat

Index

Page numbers in **bold** refer to tables, and in *italics* refer to figures.

Remote Sensing of the Cryosphere, First Edition. Edited by M. Tedesco.
© 2015 John Wiley & Sons, Ltd. Published 2015 by John Wiley & Sons, Ltd.
Companion Website: www.wiley.com/go/tedesco/cryosphere